AN INTRODUCTION TO
SEISMIC INTERPRETATION

Reflection Seismics in
Petroleum Exploration

AN INTRODUCTION TO SEISMIC INTERPRETATION

Reflection Seismics
in Petroleum Exploration

R. McQuillin, M. Bacon, W. Barclay

with contributions by

R. E. Sheriff, R. McEvoy, R. Steele

Graham & Trotman

First published in 1984
Reprinted in 1986

Graham & Trotman Limited
Sterling House
66 Wilton Road
London SW1V 1DE

British Library Cataloguing in Publication Data

McQuillin, R.
    An introduction to seismic interpretation. —
    2nd ed.
    1. Seismic reflection method
    2. Gas, natural   3. Petroleum
    I. Title II. Bacon, M. III. Barclay, W.
    622'.1828   TN271.P4

ISBN 0-86010-455-9
ISBN 0-86010-496 Pbk

ISBN 0 86010 455 9 (hardback)
ISBN 0 86010 496 6 (softcover)

Typeset in Great Britain by Herts Typesetting Services Limited, Hertford
Printed and bound in Great Britain at the Alden Press, Oxford

CONTENTS

PREFACE

The first edition of this book, work on which commenced in 1976 when all three principal authors lived in Edinburgh, was published in 1979. Plans for a new edition were conceived late in 1982, revision being completed in September 1983. The work has been substantially expanded with two main aims in view; first, to keep abreast of recent advances in the science and technology of exploration seismology whilst at the same time improving coverage of some topics we came to regard as having been given insufficient treatment in the first edition, and second, to provide a fuller treatment of basic theory making the book more useful as a student text. The book is intended for the reader needing an introduction to the seismic reflection method, especially in the oil industry, and for students of applied geophysics. We have tried to restrict the level of coverage so that no more than school (or college first year) mathematics and physics are required for the understanding of such theory as is included. It has also been our aim to structure the book in such a way that the non-mathematical reader can easily follow the main text without the need to fully comprehend the theoretical sections. To this end, we have, as in the first edition, made the fullest possible use of diagrams, illustrations and actual examples of the results of exploration surveys. In this respect we are greatly indebted to the many oil companies, exploration contractors and research institutions who have co-operated by provision of such information. Sources of data are acknowledged in the text.

The main text is a joint collaboration between the three principal authors. The chapter on 'Seismic Stratigraphy and Hydrocarbon Detection' is contributed by Robert Sheriff (University of Houston) incorporating some material from the first edition. The four case histories are individual contributions. The Moray Firth study is a revision by McQuillin of the previous study by Mike Bacon, the Rainbow Lake and Hewett studies, both revised for this new edition, are contributed by Bill Barclay, and the Kingfish Study is contributed by David McEvoy and Ron Steele of Esso Australia. Robert McQuillin undertook the task of editing material for publication.

For approving participation in the preparation of the book the authors wish to acknowledge their indebtedness to the Director of the British Geological Survey* (McQuillin) and the management of Shell UK (Bacon) and Trilogy Resource Corporation (Barclay). Many new diagrams are included in the new edition, many from the first edition have been revised and improved. We thank Angela McQuillin for her work in preparing the illustrations. Finally, we would like to thank Linda Nisbet for secretarial assistance through both editions of this book and for her work in the preparation of the manuscript, in draft, in revision and to final copy.

R. McQuillin
Edinburgh

M. Bacon
London

W. Barclay
Calgary

* The British Geological Survey was until quite recently called the Institute of Geological Sciences (IGS) to which a number of references and acknowledgements are made throughout the text.

Section 1
Theory

Chapter 1

Introduction

1.1 Role of geophysics in oil exploration

Although the seismic reflection method has its uses in several areas of the earth sciences, both pure and applied, by far its most intensive use is in the search for oil and gas. It is not hard to see why this is so. Nowadays, finding commercially valuable accumulations of hydrocarbons is rarely simple. Usually, they are to be found at depths of at least several thousand feet below the ground surface. In some areas, it is possible to infer the geology at these depths from that of the rocks exposed at the surface, but it is more common for the near-surface geology to bear little or no relation to the deeper structures. Ultimately, knowledge of this deeper geology can come only from drilling, but a deep borehole is expensive and in many areas (especially offshore) the delineation of structure by closely-spaced drilling is unthinkable. Geophysics can assist exploration in several ways.

At the reconnaissance stage, the first problem is to delineate the basins in which a thick sequence of sedimentary rocks has been deposited. Only in these basins will petroleum source rocks have been buried deeply enough to reach the temperatures and pressures needed for oil to be formed and expelled into reservoir rocks. At this stage of exploration, only a generalised picture of the subsurface is required, and the gravity and magnetic exploration methods are particularly useful. Aeromagnetic surveys can be carried out quickly and fairly cheaply even in remote areas, and interpretation of the data can, in suitable cases, reveal the general form of the igneous or metamorphic basement underlying a sedimentary sequence.

The next phase of exploration is generally to find structures within such basins that hold promise of being oil or gas traps; at least initially, this probably means searching for antiforms. The deeper the target, the less the resolution that gravity or magnetics can give. It is here that the seismic reflection method becomes useful. As we shall see, the method is capable of giving a picture of subsurface structure with a resolution of at least a few tens of metres vertically and a few hundred metres horizontally. This increases tremendously the knowledge that can be gained from a few exploration boreholes. There is usually a conflict between siting a hole so as to be a valid test of a petroleum trap (e.g. on an antiform), and locating it so as to obtain as much geological information as possible, which may mean drilling in a synform where sequences are most completely developed and preserved from erosion. In either case, however, lithological and stratigraphic information can be extrapolated away from the borehole with some confidence once seismic data have revealed the subsurface structures.

Mapping of structure remains the most

important use of the seismic tool, and is generally the most straightforward type of interpretation to carry out. Recently, however, much attention has been focused on the fine detail of seismic records. Seismostratigraphy has enabled interpreters to infer depositional environment, lithology and even chronostratigraphy in advance of drilling. Detailed studies have allowed recognition of stratigraphic traps and delineation of individual reservoir bodies. Also, in favourable cases, the presence of porosity and fluids, particularly gas, can be recognised in a reservoir directly from seismic data.

1.2 Essence of the seismic reflection method

To the horror of his geologist colleagues, the geophysicist is apt to think of the earth as a layer-cake. That is, he regards it as a series of layers with rather abrupt changes in physical properties between them vertically, but only rather gradual changes laterally, along a layer. This view, though obviously over-simplified, is helpful in understanding the principles of the method.

In essence, the reflection seismic method explores this layer-cake by bouncing sound waves off the interfaces between these various rock layers. This is analogous to the way in which a ship's echo sounder measures the depth of the sea from the time taken for a pulse of sound to return to the ship after reflection at the seabed (figure 1/1). A record can be produced automatically showing the time taken for the sound pulse to travel from the ship down to the seabed and back again (two-way-time, often abbreviated as TWT) at different points along the ship's path. By multiplying the TWT by half the velocity of sound in sea water, the time scale on the record can be marked off directly in

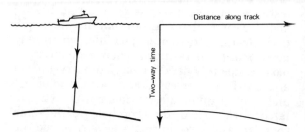

Figure 1/1 Schematic of ship's echo sounder and depth record.

depth. Even in this simple case, the reflection path is not in general vertical, so that depths found from the record are not precisely the depths of water directly under the ship at the relevant position; slopes of the seabed are generally quite gentle, however, so the difference from the true depth is negligible.

For a number of reasons, using the same technique to probe deeper into the earth is more difficult (figure 1/2). Firstly the signal received

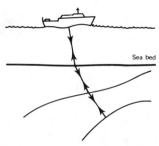

Figure 1/2 Reflections from rock layers beneath the seabed.

after reflection at a depth of several thousand feet will be very small. To get an adequate signal to (background) noise ratio, a much more powerful source will be needed. It will also be necessary to arrange for partial cancellation of such random noise by summing the signals from many source pulses received at many surface locations. Secondly, we shall see in Chapter 2 that a signal which has travelled through the earth for thousands of feet will have a quite low frequency, typically in the low tens of Hz, i.e. a single wavelength is equivalent to several hundred feet (figure 1/3). This will limit the vertical and horizontal resolution we can achieve. Thirdly, there is a good deal of layering between the source and the deep reflector. A signal can therefore bounce backwards and forwards between any two reflectors including land surface, seabed and sea surface (in marine work) a number of times, and perhaps arrive back at the receiver at nearly the same time as the deep reflector signal; these secondary signals or 'multiples' must be eliminated from our record. Fourthly, the velocity of sound varies laterally and vertically in the earth. It is therefore much harder to convert our TWTs into depths to reflectors than in the echo sounder case, where the velocity of sound in sea water is (very nearly) the same everywhere, and can be

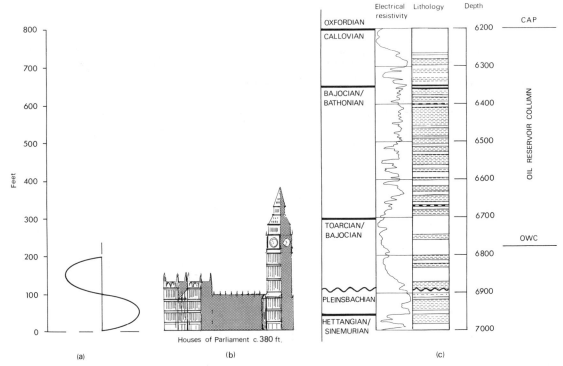

Figure 1/3 Some vertical scale comparisons. (a) A single cycle sine wave of 30 Hz in medium of velocity 6000 fps (or 60 Hz; 12 000 fps); (b) Big Ben, London; (c) Geophysical logs through the Beatrice Oil Field (see also Chapter 10).

accurately determined. Finally, the problems caused by the path not being truly vertical are going to be much worse, because structural dips of 10–20° are not uncommon at depth. The ways in which these problems are overcome are discussed in later chapters.

1.3 The seismic spectrum

Before embarking on a consideration of the theory of seismic wave propagation, it is useful to view the full range of the seismic spectrum in relation to that part of it which can be used in exploration geophysics. A logarithmic plot of the spectrum (figure 1/4) shows those parts which can be utilised in various types of study.

At the infra-sonic end we have the very low frequency seismic events associated with free oscillations of the earth. These were first recognised after the Chile earthquake of 22 May 1960 and the Alaskan earthquake of 28 March 1964 (Smith, 1966). Such events are detected by earth tide gravity meters and strain seismometers. A very large earthquake causes the earth to ring like a bell, but at very low

frequencies. The oscillations are very complex and have periods ranging from 300 to 3000 seconds.

A nuclear explosion may be thought of as a very high energy seismic source. The energy generated does not, however, match that of very large earthquakes, but nevertheless generates teleseisms (seismic events detectable at distances in excess of 1000 km from source), similar in magnitude to moderately large earthquakes. At long range sites these are detected as pressure waves in the frequency range 0.5–2 Hz and as surface waves in the range 0.01–0.2 Hz. These have been used in geological studies, particularly in North America. The US Atomic Energy Commission, by giving advance notice of underground nuclear explosions, along with source parameters and geological details of the explosion site, has allowed geophysicists to deploy specially designed experiments to study deep geological structure. The advantage of nuclear explosions over earthquakes in such studies is that in the case of earthquakes the precise location, depth and time of the seismic event must be deduced

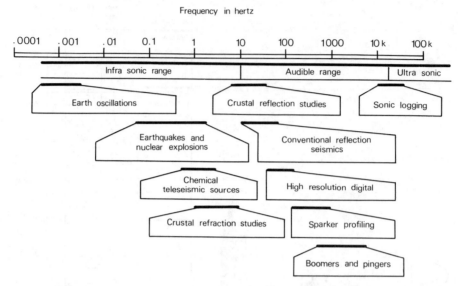

Figure 1/4 Schematic diagram of the seismic frequency spectrum.

from observations made at a number of recording stations. With an explosion, the source parameters are precisely known. Other alternative methods have been adopted for generating seismic signals which can be detected at very long range. One approach in the United States was to fill old Liberty ships with one or two kilotons of lapsed explosives and fire those at specific sites in the aseismic regions of the Atlantic and Pacific Oceans. By very careful experimental design, it is, however, possible to conduct long-range seismic experiments using only 10 tons of explosives. An experiment undertaken in the North Sea in 1972 gave an energy peak of 1.9 Hz which was detected as far away as Brasilia and Brisbane (Jacob and Willmore, 1972). Earthquakes and large explosions have provided the seismic sources for experiments which have revealed much of what is now known about the internal structure of the earth (Bott, 1982).

Although earthquake data are useful for studies of the earth's crust and upper mantle and of the boundary between them, the Mohorovicic Discontinuity (Moho), in recent years the most important advances have been made by setting up controlled experiments using both seismic refraction (see Chapter 8) and reflection methods. To obtain refracted signals from crustal layers down to the Moho it is necessary to generate a seismic pulse which can be detected at at least 200–300 km (depending on Moho depth) and ideally to about

1000 km range. The operational frequency range for such experiments is about 1–10 Hz. Crustal reflection studies can be accomplished using standard oil exploration techniques modified to operate in the 8–40 Hz range. A recently completed profile of 180 km length, offshore the northern coast of Scotland, gave good reflections from crustal layers down to and beneath the Moho (figure 1/5) with good resolution throughout the section (Smythe *et al.*, 1982).

Oil exploration is principally concerned with the definition of geological structure (and lithological variation) in the uppermost 5 km of the crust. In general, seismic reflection techniques used in such exploration are designed so that signals are detectable from the deepest target horizons in the highest frequency band attainable. There are, however, many constraints on the bandwidth of usable frequency, and, as will be described later, this is limited to the range c.12–60 Hz. This book will concentrate on the acquisition, processing and interpretation of seismic reflection data derived from surveys designed to operate in this frequency range.

In Chapter 8, conventional reflection techniques, modified for high-resolution investigations down to about 2 km depth are described ('high resolution digital' in figure 1/4). To obtain this higher resolution in a relatively shallow target zone, both seismic source and acquisition techniques are designed to operate

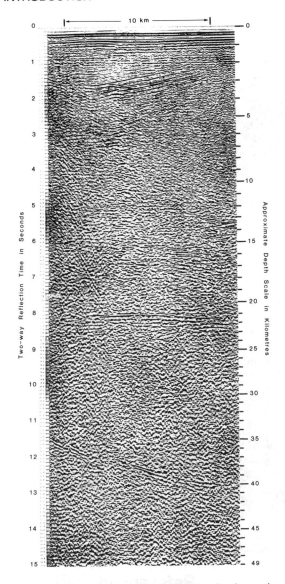

Figure 1/5 Part of an equalised stack seismic section from the Moine and Outer Isles Seismic Traverse (MOIST) surveyed across the sea area between the Orkney Islands and the Butt of Lewis, Scotland. Strong reflectors at 1.4–1.8 s TWT are from Permo-Trias redbeds beneath Jurassic shales. The Moho is at 8.0 s TWT and a major thrust zone (the Flannan Thrust) is seen as a group of reflectors dipping eastwards between 12–13 s TWT. This thrust zone intersects the Moho to the west and can be traced to mid-crustal depths (courtesy the British Institutions' Reflection Profiling Syndicate).

a further range of seismic reflection profiling tools can be employed; sparker profiling, boomer and pinger (figure 1/4), different systems operating in different frequency ranges up to a few kHz (McQuillin and Ardus, 1977). These devices naturally provide relatively low penetration, but resolution between reflecting horizons can be as good as 10–20 cm (figure 1/6).

Ultra-sonic seismic is used in some laboratory testing of rock samples and in sonic-logging during wire-line tests in boreholes. At such high frequencies detection of signals through transmission paths of up to a few metres is all that is required.

1.4 Seismic prospecting: the state of the art

Unlike an exact science, seismic prospecting for oil depends not only on high technology but also, and to a very large extent, on the interpretive powers and experience of the practising geophysicist. To a large extent, the techniques of interpretation of seismic data can only be acquired as a result of long experience translating the results of the physical experiments into geologically valid models of earth structure and then assessing the probability that any such structure might trap an accumulation of hydrocarbons. Seismic prospecting has progressed remarkably in the last two of the five decades of its regular use in petroleum exploration. Early exploration was only capable of discovering simple structures to be prospected by drilling. As techniques have improved, the geophysicist's role has extended far beyond that of simply identifying a drill site location. As a prospective area is developed and the results of exploration drilling become available, it is necessary to refine structural interpretations with information on lateral variations in the aforementioned layer-cake. Thus an element of stratigraphy is built into the geological picture revealed in seismic records.

In later chapters, as well as describing conventional exploration methods, we shall also introduce relatively new techniques, some of which are in their infancy of development: the use of P & S wave comparisons, colour displays of phase and frequency variations, modelling and pseudo-log seismic inversion, 3-D horizontal time-slicing and vertical seismic profiling. In some circumstances hydrocarbon accumulations are directly detectable by

in a higher frequency range: c.40–120 Hz. For geological mapping offshore, and for engineering studies of soil conditions close to the seabed,

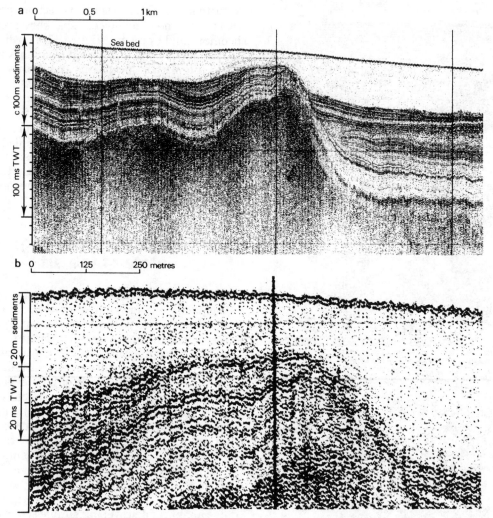

Figure 1/6 (a) Part of a Huntec DTS boomer record from the North Sea.
(b) Enlargement of record to show high resolution of bedding courtesy IGS).

seismic exploration, but even the presence of oil and/or gas is no guarantee of a find being economic. The results of drilling tests and geophysical interpretation together provide the criteria for such an evaluation.

Acquisition and processing has progressed from the use of crude analogue techniques to high technology digital systems. However, successful exploration still depends on the personal skills of the geologists and geophysicists who interpret the data. Such skills are mainly acquired by experience. This book aims to provide a broad introduction to and overview of the subject such that both practising and prospective exploration geophysicists and geologists might more quickly and more easily acquire these skills.

References

M. H. P. Bott, *The interior of the earth; its structure, constitution and evolution*. Edward Arnold, London (1982).

A. W. B. Jacob and P. L. Willmore, Teleseismic P Waves from a 10 ton explosion. *Nature*, **236** (1972), pp. 305–306.

R. McQuillin and D. A. Ardus, *Exploring the Geology of Shelf Seas*. Graham and Trotman, London (1977).

S. W. Smith, Free oscillations excited by the Alaskan earthquake. *J. Geophys. Res.*, **71** (1966), pp. 1183–1193.

D. K. Smythe, A. Dobinson, R. McQuillin, J. A. Brewer, D. H. Matthews, D. J. Blundell and B. Kelk, Deep structure of the Scottish Caledonides revealed by MOIST reflection profile. *Nature*, **299** (1982), pp. 338–40.

Chapter 2

Theoretical Background

2.1 Theoretical fundamentals

Seismic waves are elastic waves propagating through the earth. The aim of this section is briefly to develop the basic theory of these waves.

2.1.1 Basic elasticity theory

We begin by defining the concepts of stress and strain. *Stress* is the balance of internal action and reaction between different parts of a body at

Figure 2/1 Analysis of stress.

a given internal point (figure 2/1). We consider a small plane area ΔS of specified orientation; F is the force acting across this area due to the rest of the material. The stress at this point normal to ΔS is given by the three components of $\text{Lt } F/\Delta S$. The component normal to ΔS is called
$\Delta S \to 0$
the principal stress component; the components in the plane of ΔS are called shearing stress components.

By changing the orientation of ΔS, we should get different stress components. To specify the stress at a point completely, we need 3 stress components across each of 3 orthogonal surfaces (figure 2/2). It is convenient to introduce the stress tensor p_{ij} ($i, j = 1, 2, 3$) where the first suffix gives the direction of the normal to the surface, and the second suffix gives the direction of the component of the force. By using the condition that there must be no net couple acting on our infinitesimal cube, it may be shown that $p_{ij} = p_{ji}$, so there are only 6 independent stress components.

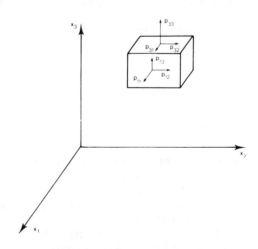

Figure 2/2 The stress tensor.

9

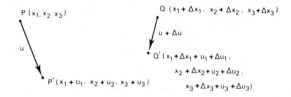

Figure 2/3 Analysis of strain.

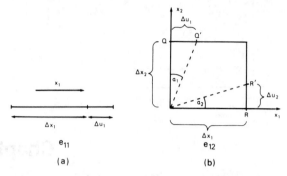

Figure 2/4 The meaning of strain components.

Strain is the deformation of a material under an applied stress. In figure 2/3 we see the effect of a small movement of the particles of a medium from an initial configuration. Thus, u is the displacement of a point P originally at (x_1, x_2, x_3), and $(u+\Delta u)$ is the displacement of a neighbouring point Q, initially at $(x_1 + \Delta x_1, x_2 + \Delta x_2, x_3 + \Delta x_3)$. Let us consider the set of quantities $\underset{\Delta x_i \to 0}{Lt} \Delta u_j / \Delta x_i = \partial u_j / \partial x_i$. These may be written as

$$\frac{\partial u_j}{\partial x_i} = \tfrac{1}{2} \left(\frac{\partial u_j}{\partial x_i} + \frac{\partial u_i}{\partial x_j} \right) - \tfrac{1}{2} \left(\frac{\partial u_i}{\partial x_j} - \frac{\partial u_j}{\partial x_i} \right)$$

$$= e_{ij} - \tfrac{1}{2} \xi_{ij}$$

where $e_{ij} = \tfrac{1}{2} \left(\frac{\partial u_j}{\partial x_i} + \frac{\partial u_i}{\partial x_j} \right)$

and $\xi_{ij} = \frac{\partial u_i}{\partial x_j} - \frac{\partial u_j}{\partial x_i}$.

It can be shown that ξ_{ij} corresponds to a simple 'rigid-body' rotation, which is of no further interest to us. The e_{ij} correspond to a pure deformation; e_{ij} is called the strain tensor.

The meaning of the e_{ij} can be seen from a diagram (figure 2/4). We know

$$e_{11} = \frac{\partial u_1}{\partial x_1}$$

$$= \underset{\Delta x_1 \to 0}{Lt} \frac{\Delta u_1}{\Delta x_1}.$$

This strain component corresponds to a pure extension in the x_1 direction of a line element originally in that direction (figure 2/4a); e_{22} and e_{33} may be interpreted similarly. If we have simultaneous strains e_{11}, e_{22} and e_{33} acting on a cube whose sides were initially Δx_1, Δx_2, Δx_3, then we should find that the volume has increased from

$$V = \Delta x_1 . \Delta x_2 . \Delta x_3$$

to

$$V + \Delta V = (\Delta x_1 + \Delta u_1).(\Delta x_2 + \Delta u_2).(\Delta x_3 + \Delta u_3)$$
$$= \Delta x_1 . \Delta x_2 . \Delta x_3 + \Delta u_1 . \Delta x_2 . \Delta x_3$$
$$+ \Delta u_2 . \Delta x_1 . \Delta x_3 + \Delta u_3 . \Delta x_1 . \Delta x_2$$

to first order.

Thus $\Delta V = \Delta u_1 . \Delta x_2 . \Delta x_3 + \Delta u_2 . \Delta x_1 . \Delta x_3 + \Delta u_3 . \Delta x_1 . \Delta x_2$.

So the fractional volume change, usually called the dilatation, θ, is given by

$$\theta = \Delta V / V$$
$$= \frac{\Delta u_1}{\Delta x_1} + \frac{\Delta u_2}{\Delta x_2} + \frac{\Delta u_3}{\Delta x_3}$$
$$= e_{11} + e_{22} + e_{33}.$$

Figure 2/4b illustrates $e_{12} = \tfrac{1}{2} \left(\frac{\partial u_1}{\partial x_2} + \frac{\partial u_2}{\partial x_1} \right)$, where as a result of deformation R moves to R' and Q to Q'. Then

$$e_{12} = \tfrac{1}{2} \left(\frac{\Delta u_1}{\Delta x_2} + \frac{\Delta u_2}{\Delta x_1} \right)$$
$$= \tfrac{1}{2} (\alpha_1 + \alpha_2)$$

where $(\alpha_1 + \alpha_2)$ is the total angular deformation. Thus e_{ij} for $i \neq j$ corresponds to pure shear, a change in shape with no volume change.

If we choose an arbitrary set of orthogonal coordinate axes, and investigate the force across a plane defined by these axes, then in general the force is at some arbitrary angle to the plane, and all 6 stress components p_{ij} are non-zero. However, for a symmetric tensor $(p_{ij} = p_{ji})$, we can always find one set of orthogonal axes such that the stress across each of the three faces defined by the axes is normal to the face, i.e. only p_{11}, p_{22} and p_{33} are non-zero. These axes are called the *principal axes* of stress. In a similar way, we can find a set of axes, the principal axes of strain, such that only e_{11}, e_{22} and e_{33} are non-zero.

The next step is to develop relationships between strain and stress. A perfectly elastic solid is one which obeys a generalised form of Hooke's Law: the components of strain are linear functions of the components of stress. It simplifies the analysis greatly if we can assume that the solid is isotropic, i.e. its properties are independent of the orientation of the coordinate axes. This is, however, only an approximation for many rocks. For an isotropic solid, the principal axes of stress and strain are coincident, and adopting this set of axes allows us to proceed as follows.

Consider a bar of material along the x_1 direction subjected to a tensional force F_1 also in the x_1 direction. Then the only non-zero stress component is p_{11}. The bar will be extended in the x_1 direction but will contract in the x_2 and x_3 directions by equal amounts. We define Young's modulus of elasticity E and Poisson's ratio σ by the relations

$$e_{11} = p_{11}/E$$

and $e_{22} = e_{33} = -\sigma p_{11}/E$ (the negative sign to indicate contraction.)

If we now introduce additional stresses p_{22} and p_{33}, then the stress–strain relations become:

$$e_{11} = p_{11}/E - \sigma p_{22}/E - \sigma p_{33}/E$$

$$e_{22} = -\sigma p_{11}/E + p_{22}/E - \sigma p_{33}/E$$

$$e_{33} = -\sigma p_{11}/E - \sigma p_{22}/E + p_{33}/E.$$

For later use, we shall need to find stress in terms of strain, which can be done as follows. Adding the above equations,

$$e_{11} + e_{22} + e_{33} = \theta$$
$$= (1-2\sigma).(p_{11} + p_{22} + p_{33})/E,$$

so $p_{11} + p_{22} + p_{33} = E\theta/(1-2\sigma).$

So, if we write

$$e_{11} = -\sigma(p_{11} + p_{22} + p_{33})/E + (1 + \sigma)p_{11}/E$$

then $e_{11} = -\dfrac{\sigma}{E} \cdot \dfrac{E\theta}{(1 - 2\sigma)} + \dfrac{(1 + \sigma)}{E} p_{11}.$

Therefore,

$$p_{11} = \frac{E\, e_{11}}{1 + \sigma} + \frac{\sigma E\theta}{(1 - 2\sigma)(1 + \sigma)}$$

with similar relations for p_{22} and p_{33}. This equation is usually written

$$p_{11} = \lambda\theta + 2\mu e_{11}$$

where λ and μ are called Lamé's elastic constants.

Note that, if there is pure shear, with no volume change, $\theta = 0$ so $p_{11} = 2\mu e_{11}$. Thus μ is the ratio of shearing stress to shearing strain (total angular deformation); it is called the *shear modulus* of the material. For a fluid, which cannot resist change of shape with no associated volume change, $\mu = 0$.

If a hydrostatic pressure p acts on a solid body then $p_{11} = p_{22} = p_{33} = -p$,

so
$$p_{11} = -p = \lambda\theta + 2\mu e_{11}$$
$$p_{22} = -p = \lambda\theta + 2\mu e_{22}$$
$$p_{33} = -p = \lambda\theta + 2\mu e_{33}$$

Adding, $-3p = 3\lambda\theta + 2\mu\theta$

so
$$-\frac{dp}{d\theta} = \lambda + \frac{2\mu}{3} = k$$

where k is called the bulk modulus of elasticity.

2.1.2 Seismic waves

The nature of the seismic wave can be illustrated by considering a plane compressional wave propagating in the x_1 direction (figure 2/5). One face of the small cube is initially in

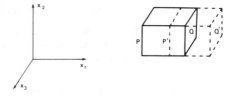

Figure 2/5 Plane compressional wave propagation.

the plane of P (x_1, x_2, x_3), the opposite face initially in the plane of Q $(x_1 + \Delta x_1, x_2, x_3)$. As a result of the passage of the wave, P and Q are displaced. At a particular instant, let P be displaced to P' $(x_1 + u_1, x_2, x_3)$, and Q be displaced to Q' $(x_1 + u_1 + \Delta x_1 + \Delta u_1, x_2, x_3)$. Then the cube is subject to an elastic strain given by

$$e_{11} = \underset{\Delta x_1 \to 0}{\mathrm{Lt}}\ \frac{\Delta u_1}{\Delta x_1} = \frac{\partial u_1}{\partial x_1}$$

The net force on the cube associated with this strain is given by

$$F_1 = \{p_{11}(Q) - p_{11}(P)\}\Delta x_2 \Delta x_3$$

$$= \frac{\partial p_{11}}{\partial x_1}\Delta x_1 \Delta x_2 \Delta x_3$$

and so the equation of motion of the cube is

$$\rho\,\Delta x_1 \Delta x_2 \Delta x_3\,\frac{\partial^2 u_1}{\partial t^2} = \frac{\partial p_{11}}{\partial x_1}\,\Delta x_1 \Delta x_2 \Delta x_3$$

where ρ is the density of the material. To proceed further, we need to consider the boundary conditions. We are interested in a wave propagating through the solid earth. In this case, lateral contraction of the earth perpendicular to the wave propagation direction will not be permitted, so $e_{22} = e_{33} = 0$. Thus, since $\theta = e_{11}$ in this case, the stress–strain relation will be

$$p_{11} = (\lambda + 2\mu)e_{11}$$

$$= (\lambda + 2\mu)\frac{\partial u_1}{\partial x_1}.$$

The equation of motion therefore becomes

$$\rho\,\frac{\partial^2 u_1}{\partial t^2} = (\lambda + 2\mu)\frac{\partial^2 u_1}{\partial x_1^{\,2}}.$$

This is the one-dimensional wave equation, whose general solution is

$$u_1 = f(x_1 - \alpha t) + g(x_1 + \alpha t)$$

where f and g are arbitrary functions and $\alpha = \left(\dfrac{\lambda + 2\mu}{\rho}\right)^{\frac{1}{2}}$. The solutions represent disturbances propagating, without change of form, in the positive and negative x_1 directions, respectively, with a propagation velocity α.

In a similar way it can be shown that elastic waves can exist in which the deformation is pure shear. The wave velocity is then $\beta = (\mu/\rho)^{\frac{1}{2}}$. The ratio of α and β is dependent only on Poisson's ratio, for

$$\frac{\alpha^2}{\beta^2} = \frac{\lambda + 2\mu}{\mu} = \frac{\lambda}{\mu} + 2 = \frac{2\,\sigma}{1-2\sigma} + 2 = \frac{2(1-\sigma)}{1-2\sigma}$$

The compressional waves are called P waves and the shear waves are called S waves. In P waves, particle motion is parallel to the direction of wave propagation. In S waves the particle motion is perpendicular to the direction

of propagation, so that these waves may have different polarisations. When the particle motion is parallel to the earth's surface they are referred to as SH waves, and when the particle motion is in a vertical plane, as SV waves (see figure 3/30).

Although there is increasing interest in the use of S waves in seismic prospecting, at present P waves are used almost exclusively. The usual seismic sources generate mainly P waves; in marine work, the source is surrounded by water, in which S waves cannot propagate. S waves can be generated when P waves strike an interface at an inclined angle, but the effectiveness of this conversion is small in most prospecting geometries.

In addition to these waves that propagate through the body of the earth, there exist waves which are confined to the vicinity of the surface. These surface waves are of two types, Rayleigh waves and Love waves. Because they carry information only about the near-surface, they are of little or no use in reflection prospecting, but are often a nuisance because they can have large amplitudes and can swamp the desired reflection signals. This 'ground-roll' has to be removed by appropriate field source and array parameter selection.

Although at large distances from the source, the plane waves discussed above are an adequate approximation for most purposes, it is of some interest to consider the spherical waves generated by a point source. It can be shown that in this case, the particle displacement in the radial direction, u_r, at distance r from the source is given by

$$u_r = \frac{\partial}{\partial r}\left[\frac{1}{r}f(r - \alpha t)\right]$$

where f is an arbitrary function. These waves are of P type; spherically symmetric S waves are, of course, geometrically impossible. Thus,

$$u_r = \frac{f'\,(r - \alpha t)}{r} - \frac{f(r - \alpha t)}{r^2}$$

This has an important consequence that the waveform varies with distance from the source. Close to the source, the $f(r - \alpha t)$ term dominates, and frequencies will be lower than at greater distances where the $f'(r - \alpha t)$ term dominates u_r. The two terms are equal in importance at a distance from the source about equal to the seismic wavelength. The implication is that the waveform of the disturbance measured close to the source will not be the same as the waveform involved in deep reflections; however, it can be shown that the pressure waveform will be the same at all distances from the source large enough for linear elastic theory to hold.

2.2 Seismic velocities

The extension of the simple theory discussed above to a seismic wave in a real rock is not simple. The wave is then propagating in a solid formed from grains of different minerals, with pore spaces between the grains which may be filled with liquid (brine or oil) or gas.

Let us look first at the experimental evidence. We can start by considering the velocities in various minerals individually. Mineral crystals exhibit anisotropy; the elastic moduli, and therefore seismic velocities, are dependent on the orientation of the crystal lattice relative to the direction of seismic wave propagation. Average values approximating measurements on polycrystalline bodies can be calculated from a compilation by Anderson and Lieberman (1966):

Mineral	Density, g/cc	Velocities, ft/s	
		P	S
Pyrite	4.93	26610	16990
Magnetite	5.20	24220	13750
Haematite	5.12	21800	14120
Quartz (β)	2.53	21770	13240
Quartz (∝)	2.65	19870	13620
Calcite	2.71	20530	10640
Barite	4.50	14060	7300
Halite	2.16	14850	8580

Logically, the next step would be to look at velocities in non-porous sedimentary rocks. These are difficult to establish, because almost all sedimentary rocks are porous to some extent. Values estimated from the work of Anderson and Lieberman are as follows:

Rock type	Density, g/cc	Velocities, ft/s	
		P	S
Dolomite	2.84	23000	13000
Limestone	2.73	21000	11000
Sandstone	2.65	18500	11500

Real rocks will be porous, and the pore space will be filled by fluids of much lower seismic velocity:

Liquid	Density, g/cc	P wave velocity, ft/s
Water (20% NaCl)	1.14	5290
Water (fresh)	1.00	4590
Oil	0.80–0.85	4200

It is not surprising that real rocks show a wide variation of seismic velocity. Figure 2/6 is a histogram of measurements on various classes of rocks, after Grant and West (1965). It is clear that, in general, the range of seismic velocity associated with a given lithology is too large for lithology to be inferred from velocity. However, it may be possible to predict, for example, sand–shale ratio in a specific environment, where borehole control is sufficient to allow one to study the variation of velocity in a restricted area, stratigraphic interval, and depth of burial.

A cross-plot of velocity and density shows some consistency between most rock types (figure 2/7, after Gardner et al., 1974). An empirical fit to the data is

$$\rho = 0.23 \, \alpha^{\frac{1}{4}}$$

where ρ is density in g/cc and α is P wave velocity in ft/s.

Various empirical relations between velocity, geological age and burial depth have been reported, such as that of Faust (1951):

$$\alpha = 125.3 \, (ZT)^{\frac{1}{6}}$$

where α is P wave velocity in ft/s, Z is depth in ft, and T is geological age in years.

From both the theoretical and practical points of view, an important problem is to describe how seismic velocity varies with porosity and with the nature of the pore-fill (liquid or gas). A simple empirical equation has been proposed by Wyllie et al. (1956):

$$\frac{1}{\alpha} = \frac{\phi}{\alpha_w} + \frac{1-\phi}{\alpha_m}$$

where ϕ = porosity, α_w = P wave velocity in water, and α_m = P wave velocity in solid matrix material. This 'time-average' equation is a reasonable approximation for water-saturated

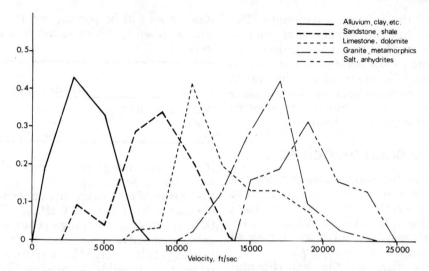

Figure 2/6 Histogram of seismic wave velocities of various classes of rocks (after Grant and West, 1965).

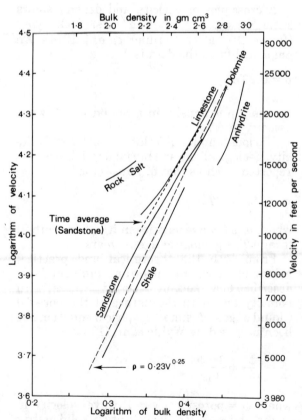

Figure 2/7 Velocity–density relationships in rocks of different lithology (after Gardner *et al*., 1974).

sands at depths of burial greater than about 6000 ft.

A simple theoretical approach to unconsolidated sediments is provided by the Wood equation (Wood, 1949). This equation is strictly applicable to a suspension of solid particles in a continuous liquid phase. In this case,

$$\alpha = \sqrt{\left(\frac{\lambda + 2\mu}{\rho}\right)} = \sqrt{\left(\frac{k + 4\mu/3}{\rho}\right)} = \sqrt{(k/\rho)}$$

since the bulk modulus $k = \lambda + 2\mu/3$ and $\mu = 0$ for a continuous liquid phase. k for the mixture is readily found because an incremental change ΔP in pressure will cause a fractional volume change $\Delta V/V$ in each constituent:

$$\frac{\Delta V_{\text{fluid}}}{V_{\text{fluid}}} = \frac{-\Delta P}{k_{\text{fluid}}} \text{ and } \frac{\Delta V_{\text{matrix}}}{V_{\text{matrix}}} = \frac{-\Delta P}{k_{\text{matrix}}}$$

so

$$\frac{\Delta V_{\text{total}}}{V_{\text{total}}} = \frac{-\Delta P}{k_{\text{mixture}}}$$

$$\frac{-\Delta P}{k_{\text{fluid}}} \cdot \frac{V_{\text{fluid}}}{V_{\text{total}}} - \frac{\Delta P}{k_{\text{matrix}}} \cdot \frac{V_{\text{matrix}}}{V_{\text{total}}}$$

Thus

$$\frac{1}{k_{\text{mixture}}} = \frac{\phi}{k_{\text{fluid}}} + \frac{1 - \phi}{k_{\text{matrix}}}$$

Since

$$\rho_{\text{mixture}} = \phi\rho_{\text{fluid}} + (1 - \phi)\rho_{\text{matrix}},$$

α can be calculated by $\alpha = \left(\dfrac{k_{\text{mixture}}}{\rho_{\text{mixture}}}\right)$.

This equation correctly predicts that very unconsolidated sediments such as those found immediately below the sea floor may have velocities less than that in sea water. However, the equation is not useful for consolidated sediments, where μ of the granular skeleton is not negligible and velocities will be higher than the Wood equation predicts.

A more generally useful relationship is that given by Gassman (1951):

$$\alpha = \frac{1}{\rho^{\frac{1}{2}}}\left[\left(k_b + \frac{4\mu_b}{3}\right) + \frac{(1-k_b/k_s)^2}{(1-\phi-k_b/k_s)/k_s+\phi/k_f}\right]^{\frac{1}{2}}$$

where k_s = bulk modulus of matrix material
          k_b = bulk modulus of empty reservoir
          μ_b = shear modulus of reservoir bulk
              material
and k_f = bulk modulus of fluid.

This equation holds in the low-frequency limit, when fluid-matrix coupling is perfect; this is a reasonable assumption for frequencies used in seismic prospecting. The equation is not easy to use to calculate velocites *ab initio*, for it is not clear how to estimate k_b. However, it is well suited to calculating the effect on α of a change in pore-fill from liquid to gas. Intuitively, one might expect a small amount of gas to produce a large drop in k_f and thus a sharp decrease in velocity. Curves calculated by Domenico (1974) show just such an effect (figure 2/8). As we shall see, the change in seismic velocity can cause a marked change in amplitude of the seismic reflection on passing laterally from a gas-filled to a brine- or oil-filled sandstone. However, these amplitude anomalies allow an inference only of the presence of gas, and not of its quantity; the curves show that the largest velocity effects can be produced by small (and possibly sub-commercial) gas saturations.

By comparison, the effect of water saturation on shear wave velocity is small. The fluid, whether water or gas, cannot support shear wave motion ($\mu = 0$). Therefore the effect on shear wave velocity is essentially through the bulk density. The decrease of bulk density with decreasing water saturation leads to an increase in S wave velocity. For the sandstone of figure 2/6 the difference between the velocities at $S_w = 0$ and $S_w = 1$ is calculated to be only 11% at a depth of 2000 ft.

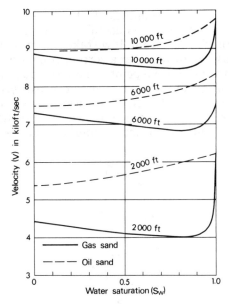

Figure 2/8 Longitudinal velocity V as a function of water saturation S_w for gas and oil sands at depths of 2000, 6000 and 10 000 ft (adapted from S.N. Domenico, *Geophysics*, **39** (1974), p. 764).

2.3 Elementary wave propagation theory

In Section 2.1 we derived an expression for the displacement in a one-dimensional wave at time t:

$$u = f(x_1 - \alpha t).$$

The wave is here propagating in the x_1 direction. Figure 2/9 shows u_1 as a function of x_1 at two different times. If we concentrate on any particular point of the displacement graph, say the peak which is at A at time t_1, we see it move steadily to the right as time increases, reaching A' at time t_2.

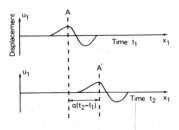

Figure 2/9 Propagation of a one-dimensional wave.

In a more complicated, two- or three-dimensional case, it is useful to introduce the concept of a wavefront. If we take a snapshot of the wave disturbance at a certain time, we can join together all the points where the displacement is about to start to define a *wavefront* (figure 2/10). Given the position of the wavefront at a certain time, we can calculate the position at a later time by means of

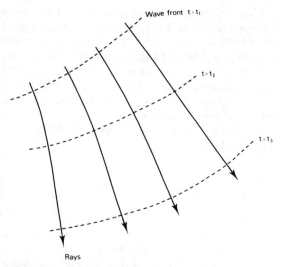

Figure 2/12 Schematic relationship between rays and wavefronts.

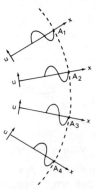

Figure 2/10 Wavefront joining points A₁–A₄ where disturbance is about to start.

Huygens' principle (figure 2/11): each point on the wavefront is regarded as a new source point, from which a spherical wavefront diverges. The new wavefront is the envelope of these subsidiary wavefronts.

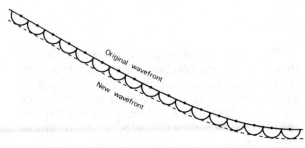

Figure 2/11 Wavefront propagation according to Huygens' principle.

In a homogeneous medium, the wavefront from a point seismic source will be spherical. As the wavefront expands, the energy of the disturbance is spread over a larger and larger area, so the amplitude of the seismic wave decreases. As we saw in Section 2.1, the amplitude is in fact proportional to the reciprocal of the distance from the wavefront to the source. This effect is called spherical divergence, and is the main reason why reflections from deep interfaces tend to have small amplitudes.

For many purposes, it is easier to describe the geometry of wavefront propagation by considering the behaviour of *rays*. These are perpendiculars to the instantaneous wavefront (figure 2/12). The ray is the path along which energy travels. In the case of a plane wave, the wavefront is planar and the rays are parallel straight lines.

When a wave strikes a boundary between different media, part of the energy is reflected and part is transmitted into the second medium. Let us, for simplicity, consider a plane wave. The geometry of the ray paths is governed by Snell's Laws (figure 2/13). For the reflected wave, $A'B' = AB = \alpha_1 t$, so triangles $AA'B'$ and $B'BA$ are congruent, so $\phi_1 = \phi_2$; the angle of incidence equals the angle of reflection. For the refracted wave, $A'D' = \alpha_1 t$ and $AB = \alpha_2 t$, so $AB' = \alpha_1 t/\sin \phi_1 = \alpha_2 t/\sin \phi_3$. Thus we arrive at the usual form of Snell's Laws in Figure 2/14.

Note that if $\alpha_2 > \alpha_1$, then for some particular value of $\phi_1, \phi_3 = 90°$ and the refracted ray travels along the interface between the two media. This value of ϕ_1 is called the critical angle. For angles of incidence greater than the critical angle, there is no transmitted wave and all the energy is reflected. The wave which travels along the interface when the incident

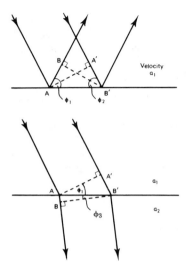

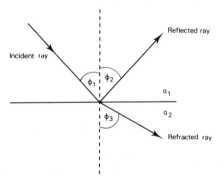

Figure 2/13 Snell's Laws. AA′ and BB′ are wavefronts time *t* apart. On the left, the reflected wavefront; on the right, the refracted wavefront.

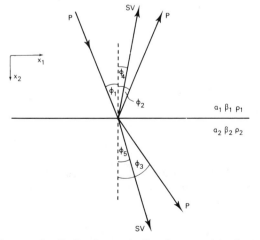

Figure 2/14 Snell's Laws: $\phi_1 = \phi_2$ and

$$\frac{\sin \phi_1}{\alpha_1} = \frac{\sin \phi_3}{\alpha_3}$$

ray is at the critical angle is called the head wave. The precise theory of this wave is not simple; for an incident plane wave, the amplitude of the head wave is theoretically zero, but a finite-amplitude head wave is generated when the incident wavefront is curved. The head wave travelling along the interface continuously feeds energy back into the upper medium; the wave in the upper medium behaves as a plane wave leaving the interface at the critical angle. A signal from the surface source may therefore travel as a head wave along a subsurface interface, and be received at the surface again. This is possible whenever there is an increase in velocity downwards at the interface concerned. The

seismic refraction exploration method makes use of these head waves to investigate subsurface structure (see Chapter 8).

So far we have considered only the geometry of reflection; we shall now investigate the relative amplitudes of the reflected and transmitted waves.

2.4 Reflection of seismic waves

When a plane SH wave strikes a boundary between two different media, reflected and refracted SH waves are generated. The case of most interest in seismic prospecting, however, is when a plane P wave strikes a boundary. This is more complicated, because in general this results in both P and SV waves, both reflected and transmitted (figure 2/15). The geometry is governed by a generalisation of Snell's Laws:

$$\phi_2 = \phi_1$$
$$\alpha_1/\sin \phi_1 = \beta_1/\sin \phi_4 = \alpha_2/\sin \phi_3 = \beta_2/\sin \phi_5.$$

To calculate the relative amplitudes of the various waves, we need to consider the conditions that must be satisfied at the boundary. These are:

(i) continuity of displacement (no slip at boundary). Thus u_1 and u_2 must be continuous across the boundary.
(ii) continuity of stress (by Newton's Third Law). Thus p_{21} and p_{22} must be continuous across the boundary.

Unfortunately, the detailed calculation for the

Figure 2/15 Reflection and refraction at an interface. α and β are P and S wave velocities, ρ is density.

general case is rather laborious; it may be found, for example, in Ewing *et al.* (1957). In many cases of prospecting interest, the incident P wave is almost normal to the surface ($\phi_1 = 0$). At normal incidence, the calculations are much simplified, because in the incident P wave $p_{21} = 0$ so no SV waves can be generated. Thus we shall have only reflected and transmitted P waves, and the boundary conditions become:

(i) continuity of u_2, and
(ii) continuity of p_{22}.

It is convenient to consider the incident wave to be a sine wave. This is not a particularly restrictive assumption, for we shall see in Chapter 4 that the methods of Fourier analysis allow us to describe any physically realisable waveform as the sum of a series of sine waves of different frequencies. We therefore consider an incident wave of the form:

$$u_2 = A \sin 2\pi(x_2/\lambda_1 - \nu t).$$

This is a sine wave of amplitude A, wavelength λ_1 and frequency ν; wavelength and frequency are related by $\alpha = \lambda\nu$ (figure 2/16). To

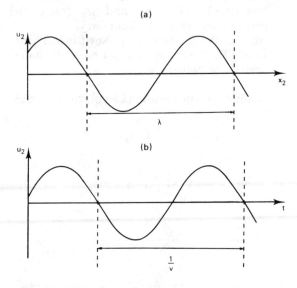

(a)

(b)

Figure 2/16 Sine waves: (a) as a function of x_2 at a particular time and (b) as a function of t at a particular value of x_2.

satisfy the conditions at the boundary, which we shall take at $x_2 = 0$, it is clear that the time

behaviour of all the waves at the boundary must be the same. We therefore have reflected and transmitted waves of the same frequency ν:

Reflected wave $u_2 = B \sin 2\pi(-x_2/\lambda_1 - \nu t)$

Transmitted wave $u_2 = C \sin 2\pi(x_2/\lambda_2 - \nu t)$.

The frequency of the transmitted wave is the same as that of the incident wave, so its wavelength must be different. The first boundary condition then gives, at $x_2 = 0$:

$$A \sin 2\pi(-\nu t) + B \sin 2\pi(-\nu t)$$
$$= C \sin 2\pi(-\nu t)$$
$$\therefore \quad A + B = C \tag{i}$$

We recall from Section 2.1 that $p_{22} = \rho\alpha^2 \dfrac{\partial u_2}{\partial x_2}$, so the second boundary condition gives

$$\frac{\rho_1\alpha_1{}^2}{\lambda_1} \cdot A \cos 2\pi(-\nu t) - \frac{\rho_1\alpha_1{}^2}{\lambda_1} \cdot B \cos 2\pi(-\nu t)$$
$$= \frac{\rho_2\alpha_2{}^2}{\lambda_2} \cdot C \cos 2\pi(-\nu t)$$
$$\therefore \quad \frac{\rho_1\alpha_1{}^2}{\lambda_1} A - \frac{\rho_1\alpha_1{}^2}{\lambda_1} B = \frac{\rho_2\alpha_2{}^2}{\lambda_2} C \tag{ii}$$

Substituting (i) in (ii):

$$\frac{\rho_1\alpha_1{}^2}{\lambda_1} A - \frac{\rho_1\alpha_1{}^2}{\lambda_1} B = \frac{\rho_2\alpha_2{}^2}{\lambda_2} (A + B)$$

$$\text{so } A \left\{ \frac{\rho_1\alpha_1{}^2}{\lambda_1} - \frac{\rho_2\alpha_2{}^2}{\lambda_2} \right\} = B \left\{ \frac{\rho_2\alpha_2{}^2}{\lambda_2} + \frac{\rho_1\alpha_1{}^2}{\lambda_1} \right\}$$

Recalling that $\lambda = \alpha/\nu$,

$$A (\rho_1\alpha_1 - \rho_2\alpha_2) = B (\rho_2\alpha_2 + \rho_1\alpha_1).$$

We may define a reflection coefficient R by

$$R = -\frac{B}{A} = \frac{\rho_2\alpha_2 - \rho_1\alpha_1}{\rho_2\alpha_2 + \rho_1\alpha_1}$$

and the amplitude of the transmitted wave is given by $C = A(1 - R)$. $(1 - R)$ is generally called the transmission coefficient.

If R is positive, then the displacement in the reflected wave is in the opposite direction to that of the incident wave. Thus, if the signal was originally a compression (displacement in the direction of the wave's progress), it will be reflected as a compression, because the reflected wave is travelling in the opposite direction to the incident wave. If R is negative, a compression is reflected as a dilatation; we say that the negative reflection coefficient causes a 180° phase change in the reflected signal.

The quantity $\rho\alpha$ is called the *acoustic impedance* of a medium; the amplitude and sign of the reflection coefficient is thus dependent on the acoustic impedance contrast across an interface. The negative reflection coefficient arises when the incident ray is in the medium of higher acoustic impedance.

Typical interfaces giving rise to reflections on seismic records may have reflection coefficients of the order of 0.1. In marine work, it is important that the sea surface is an almost perfect reflector of seismic waves (incident from below the surface). The seabed is also often a strong reflector (perhaps $R = 0.3$), so the water-layer is apt to be a strong source of reverberation.

Eventually, the transmitted signal will be reflected at a deeper interface and will be incident on the same intermediate interface again, this time from below. The transmission coefficient on this upward journey will be $(1 + R)$, for the reflection coefficient will have the opposite sign when the interface is crossed in the reverse direction. Thus the two-way transmission coefficient at the interface will be $(1 - R)(1 + R) = 1 - R^2$. Even for a strong reflector, this two-way coefficient will be little different from 1 (e.g. 0.99 for $R = 0.1$). Thus single interfaces, even where there is a large impedance contrast, have little effect on the transmission of the signal to deeper levels. However, a large number of interfaces will have a cumulative effect in decreasing the signal from deeper reflections. For a practical analysis of reflection coefficients, see Chapter 13, the Hewett Gas Field Case History.

2.5 The earth as a low-pass filter

In general, the frequency of reflected signals decreases with increasing TWT. Two causes of this effect are absorption and the effect of short-period multiples.

Absorption is essentially a loss of energy by conversion into heat. We can picture this happening on the microscopic scale: particle motion during passage of the seismic wave causes grains to slide against one another, and friction between the grains causes energy loss. In a given rock, there is generally a constant fractional energy loss per cycle of the seismic wave. This can conveniently be expressed in decibels (dB) per wavelength. The decibel is defined by the relation that a difference of N dB between two signals of amplitudes a and b implies that $N = 20 \log_{10} a/b$; thus, for example, 6dB corresponds to a factor 2 in amplitude, 20 dB to a factor of 10, 40 dB to a factor of 100. There is considerable variation in attenuation between different rock types, in the range 0.1 to 3 dB per wavelength.

Since there is a constant fractional loss per wavelength (see figure 3/12), higher frequencies are attenuated more than lower ones over a given path. Consider, for example, a path 2 km long in a rock of velocity 3 km/s and absorption 0.5 dB per wavelength. Then different frequencies are affected as follows:

Frequency Hz	Wavelength m	No. of wavelengths in path	Loss dB
10	300	6.7	3
20	150	13.3	7
40	75	26.7	13
80	37.5	53.3	27

It is clear that the higher frequencies will be quite strongly absorbed in traversing a path typical of reflection prospecting, leaving a signal whose dominant frequencies are a few tens of Hz.

Short-period multiples also cause a decrease of frequency with travel-time (O'Doherty and Anstey, 1971). Briefly, part of the seismic energy is delayed on its path by reverberation between closely-spaced reflecting interfaces. Thus an initially sharp pulse (even if we could produce one) will be smeared out in its passage through the earth, in effect leading to a slow decrease of frequency with depth.

2.6 Geometry of reflection prospecting

Usually, we place our source of seismic waves at or near the earth's surface. Seismic waves travel down into the earth and are reflected back from interfaces to a receiver, also at or near the surface. The simplest case is that of a plane horizontal reflector overlain by a medium of constant velocity (figure 2/17). The source–receiver distance x is called the offset. Clearly, the travel time t will be

$$t = \frac{2\sqrt{[h^2 + (x/2)^2]}}{v}.$$

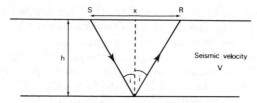

Figure 2/17 Geometry of reflection.

At normal incidence (zero offset), the travel time would be

$$t_0 = 2h/v.$$

This zero-offset travel time is often called simply the 'two-way time' (TWT) to the reflector. The difference between t_0 and the travel time at finite offset is called the normal moveout (NMO).

In this simple case,

$$t^2 = t_0{}^2 + x^2/v^2.$$

There is thus a hyperbolic relationship between travel-time and offset. A case more closely resembling the real earth is a 'layer-cake' of n horizontal layers of different thicknesses and velocities. For a reflector at the base of such a sequence,

$$t^2 = t_0{}^2 + x^2/v_{\text{rms}}^2 + c_1 x^4 + c_2 x^6 + \ldots$$

where the c_i are functions of the thicknesses and velocities of the n layers (and are generally small). v_{rms} is the root-mean-square velocity along the zero-offset trajectory, defined by

$$v_{\text{rms}}^2 = \sum_{k=1}^{n} v_k^2 t_k / t_0$$

where v_k and t_k are the velocity and TWT in the kth layer (Taner and Koehler, 1969). In the analysis of seismic data, it is usual to approximate the above series by

$$t^2 = t_0'^2 + x^2/v'^2$$

where t_0' will be slightly different from t_0 and v' is a 'stacking velocity' slightly different from v_{rms}.

If we consider the case of a dipping reflector beneath an overburden of uniform velocity v, then it can be shown (Levin, 1971) that

$$t^2 = t_0{}^2 + x^2 \cos^2\alpha/v^2,$$

where α is the component of dip along the profile. Thus the NMO has the same form as in the horizontal layer case, but the apparent velocity is $v/\cos\alpha$.

The simplest possible method of reflection shooting would be to collect a number of zero-offset records along a profile. This is essentially the technique of the ship's echo-sounder, and may be applied in practice where only very shallow penetration is needed. However, for deep reflection work it is usual to obtain reflections at a series of different offsets (figure 2/18). The rays involved are reflected at a common subsurface point, or *common depth point* (CDP). By adding together these records, known as a CDP gather, we can improve the signal to noise ratio of the data. This addition process is called the CDP stack. Obviously, before stacking, the records must be corrected for NMO, i.e. converted to the form they would have had at zero offset. Stacking reduces random noise, and also attenuates multiples. Multiples are reduced because the dependence of NMO on offset is different from that of primaries; thus when the records are shifted so as to line up the primary reflections across the gather, the multiples will not line up properly and so will be smeared out and (at least partly) destroyed by stacking.

Derivation of the NMO corrections to be applied prior to stack is equivalent to determining the velocity structure of the section. Knowledge of these velocities is useful for several purposes: conversion of TWT to depth, as an input to the migration process, and perhaps to infer the lithologies present.

The stacked section approximates a zero-offset section. If the layering in the earth is horizontal and planar, a seismic section thus has a similar appearance to a geological section, except for a varying vertical scale due to changes in seismic velocity with depth. However, if the geological structure is complex, the seismic section may look very different from a geological section. The problem is that the

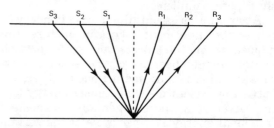

Figure 2/18 Geometry of the CDP gather.

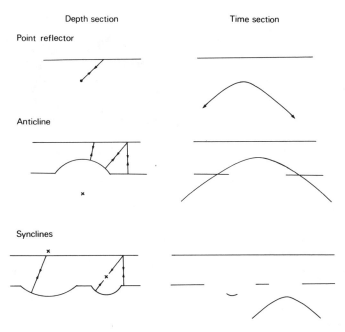

Figure 2/19 Some geometrical effects on reflections. Crosses denote centres of arcs. On the left are shown model structures and on the right the equivalent segments of reflector events as seen on time sections.

reflected raypath (at zero offset) is perpendicular to the reflector, but the reflection record is plotted vertically below the source-receiver point. Some simple examples are shown in figure 2/19.

Point reflectors give rise to hyperbolae, theoretically of indefinite extent but in practice falling eventually below the noise level. Anticlines appear broadened. Synclines can show a surprising effect; if their curvature is greater than that of the incoming wavefront, they appear inverted and resemble anticlines. Thus, regions of complex structure can produce confused seismic sections from which it will be difficult to deduce the geometry of the corresponding geological section. The process of migration (Chapter 4) attempts to place reflectors in their true positions on the section.

2.7 Interference

When we have closely spaced reflecting interfaces, the reflected signal from one of them may overlap in time with the reflected signal from another. We shall observe the sum of the reflected waveforms; the reflections are said to interfere with one another.

To make this idea more concrete, suppose we

have a thin bed of material enclosed in a substance of different acoustic impedance (figure 2/20). At normal incidence, the reflection coefficient will be the same except for sign at the two interfaces. If we neglect the small loss due to two-way transmission at the upper interface, we shall observe the same reflected signal from the two interfaces, except for a polarity reversal. Figure 2/21 shows how the total reflected signal changes as the thickness of the bed varies. If the TWT in the bed is long compared with the input signal length, the top and bottom reflections are well separated. As the bed thickness is reduced, the top and bottom reflections begin to merge, and it is no longer possible to resolve the top and bottom of the bed separately. If the bed is thin enough, we get constructive interference or 'thin bed tuning': a particularly large amplitude is observed where the second cycle of the reflection from the top is

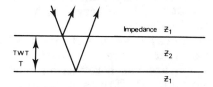

Figure 2/20 Reflection from a thin bed.

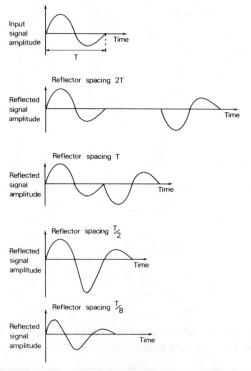

Figure 2/21 Reflected signals from beds of different thicknesses. The input signal has a pulse width of T and the beds have reflector spacings between top and bottom equivalent to seismic wave TWTs of $2T$, T, $T/2$ and $T/8$. This model depicts the case when the acoustic impedance of the reflecting bed is higher than the enclosing media and the impedances of overlying and underlying beds are equal.

added to the first cycle of the reflection from the base. Finally, when the bed is very thin, we get destructive interference, and the signal amplitude becomes small as the top and base reflections tend to cancel one another.

This effect will limit the vertical resolution we can hope to achieve. To improve our resolution, we need to have a sharper, higher frequency input signal. An ideally sharp pulse is not physically attainable, especially in view of the low-pass filter effect of the passage of the signal through the earth. Deconvolution (Chapter 4) of the seismic trace can try to convert it to the form that would be observed with an ideally sharp input pulse.

Interference has an important effect on the marine seismic source signal (figure 2/22). The sea surface has a reflection coefficient nearly equal to -1. If the source emits a sine wave of wavelength λ, then the vertically downgoing signal is the sum of two waves with a phase difference of $(\pi + 2\pi \frac{2D}{\lambda})$, where the π (180°) phase shift is due to the negative reflection coefficient at the sea surface. Therefore, we shall get constructive interference and a maximum downgoing signal when

$$\pi + \frac{4\pi D}{\lambda} = 2\pi, 4\pi, 6\pi \ldots$$

i.e. $4D/\lambda = 1, 3, 5 \ldots$

or $\lambda = 4D, 4D/3, 4D/5 \ldots$

and destructive interference and no downgoing signal when

$$\pi + \frac{4\pi D}{\lambda} = \pi, 3\pi, 5\pi \ldots$$

i.e. $4D/\lambda = 0, 2, 4 \ldots$

or $\lambda = \infty, 2D, D \ldots$

The effect is to filter the downgoing signal. Thus, for a tow depth of 10 m, the maximum signal strength would be at a wavelength of 40 m (frequency approximately 37 Hz) and there would be no signal at all at $\lambda = 20$ m (frequency approximately 75 Hz). A similar effect will occur at the receiver. These effects may limit the frequency range available in the data.

2.8 Diffraction

The simple ray geometrical model of wave propagation which we have used so far is adequate for many purposes. However, the theory is not adequate to describe what happens in the vicinity of sharp discontinuities, such as the abrupt termination of a reflector (e.g. at a fault plane). A more precise description is provided by the theory of diffraction, which predicts important limitations on the horizontal resolution we can hope to achieve in seismic data.

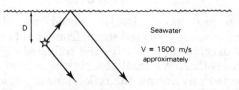

Figure 2/22 The effect of tow depth on output of a marine seismic source.

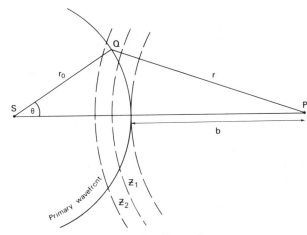

Figure 2/23 Fresnel zones.

We begin by considering a little more closely the implications of Huygens' principle. Suppose a spherical wavefront is emitted from a source S (figure 2/23); we want to evaluate the disturbance at a point P. The wavefront is shown at some intermediate position. Each element of the wavefront acts as the centre of a secondary disturbance in the form of a spherical wavelet. Suppose that the disturbance at Q is a sine wave of frequency v, wavelength λ and amplitude A/r_0 (in the far-field approximation). Then the effect at P of the secondary wavelet due to an element dS of the wavefront at Q will be

$$du = K \cdot \frac{A}{r_0} \cdot \frac{1}{r} \cdot \sin 2\pi \left(\frac{r_0+r}{\lambda} - vt + \phi\right) \cdot dS$$

where K is a factor which relates the amplitude of the secondary wavelet to that of the primary disturbance, and $2\pi\phi$ is the phase shift between secondary and primary waves. We shall assume that K is dependent on direction, being a maximum in the direction of propagation of the primary wave, and decreasing as QP becomes more nearly tangential to the wavefront.

The total effect at P can be evaluated by constructing *Fresnel zones*, as follows. With centre at P, we construct spheres of radii b, $b + \lambda/2$, $b + 2\lambda/2$, $b + 3\lambda/2$ .... These spheres divide the wavefront into a number of zones z_1, z_2, etc. Within a zone, K may be assumed constant provided r_0 and b are much greater than a wavelength. All the secondary wavelets produced by a zone will have phases close enough together to interfere constructively.

The contribution of the jth zone to the disturbance at P is

$$u_j = K_j \cdot \frac{A}{r_0} \int_s \frac{1}{r} \cdot \sin 2\pi \left(\frac{r_0+r}{\lambda} - vt + \phi\right) \cdot dS$$

where dS is the area of the primary wavefront cut off between the zone boundaries, given by

$$dS = 2\pi r_0^2 \sin\theta \, d\theta.$$

But since $r^2 = r_0^2 + (r_0 + b)^2 - 2r_0(r_0 + b) \cos\theta$

then $r \, dr = r_0(r_0 + b) \sin\theta \, d\theta$

$$\therefore dS = \frac{2\pi r_0 r \, dr}{(r_0 + b)}$$

$$\therefore u_j = K_j \frac{A}{r_0} \cdot \frac{2\pi r_0}{(r_0+b)} \times$$

$$\int_{r = b + \frac{(j-1)\lambda}{2}}^{b + j\lambda/2} \sin 2\pi \left(\frac{r_0+r}{\lambda} - vt + \phi\right) \cdot dr$$

$$= K_j A \cdot \frac{2\pi}{(r_0+b)} \cdot \frac{\lambda}{2\pi} \times$$

$$\left\{\cos 2\pi \left(\frac{r_0+b+(j-1)\lambda/2}{\lambda} - vt + \phi\right)\right.$$

$$\left. - \cos 2\pi \left(\frac{r_0+b+j\lambda/2}{\lambda} - vt + \phi\right)\right\}$$

$$= \frac{\lambda K_j A}{(r_0+b)} \cdot 2 \sin 2\pi \times$$

$$\left(\frac{r_0+b+(2j-1)\lambda/4}{\lambda} - vt + \phi\right) \sin 2\pi/4.$$

We expect the total disturbance to vary as $\sin 2\pi ((r_0 + b)/\lambda - vt)$, and this will be the case if we assume $\phi = \frac{1}{4}$ (i.e. secondary wavelets 90° different in phase from the primary wavefront). We then obtain

$$u_j = \frac{2\lambda K_j A}{(r_0+b)} \cdot \sin 2\pi \left(\frac{r_0+b}{\lambda} - vt + j/2\right)$$

$$= \frac{2\lambda K_j A}{(r_0+b)} \cdot \sin 2\pi \left(\frac{r_0+b}{\lambda} - vt\right) \cdot (-1)^{j-1}.$$

Thus the total disturbance at P will be given by

$$u = \frac{2\lambda A}{(r_0+b)} \cdot \sin 2\pi \left(\frac{r_0+b}{\lambda} - vt\right)$$

$$\times (K_1 - K_2 + K_3 - K_4 \ldots)$$

Each Fresnel zone can therefore be thought of as making a contribution to the total disturbance; alternate zones make contributions of opposite sign, and the contributions of successive zones are successively smaller, because K_j decreases as the angle SQP becomes smaller.

The most important contribution to the observed disturbance is the effect of the first Fresnel zone. When a wavefront hits a reflector, the reflected signal is related to the behaviour of the first Fresnel zone (figure 2/24). Thus, in the vertical incidence case, we obtain information not about the reflector at the point P, the reflection point of the geometrical ray, but rather an average over the whole Fresnel zone. (Note that in this case we construct zones a distance $\lambda/4$ apart rather than $\lambda/2$ to allow for the two-way-path.)

The radius f of the first Fresnel zone is given by

$$f^2 = (h + \lambda/4)^2 - h^2$$

$$= \frac{\lambda h}{2} + \frac{\lambda^2}{16} \doteqdot \frac{\lambda h}{2}.$$

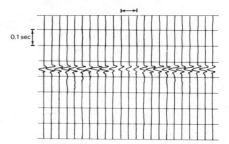

Figure 2/25 Reflection from reflector containing a hole. Reflection is observed at hole because hole is smaller than Fresnel zone. Location of hole and its dimensions are indicated by arrow (adapted from R. E. Sheriff, *AAPG Mem*., **26** (1977), p. 12).

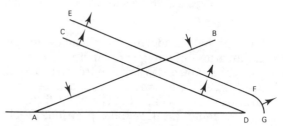

Figure 2/26 Reflection from a plane terminated at D.

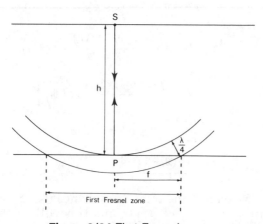

Figure 2/24 First Fresnel zone.

Thus, for example, for a reflector at a depth of 2000 m, with a seismic frequency of 30 Hz and seismic velocity 3000 m/s, the wavelength is 100 m and $f = 316$ m.

This effect limits the horizontal resolution we can expect from the reflection seismic method. In broad terms, we can perceive only the average structure over an area the size of the first Fresnel zone. Thus, small-scale changes in structure will not be observable (figure 2/25).

A related diffraction effect is observed at discontinuities of a reflector, for example where a reflector is sharply terminated by a fault (figure 2/26).

The incident wavefront AB is reflected as wavefront CD and, later, as wavefront EF, but Huygens' principle implies that there is also a circular wavefront FG with centre at D. This scattered energy, which appears to come from the edge of the reflector, is called the diffracted wave. Detailed calculation of the effect is laborious; the important result is that the edge behaves like a line reflector, but with several important differences (Trorey, 1970) (figure 2/27).

The polarity of the diffracted wave changes in going from one side of the edge to the other. The amplitude decays faster with increasing travel path than would be expected for a line reflector. Finally, the waveform is different from the incident waveform and becomes richer in low frequencies at greater distances from the diffracting edge.

Termination of a reflector by a fault thus gives rise to characteristic quasi-hyperbolic diffraction patterns on a seismic section (figure 2/28).

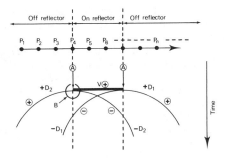

Figure 2/27 Illustration that a diffraction must undergo a 180° phase change on either side of a diffracting edge. P_1, P_2 etc. are successive source receiver locations along a line. The diffractor is a phase strip with edges at A–A′. D_1 and D_2 are diffractions and V is the reflection. The circled signs show the phase of the various parts of the response (adapted from A. W. Trorey, *Geophysics*, **35** (1970), p. 770).

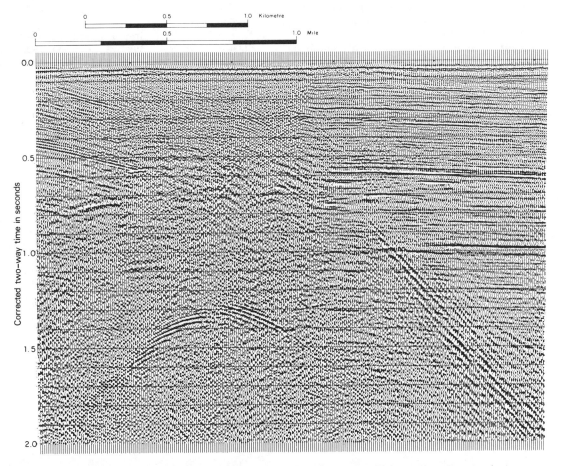

Figure 2/28 Hyperbolic diffraction patterns caused by faulting (courtesy: IGS, Horizon Geophysical record).

References

O. L. Anderson and R. C. Lieberman, Sound velocities in rocks and minerals: *VESIAC state-of-the-art Report: 7885-4-X*, Univ. of Mich., Ann. Arbor. (1966).

S. N. Domenico, Effect of water saturation on seismic reflectivity of sand reservoirs encased in shale. *Geophysics*, **39**, (1974), pp. 759–769.

W. M. Ewing, W. S. Jardetsky and F. Press, *Elastic waves in layered media*. McGraw Hill, New York (1957).

L. T. Faust, Seismic velocity as a function of depth and geologic time. *Geophysics*, **16**, (1951), pp. 192–206.

G. H. F. Gardner, L. W. Gardner and A. R. Gregory, Formation velocity and density – the diagnostic basis for stratigraphic traps. *Geophysics*, **39**, (1974), pp. 770–780.

F. Gassman, Elastic waves through a packing of spheres. *Geophysics*, **16**, (1951), pp. 673–685.

F. S. Grant and G. F. West, *Interpretation theory in applied geophysics*. McGraw Hill, New York (1965).

F. K. Levin, Apparent velocity from dipping interface reflections. *Geophysics*, **36**, (1971), pp. 510–516.

R. F. O'Doherty and N. A. Anstey, Reflections on amplitudes. *Geophysical Prospecting*, **19**, (1971), pp. 430–458.

R. E. Sheriff, Limitations on resolution of seismic reflections and geological detail derivable from them. *AAPG Mem.*, **26** (1977), pp. 3–14.

M. F. Taner and F. Koehler, Velocity spectra – digital computer derivation and application of velocity functions. *Geophysics*, **34** (1969), pp. 859–881.

A. W. Trorey, A simple theory for seismic diffractions. *Geophysics*, **35** (1970), pp. 762–784.

A. B. Wood, *A textbook of sound*. G. Bell and Sons, London (1949).

M. R. J. Wyllie, A. R. Gregory and L. W. Gardner, Elastic wave velocities in heterogeneous and porous media. *Geophysics*, **21** (1956), pp. 41–70.

Chapter 3

Data Acquisition

In this chapter we consider the application of seismic theory, as described in Chapter 2, to the design of equipment for generating seismic pulses, and detecting and recording the earth's response to the passage of seismic waves through it.

Exploration seismology is a remote-sensing technique in which the aim is to record as detailed a picture as possible of subsurface geology. The product of a seismic investigation is a geological model which can be described as the sum of a finite series of layers of varying thickness, physical properties (density and seismic velocity) and structural attitude. Interpretation of this model is in terms of geological structure, lithological variation, stratigraphy and, in oil exploration, hydrocarbon prospectivity.

Seismic data are acquired using a system consisting of three main components: an input source, an array of detectors and a recording instrument. The input source is designed to generate a pulse of sound which meets, as near as possible, certain predefined requirements of total energy, duration, frequency content, maximum amplitude and phase. Reflected and refracted seismic pulses (the output from the earth) are detected by an array of geophones or a hydrophone array, then recorded by a recording instrument, and in both cases these output signals will be modified by the response characteristics of that part of the system. Each

seismic record is thus a time record of the output signals which are generated at interfaces in a series of stratigraphic layers because of the changes in acoustic impedance which occur at such boundaries, modified firstly by transmission decay and noise interference in the earth and then by detector and recorder response characteristics. This can be summarised as follows:

Recorded signal = Source pulse * [Reflectivity * (Earth filter + Noise)] * Detector response
* Recording instrument response

where * represents convolution (see p. 72).

Assuming that we know the signal characteristics of the seismic pulse and the response characteristics of detector and recording instrument, then we can separate that part of the function contained in square brackets, and this is the earth's impulse response. The earth's reflectivity is what we wish to measure. The earth's filter is a variable function of absorption and attenuation which can be compensated for in data processing. Noise cannot be so adequately treated by data processing and, as far as possible, must be measured and compensated for during data acquisition. This is mainly achieved by layout design; on land by proper design of geophone spreads and arrays and at sea by use of well designed hydrophone arrays. Display of the recorded signal will not, in most situations, give an easily interpretable picture of geological structure. This record needs further processing to achieve such clarification, and these processing techniques are the subject of the following chapter.

3.1 Recording spread design

The quantity and quality of the primary seismic signals recorded in the field are affected by many different types of background noise which can be conveniently described in two main categories: random and coherent. Much depends on field analysis, the ultimate objective of which is concise layout design of source and recording spread parameters.

3.1.1 Spatially random noise

In the recording of a single seismic trace, the weaker the source energy is, not only will the depth of penetration be limited, but so also will the strength of the primary signals be limited.

The use of multiple sources, multiple detectors per trace and the summing of common reflection point traces (see figures 3/1 and 4/5) brings a distinct improvement in the signal to noise ratio of even the weakest source energy. Indeed, in the period preceding the inception of multi-fold coverage, adjacent traces were often successfully summed or mixed in various proportions. The improvement is proportional to $\sqrt{n}$ where n is the number of detecting elements in the acquisition system, the signals from which are added together to provide the final record. For example, the summing of eight separate seismic signals (eight shots at same shot-point location), detected by geophone spreads of twenty geophones per trace then subject, during processing, to twenty-four fold stacking, will provide an improvement ratio of $\sqrt{(8 \times 20 \times 24)} = 62$ or 36dB. This may be compared with a single shot record, single-fold processing with, as before, 20 geophones per

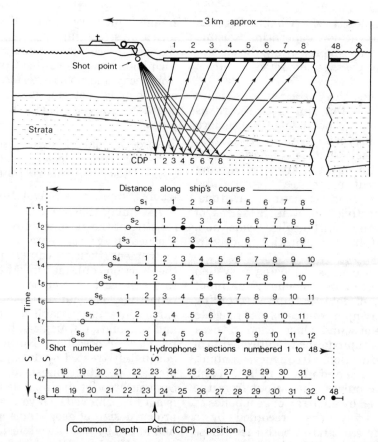

Figure 3/1 Schematic diagram showing the use of multi-channel hydrophone streamers to acquire data which can be common depth point (CDP) stacked.

spread, in which case the improvement ratio is $\sqrt{20} = 4.5$ or 13dB. The former acquisition method shows a relative improvement of 23dB over the latter in signal to noise ratio enhancement.

3.1.2 Multi-fold acquisition

Figure 3/1 shows a typical marine multi-channel acquisition system and illustrates how data are acquired in a way which allows stacking during processing. Although this illustration shows only a marine system, acquisition of land data is based on identical principles. At sea, a survey ship tows a hydrophone streamer made up of a number of sections, numbered one to forty-eight in the figure. Modern streamers are fitted with 24, 48, 96 or even more such sections and each section consists of a group of hydrophones which are pressure sensitive sound detectors (see p. 51). Signals received by the hydrophones in each section are summed so that each section is considered to be an independent single detector. In figure 3/1 the reflections from a single horizon are schematically portrayed as received in the first eight sections of the 48-section streamer. Let us assume that the distance between sections is 50 m and that the ship is travelling at 8 km/h (approximately 4knots). If the first shot S_1 occurs at time t_1, a reflection from depth point no. 1 is received in section no. 1 of the streamer and thence recorded in channel no. 1 of the seismic recording system onboard ship. The common depth point (CDP) position is located midway between the locations of S_1 and section no. 1; in the lower half of the diagram at t_1 the section no. 1 location is highlighted as a large dot. The next shot S_2 is timed so that the ship has progressed to a position such that the location of the midway point between source and section no. 2 of the streamer is the same CDP location as for S_1. This occurs at t_2, see lower half of diagram, and it can be seen that the distance between S_1 and S_2 is half the distance between sections, that is 25 m. The interval between shots (t_2-t_1) should be set therefore at 11.15s.

At t_2, the signals recorded on channel no. 2 of the recording system are therefore those associated with CDP no. 1. Shots S_3 to S_{48} follow at the same interval and successive records are obtained on ship from CDP 1, until data from shot S_{48} is recorded on channel no. 48 of the recording system. It should be noted that as the ship progresses along course, the seismic signal reflected from CDP 1 will have travelled an ever increasing distance between shot-point and receiver streamer sections. The change in geometry is corrected for during processing, and it is possible to add together (stack) all 48 records pertaining to CDP 1. Obviously the same is true for the locations CDP 2, 3 etc. CDP stacking is valuable not only as a means of increasing signal to noise ratio but also, during processing (see Chapter 4), of allowing differentiation between primary reflections from geological structure, and multiple reflections in sea and rock layers. Multiple reflections (see following section) can then be suppressed to improve the quality of the final seismic section display.

In land surveys, shot-point locations are surveyed at fixed intervals and groups of geophones are pegged into the ground with a group interval which is equivalent to the section interval of the marine streamer. After the initial layout of a spread, the speed of continuous acquisition is limited only by source-firing and the leap-frogging of geophones from already surveyed locations ahead to the front end of the spread. A roll-along switch redesignates the channel numbers of each recording station. Otherwise, the geometrical principles involved in continuous common depth point coverage are identical for both land and marine acquisition.

3.1.3 Coherent noise

Noise which is organised, predictable or repetitive in pattern is described as coherent and can be further categorised as follows:

(a) Multiples:

    (i) simple surface,
    (ii) peg-leg, inter-bed.

(b) Direct and near-surface waves:

    (i) low-velocity, marine or land traffic and source generated noise,
    (ii) medium to high velocity near-surface refractions.

(c) Diffractions (point source).

(d) Compressional converted (vertical) shear waves.

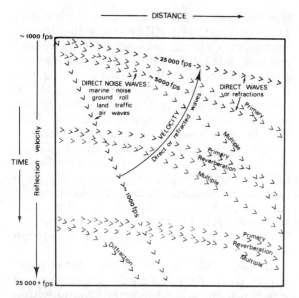

Figure 3/2 Coherent seismic noise types. For primary and multiple reflections the average velocities generally increase and normal moveouts decrease with depth and thus reflection time. The relationship between primary and multiple is shown for each at three different reflection times, and for two of these the interference between primary, multiple and a reverberation of the primary. A diffracted deep event is shown as well as the zone affected by direct waves of differing velocity (after S. D. Brasel in unpublished report, *Design of seismic field techniques.* Atlantic Richfield, 1971).

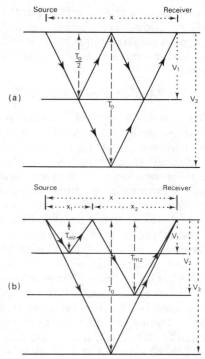

Figure 3/3 Ray path multiple reflection geometry (not adjusted for refraction according to Snell's Law). (a) Simple multiple. (b) Peg-leg multiple, that is, $T_0 = T_{M1} + T_{M2}$. Such multiples are common in marine survey data due to repeated reflections between seabed and sea surface.

These are illustrated schematically in figure 3/2. Removal of diffractions is attempted during processing by migration and is fully described in Chapter 4. Near-surface medium to high velocity refracted waves only interfere with the early part of the record, and are also removed during processing, by muting. Converted shear waves with slower velocities than compressional waves and therefore capable of interfering with later arrivals of the latter are confined to land seismic and are not considered a serious problem because of their generally higher rate of absorption. As yet there is no standard method of processing them out of the record.

The remaining two types of coherent noise can be reduced in two different ways depending on whether they are direct near-horizontal waves originating immediately below ground level, seabed or in the sea, or are reflected near-vertical (multiple) travelling waves.

In general, lowest velocity direct waves arrive last, a factor which can be utilised in design criteria for detector arrays. With reflected noise waves, lowest velocity waves arrive earliest and this can be utilised in the design of optimum trace spacing for attenuation stacking.

Multiple reflections can be attenuated or even effectively eliminated by common reflection point (or common depth point, CDP) stacking. The principle is to design the trace spacing such that these reflections have the appropriate residual normal moveout to be stacked out of phase and consequently much reduced in amplitude, while the primary reflections are stacked in phase by application of the correct normal moveout velocity (see Chapter 4). The formulae to be utilised for the simplest cases (see figure 3/3) are as follows:

(i) The symmetrical or surface multiple moveout,

$$\Delta t = \Delta T_{\text{multiple}} - \Delta T_{\text{primary}}$$
$$= 2\left[\sqrt{\left(\frac{T_0^2}{4} + \frac{x^2}{4V_1^2}\right)} - \frac{T_0}{2}\right]$$

$$- \left[\sqrt{\left(T_0{}^2 + \frac{x^2}{V_2{}^2} \right)} - T_0 \right]$$

$$= \sqrt{\left(T_0{}^2 + \frac{x^2}{V_1{}^2} \right)} - \sqrt{\left(T_0{}^2 + \frac{x^2}{V_2{}^2} \right)} .$$

(ii) The asymmetrical or peg-leg multiple residual moveout,

$$\Delta t = \Delta T_{M1} + \Delta T_{M2} - \Delta T_{primary}$$

$$= \sqrt{\left(T_{M1}{}^2 + \frac{x_1{}^2}{V_2{}^2} \right)} - T_{M1}$$

$$+ \sqrt{\left(T_{M2}{}^2 + \frac{x_2{}^2}{V_2{}^2} \right)} - T_{M2}$$

$$- \sqrt{\left(T_0{}^2 + \frac{x^2}{V_3{}^2} \right)} + T_0$$

$$= \sqrt{\left(T_{M1}{}^2 + \frac{x_1{}^2}{V_1{}^2} \right)}$$

$$+ \sqrt{\left[(T_0 - T_{M1})^2 + \left(\frac{x - x_1}{V_2{}^2} \right)^2 \right]}$$

$$- \sqrt{\left(T_0{}^2 + \frac{x^2}{V_3{}^2} \right)} .$$

Examination and time/distance analysis of a NMO corrected and uncorrected expanded spread (figure 3/4) or its equivalent common offset gather (figure 4/18) will identify the multiple with the most damaging interference effect and provide relevant velocities. Figure 3/5 is a plot of the NMO for both primary and multiple events as well as the Δt difference; it is adapted from the original paper on stacking by W. H. Mayne. From the above equations and the graph, it can be seen that the Δt differences are hyperbolic and that ideal attenuation would be achieved by the stacking out of phase of the multiple wavelets from a hyperbolic-function trace spacing arrangement: this is impracticable and a more limited attenuation is obtained by choosing the uniform trace spacing such that the average Δt differences are approximately equal to the multiple's period divided by the number (fold) of stack less one (-1). Elimination of remaining residual multiple effects, especially at other time levels, in the records would have to be handled by processing methods such as F/K filtering or deconvolution, described in the following chapter. Figure 3/6 is

an illustration of a single-fold as against multi-fold comparison; the distinct multiple suppression and improvement in the signal to noise ratio of the primary reflections is apparent.

Suppression of direct noise waves is attained by using an appropriate number of detectors, spaced areally or linearly at pre-determined intervals. The problem of surface noise is greater in land surveys than marine surveys and usually it is necessary to conduct noise spread or 'walk-away' surveys (see figure 3/7) in each new area using single phones per trace over the total length of the planned spread. Arrays of detectors can discriminate against waves travelling in nearly horizontal directions, in favour of those (desired) travelling nearly vertically. This is due to interference between the individual receivers of the array (figure 3/8). The wavefront arrives successively later at the more distant receivers, the time difference between adjacent receivers being $d \sin\theta/v$. If the wavelength of the seismic wave is λ, this corresponds to a phase difference $2\pi d\sin\theta/\lambda$ between the arrivals at successive receivers. Just as in the theory of the diffraction grating, the summed amplitude response for all N receivers can be shown to be

$$A \frac{\sin (N\pi d \sin\theta/\lambda)}{\sin (\pi d \sin\theta/\lambda)}$$

where A is the amplitude of the incident seismic wave. The response thus depends on the ratio of detector spacing d to $\lambda/\sin\theta$, the 'apparent wavelength' of the seismic disturbance travelling across the array. When θ is very small (reflections close to vertical incidence), the apparent wavelength is very large, and the response approximates to NA. For $\theta = 90°$, a horizontally travelling wave, the response will be zero for $N\pi d/\lambda = n\pi$ where n is an integer, i.e. for

$$\lambda = Nd, Nd/2, Nd/3 \ldots$$

and if $\lambda > d$ the intervening maxima will have lower amplitudes than NA.

By giving different response weights to the signals from different receivers, or by varying the receiver spacing, it is possible to tailor the array response so as to reject much of the horizontally travelling energy. The periods and velocities of noise waves (such as those

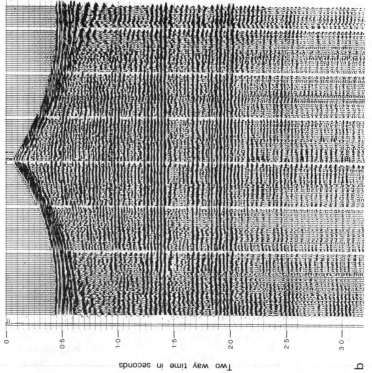

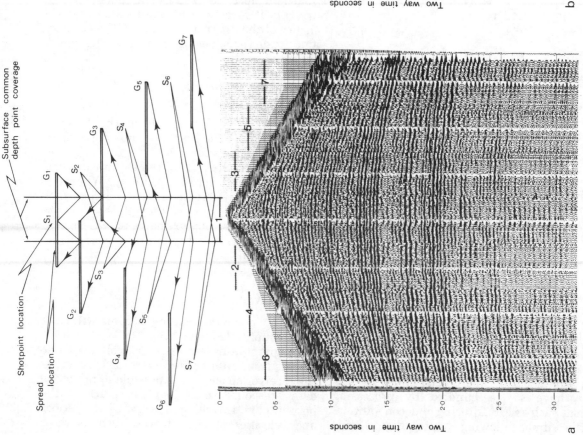

Figure 3/4 Expanded spread shooting for velocity determination and identification of multiples. (a) The same zone of common depth point (CDP) coverage is tested seven times by shooting at locations S_1 to S_7; the geophone spreads G_1 to G_7 being located with varying offsets. (b) Application of normal moveout and datum reflection flattening makes identification of multiples easier because of their residual moveout. (Courtesy: Mobil Oil Canada Ltd.)

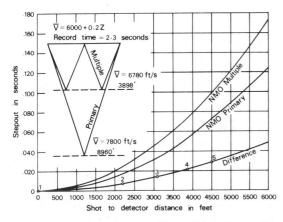

Figure 3/5 Normal moveout differentials between time co-incident primary and multiple reflections plotted against length of spread. Detector spacing across the spread can be designed to stack the multiple signals out of phase after applying normal moveout corrections to the primary signals (after W. H. Mayne, 1962).

within the time triangle of 0.0, 0.8 and 2.6 s in figure 3/7) are measured from the noise-spread; these are plotted on standard amplitude/wavelength geophone spread response curves and the most suitable design selected. Figure 3/9 is an example of such a response curve involving a spread of eight weighted elements spaced uniformly at 25 ft and designed to attenuate wavelengths ranging from 30 ft to 150 ft by a minimum of 13.69 dB. Although complex areal and non-uniformly spaced linear arrays are also used for specially severe noise problems, in practice, the economics of equipment and time-efficiency often

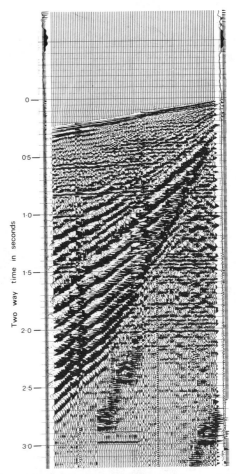

Figure 3/7 Example of a noise spread record. Each trace is derived from a single geophone. Spacing is usually 3–9 m. Severe interference is seen between flat-lying (primary and multiple) events and sloping events (refractions, ground roll and air waves). (Courtesy: Mobil Oil Canada Ltd.)

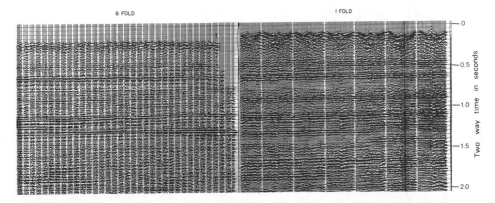

Figure 3/6 Single-fold and multi-fold seismic section comparison. On the single-fold section pronounced multiples interfere with reflections from a target horizon at between 1.05 and 1.15 seconds two-way time. (Courtesy: Mobil Oil Canada Ltd.)

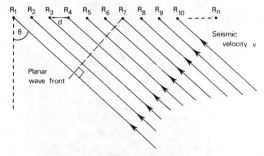

Figure 3/8 Seismic wave impinging on an array of receivers.

$$N > \frac{2\lambda_{\mathrm{L}}}{\lambda_{\mathrm{S}}},$$

$$d = 1.5\,\frac{\lambda_{\mathrm{L}}}{N}$$

where λ_{L} and λ_{S} are, respectively, the wavelengths of the longest and shortest waves to be suppressed.

It should be noted here that the first direct arrivals, or first-breaks, provide valuable near-surface refraction information, and in both land and marine work these are not suppressed in the field. Where they interfere with desired shallow reflections they are muted during the data processing stages as described in the following chapter.

dictate uniformly weighted and spaced arrays, especially for average and consistent noise problems. In those cases, from the above equations, the following relationships can be established:

Source arrays can be used for coherent noise suppression, either as an alternative to geophone arrays, or, more commonly, as a

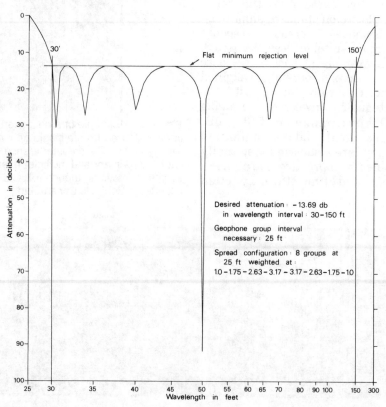

Figure 3/9 Computer-generated amplitude/wavelength response of an idealised geophone spread designed to attenuate unwanted direct noise waves in the wavelength band from 30 to 150 ft using different weighted geophone output levels. (Courtesy Digitech).

complement. The same principles of design as for detector arrays are involved and this is most important for proper utilisation of low energy surface sources such as dinoseis, vibroseis and thumper.

So far we have discussed design criteria in terms of the spacing of detectors, the number of groups (or sections in marine work) to be used and the level of fold or stacking multiplicity. A final consideration is that of the length of spread. Design criteria here are not so specific, but the length of spread used is usually related to the depth of geological objective; long spreads are used to investigate deep structure, short spreads to obtain highly resolved data on shallower objectives.

In marine work, the spread is towed in an 'end-on' configuration whereas on land it may be arranged end-on or either side of the source location (shot-point) as in 'split' or 'straddle' spreads. It should be noted, however, that with symmetrical split spreads the stacking multiplicity, which is defined as:

$$\frac{\text{Number of channels} \times \text{Interval between detector stations}}{2 \times \text{Interval between source locations}}$$

while being fully effective in signal to random noise enhancement, is however less effective than an ideal end-on spread in attenuating multiples. The duplication of common offset traces with symmetrical split spreads means that the records provide equal multiplicity (to end-on spreads) as far as random noise attenuation is concerned, but only half-fold multiplicity for multiple attenuation. In most situations, both on land and sea, it is common to leave a gap between the source and the first detector station (or even within a spread) to avoid interference from low velocity noise trains.

3.2 Line grid design, 2D and 3D

In an area where operational conditions and geological problems are well defined it is a relatively simple task to define the required layout of a seismic grid. Where shooting is to be undertaken in a new area, several factors need to be considered of which the more important are orientation and line-spacing of the seismic grid, field operation logistics, and budget.

Where operational conditions allow, it is generally advantageous to orientate a seismic grid such that lines are closely spaced parallel to the regional dip, or normal to the trend of major faults, anticlinal and synclinal axes, with only an open spacing of tie-lines at approximately 90°.

In the absence of information on the size of potential targets, definition of line-spacing will vary according to circumstances. Offshore, the high cost of exploration drilling and development suggest that only large scale structures are likely to be prospective. On the other hand, the costs of seismic surveys at sea are less than on land. As a consequence, it is common for marine investigations to commence with a survey of a widely spaced reconnaissance grid which is infilled over identified target structures at a later stage. If the size of potential target structures can be estimated, then the grid spacing should directly relate to this estimate. A spacing of half the width of such a structure would have a high probability of detecting its presence.

Field operational logistics will always have some influence on the layout of seismic programmes. Offshore, shallow water may prevent complete coverage of a prospect area, or it may necessitate splitting the programme into shallow and deep water surveys with use of appropriate vessels and equipment. Tidal conditions may influence orientation of lines to avoid excessive feathering of the streamer (the feather angle at any location is the angle between the line of the streamer and the line of profile and for good CDP stacking this is kept to a minimum except when using specialised acquisition techniques to obtain three dimensional coverage). On land, the distribution of access roads may influence the layout design, so might the existence of difficult terrain, bogs and marshes, etc.

Budgetary constraints show marked contrast between onshore and offshore operations, both in respect of seismic survey and drilling costs. Offshore the cost of a 3000 m exploratory well may be as high as $10 000/m whereas on land, a similar well might be drilled for $300/m. Proprietary seismic data offshore may cost approximately $400/km, whereas on land costs are very variable depending on method used and local conditions, but they are always likely to be higher by a large factor (see Appendix 2). If a budget is severely limited then it will be more important to eliminate areas of low target probability. This can be done through a careful

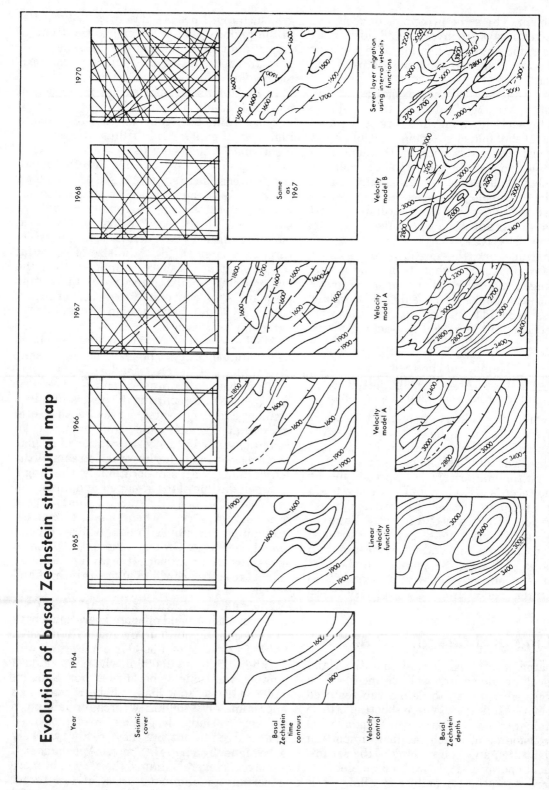

Figure 3/10 Seismic grid survey history from the North Sea; the West Sole gas field. Shooting dip lines in 1966 identified faulting after which all follow-up survey lines were orientated to investigate the nature of these faults (after J. T. Hornabrook 1974).

evaluation of known geology, and the interpretation of gravity, magnetic and shallow geophysical data prior to definition of the seismic programme layout (see Chapter 8).

A case history involving reorientation of successive surveys as knowledge of first the regional dip, then the strike of controlling faults became known is illustrated in figure 3/10.

Although it had long been recognised that seismic sections sometimes registered events from out of their planes, until recent years, the interpreter's 3D view of his prospect was limited to isometric projections of his 2D contour maps (figure 6/9) or sections (figure 6/8). Now, due originally to an uncontrollable quirk of nature and the discovery of its potential, almost by accident, three dimensional acquisition, processing and presentation is readily available. The quirk was the tendency of cables to feather or stray off the line of profile due to excessive wind and tide conditions; its potential lay in the recognition that a feathered cable was collecting data along a swath or two-dimensional area and not a one-dimensional line (figure 3/11). The data collecting procedure then became a matrix exercise with two outstanding problems – one an analogue one, position locating of the recording points and a digital one, the complex three dimensional processing requirements.

The positional problem was solved by installing location devices for each recording group, designing the parallel line grid such that the swaths butted in to each other and also arranging the sequence of line-shooting in conjunction with prognosed daily and hourly changes in tides and winds (figure 3/11). As in all new techniques, the software came slowly and both processing capabilities and method of presentation of results were initially somewhat limited. However, through an extension of the downward continuation wave equation migration method into the third dimension, improved interactive graphics and iso-time (horizontal) and iso-distance (vertical) slicing, even motion picture presentation became available. A side benefit has been the application of 3D migration for 2D data. Three-dimensional shooting is at present mainly a development method not only because of cost, but also because geological well control ties are essential. However, the cost is a small fraction of offshore appraisal wells or actual field

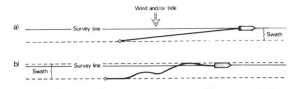

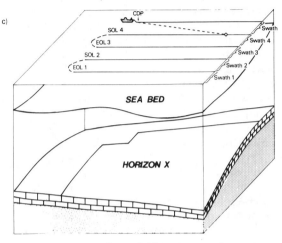

Figure 3/11 3D data acquisition at sea using a feathered cable. (a) Streamer disposition in relation to survey line in normal weather/tide conditions. (b) Effect of severe weather/tide variations on cable disposition. (c) Schematic oblique representation of the 3D coverage attained at horizon X by a survey conducted using a feathered cable along the indicated ship tracks.

development. Because of the high seismic to well cost ratio onshore mentioned earlier (and inversely proportional to the offshore ratio) and some technical limitations, 3D land seismic is increasing at an apparently slow rate. With improved operating and processing techniques it will probably join with marine seismic in being the acquisition tool of the future.

Interesting results of a 3D survey over a prospect in the Gulf of Thailand are shown in figure 4/45.

3.3 The seismic pulse

At the beginning of this chapter we referred to the earth's filter as one of the convolution factors of the seismic pulse which contributed to the final signal output. The earth's filter attenuates the seismic pulse in three ways: absorption of energy through conversion into heat; spherical divergence of the wave front;

and reflectivity losses at acoustic interfaces. The attenuation involves amplitude and frequency and these are interrelated. In normal seismic prospecting, the bandwidth of useful frequencies is extremely narrow and except for special shallow, high frequency input, high resolution surveys, it is limited to around 5–100 Hz. Commonly, at exploration depths around 10 000 ft or deeper, frequencies higher than 40 Hz are rare. This means that thin bed resolution becomes increasingly difficult with increased depths of exploration. For a 40 Hz wavelet, and an interval velocity of 15 000ft/s, beds thinner than 94ft (the quarter-wavelength one-way width) would be irresolvable. The band-limiting nature of the earth's filter and its frequency attenuation with depth is graphically illustrated in figure 3/12. At 12 000ft the 40 Hz amplitude factor of a waveform is one third of the 20 Hz factor, while that for a 100 Hz waveform is about one fortieth. It should be appreciated that the ideal seismic pulse from an implosive or explosive source should be as close to a spike as possible, which implies an instantaneous, short-period build up of the high frequency energy, decaying rapidly with minimum reverberation or development of source-generated interference. In practice this ideal is impossible to attain, but the closer it can be approached the greater the high frequency content, and quality of the recorded signal, and consequently the reflection seismic detail. The signatures of eight different types of marine sources in figure 3/21 show a complete range of attainment towards this ideal.

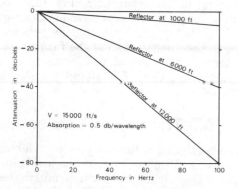

Figure 3/12 The effect of absorption in reducing seismic bandwidth with depth of a reflector. High frequencies are severely attenuated in the spectrum of deep reflections (after Evenden et al., 1971).

3.4 Land sources

3.4.1 Dynamite

On land dynamite is the traditional material used to generate pulses and in 1982 it was utilised in over one half of worldwide surveys. Normally, charges tailored to the required depth of penetration are detonated in bedrock beneath the highly absorptive weathered rock layer, and in a hole which is tamped with water or mud above the charge. The quality and quantity of the input acoustic energy is related to the size and shape of charge and the depth at which it is exploded as well as to the type of surrounding rock material. Experimentation is required in each new area and in poor record areas, multiple charges may be required.

3.4.2 Vibroseis*

This mechanical wave generator method is rapidly increasing its share of the acquisition market. In 1982, over 40% of worldwide surveys employed this source. Its growing popularity is due to a number of factors: costs per mile of survey are generally less and considerations of safety and minimal environmental disturbance are among the more important. A truck-mounted vibrator (figure 3/13) is coupled to the land surface and a long train of waves of progressively varying frequency is generated over a period of around seven seconds. The outputs from either an upsweep (increasing frequency input) or a downsweep (decreasing frequency input) are summed and correlated with the input sweeps to provide a 'conventional' field record (figure 3/14). Usually several vibrators sweep simultaneously in source arrays appropriately designed to attenuate surface noise and to improve the weak signal to noise ratio intrinsic to land-surface sources. Static corrections are obtained by shooting up-hole surveys and short refraction lines.

3.4.3 Dinoseis

Like Vibroseis, this source has advantages over dynamite sources. Basically it is a gas exploding device with the energy being imparted to the land surface via a heavy plate rammed downwards by the force of the explosion. Like Vibroseis, the signal to noise

* Trademark of Continental Oil Co.

VVDA vibrator

1 Internal combustion engine
2 Adapting gear
3 Axial-flow pump (vibrator unit)
4 Lift system
5 Vibrator control unit
6 Vibrator
7 Hydraulic accumulators
8 Oil tank (main oil circuit)
9 Oil cooler

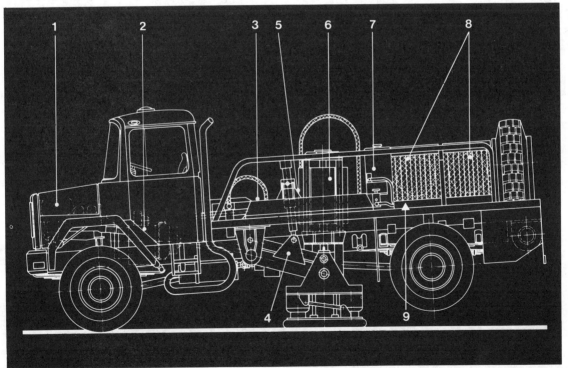

Side view of the vibrator system VVDA

Figure 3/13 Truck mounted vibrators used as a seismic source in vibroseis surveying. (Courtesy: Prakla-Seismos.)

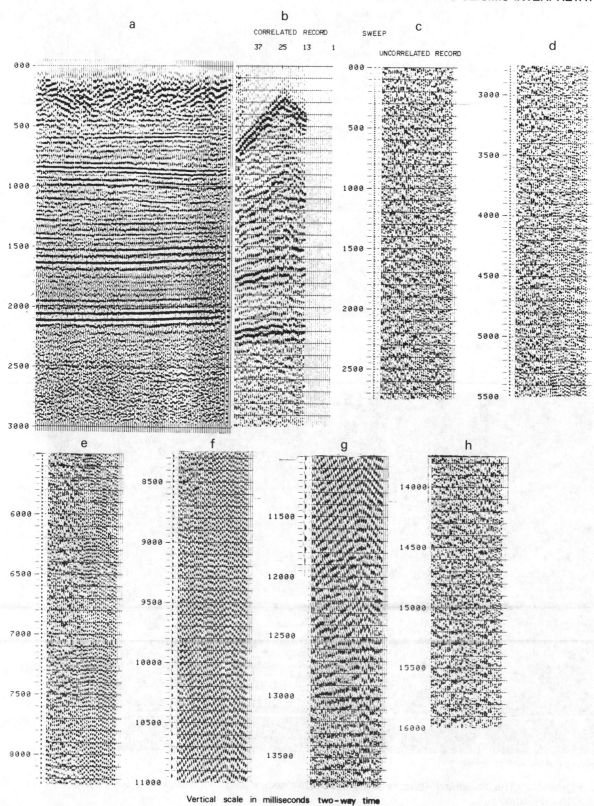

Vertical   scale   in   milliseconds   two - way   time

Figure 3/14 Vibroseis: (a) 2 × 6 fold stack. (b) Correlation corrected record. (c) to (h) Uncorrelated record traces for full 16 second sweep. Horizontal scale 30 traces/km. At this shot-point, sixteen 12 s downsweeps ranging from 56–10 Hz were applied. The 2 × 6 fold stack shows good quality down to economic basement at approximately 2.10 s. (Courtesy: Grant Geophysical.)

ratio is improved by the use of several units and/or the recording and stacking of several 'pops' at each station. Again it is necessary to run a supporting weathering correction survey.

3.4.4 Thumper or weight-dropping

This surface device is not in widespread use but is a leading seismic energy source in the desert areas of North Africa and the Middle East. Typically a 2–3 ton plate is dropped about 10ft. Summing of several drops is required and weathering correction data are collected separately as for Dinoseis and Vibroseis.

3.4.5 Air guns

This type of source, originally developed for marine surveys (see below) has been modified for use on land and is being increasingly used

Figure 3/15 Mini-Sosie seismic source. A signature recording sensor is mounted on the top side of the vibrator base-plate. (Courtesy: Société Nationale Elf Aquitaine (Production): SNEA (P).)

for work either in fresh water surveys or on dry land where they can be exploded in a suitably contrived liquid environment.

3.4.6 Mini-Sosie seismic source

A recently developed method provides a variation on the Vibroseis theme; a modified manually operated pneumatic hammer (figure 3/15) vibrates a plate coupled with the ground in a non-controlled random sequence. The signature of the random source energy is recorded by a sensor on the vibrating plate and is correlated with the recorded output to provide a conventional record. Again, summing of several signals is required. Indeed, the source is capable of producing 600 pops/minute and the methods depend on the summing of the results of a few hundred pops for each shot-point location. It is estimated that 8 to 12 records per hour can be obtained during smooth operations allowing approximately one mile of seismic survey per production day per seismic team. Use of this system is confined to shallow, high resolution seismic investigations more generally associated with engineering and mining studies than with hydrocarbon evaluation. In figure 3/16 is shown a comparison of a Mini-Sosie record with a dynamite record and in figure 3/17 a section obtained during a coal exploration programme in which good resolution of near surface faulting is clearly evident.

3.5 Marine seismic sources

As more and more petroleum exploration is conducted offshore in new sedimentary basins in search of large scale hydrocarbon reserve traps, recently well over twice as many miles of marine seismic were being surveyed as on land (see figure 3/18 and Appendix 2). Substantial technological advances have been made over the past two decades in both streamer and source design.

The major limiting factor in obtaining an ideal spike-type seismic pulse from marine sources is due to the effect of what is known as the bubble pulse. Within a water body, simple explosive sources generate an unwanted train of secondary pulses produced by multiple oscillations of a gaseous bubble. Immediately succeeding the original impulse or main seismic signal, little new energy is being generated as the bubble expands, but as it implodes under the ambient hydrostatic pressure, water surrounding the bubble compresses and a new

COMPARISON TEST

Mini SOSIE ←——→ DYNAMITE

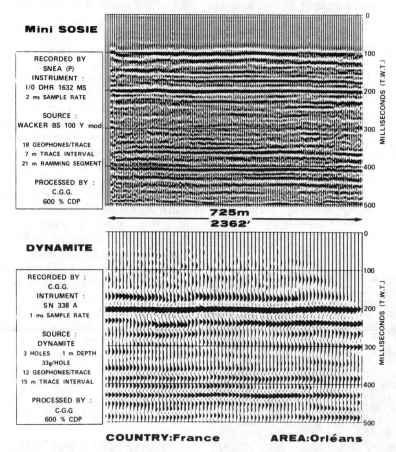

Figure 3/16 Comparison test between use of Mini-Sosie and dynamite as a seismic source.

pressure wave is radiated. Eventually the bubble breaks up due to either condensation or solution of the gases, or instability as it rises towards sea surface. Until this happens multiple periodic oscillations will occur, dependent on the bubble-size/energy and water depth/pressure. The far field seismic signature for a single Par airgun in figure 3/21 graphically illustrates such oscillations.

Lord Rayleigh's equation, originally describing the effect of steam bubbles in boiling water, was amended by Willis to define the bubble-pulse period T, as:

$$T = \frac{0.0452\ Q^{1/3}}{(D + 33)^{5/6}}$$

where Q is the source energy in joules and D the water depth in feet to the centre of the bubble. Figures 3/19 (a) and (b) illustrate comparisons, respectively, of source energy and water depth bubble oscillation periods to the Rayleigh–Willis curve. High energy sources which have larger bubble periods conform more closely to the Rayleigh–Willis curve, and in seismic processing, it is preferable that the period should be so large that there is no interference with the primary source signature. In such a case the bubble oscillation signature can be removed using specifically designed deconvolution operators.

Several solutions have evolved to combat the bubble pulse problem. With conventional explosives, the solution is to release the energy at such a shallow depth that the gas bubble is

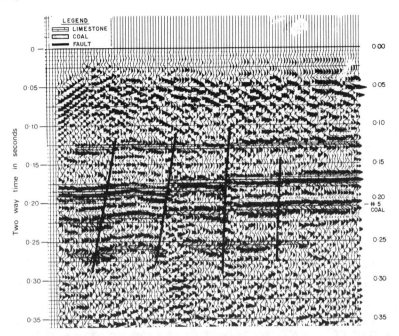

Figure 3/17 Mini-Sosie record showing high-resolution of faulting which affects a group of coal seams. (Courtesy: Geotemex).

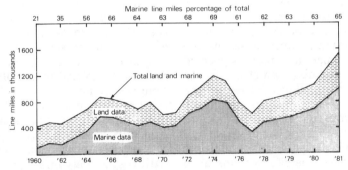

Figure 3/18 World wide seismic activity as measured in line miles, 1960–1981; includes refraction and both P and S wave acquisition (source: *The Leading Edge*, September 1982).

vented into the air. The results are spectacular, but grossly inefficient as a large proportion of the energy is wasted. Thus in early days of offshore shooting, it was necessary to utilise large explosive charges, typically 50lb TNT, a costly operation of benefit principally to the explosive manufacturing companies. Such large explosive charges are now seldom used. In one system, Maxipulse (a Western Geophysical system) a small explosive charge is used, about half a pound in weight, and de-bubble processing is used to remove bubble effects. To

do this it is necessary to embody a gun-channel record in the system as an auxiliary to the streamer records, and this is used to produce a filter operator which collapses the bubble pulse sequence into the equivalent of a simple explosion pulse. An alternative way of counteracting the bubble effect is to dissipate the bubble energy. This technique is used in the Flexotir (Institut Français du Pétrole) method. A small charge is exploded within a perforated steel chamber. Following the initial explosion, flow of fluids through these perforations to a

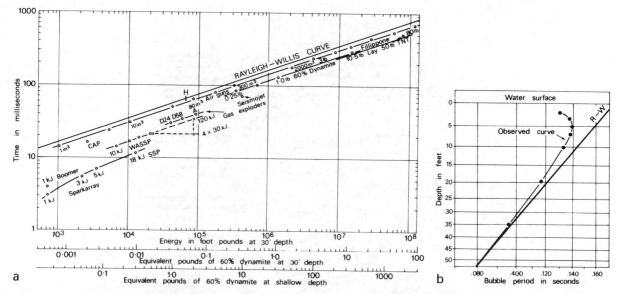

Figure 3/19 (a) Bubble oscillation periods of various seismic sources compared with the Rayleigh–Willis curve. (b) Plot of an observed bubble oscillation period versus depth; 300 in³ bubble at 2000 psi air. R–W is the Rayleigh–Willis curve. Near sea surface, the bubble period shows a divergent decrease due to the mass unloading effect which occurs as the bubble approaches the air/water interface (after Kramer *et al.*, 1968).

large extent attenuates the bubble pulse. Implosion sources have been developed, for example Vaporchoc and Starjet (Compagnie Générale de Géophysique), which do not have a bubble effect. With Vaporchoc, steam is injected into the water to form a bubble which condenses and implodes; bubble collapse provides the seismic pulse and there are no secondary impulses. An important development of airgun technology which converts the basic system into an implosive source is the water gun, and this is described in more detail below. Systems have also been developed which generate explosions within a flexible container and vent the exhaust gases through hoses to the atmosphere. Conventional airguns have been modified so that air is blod into the air bubble to increase pressure during collapse and inhibit oscillation; such a modification is termed a wave-shaping kit. Perhaps, however, the most widely used method of countering the effects of bubble oscillation is to design a tuned array of sources, particularly when air guns are used. Two techniques can be used. Firstly, a number of sources may be fired simultaneously with different energy levels and hence different bubble periods so that the summation of the wavelets produces the desired source signature.

Secondly, a number of sources may be fired with pre-set delays. With proper design, in either case, the end result is destructive interference of the bubble pulses.

The latest worldwide annual survey of the Society of Exploration Geophysicists shows that compressed air sources are by far the most favoured marine source, being used in two thirds of all data acquisition (in line miles; Appendix 2). Gas/sleeve exploders, and implosive devices (other than water guns) followed with respectively one-fifth and one-fourteenth shares. In actual number of marine vessels, a 1976 'Offshore' review indicated a total of 77 working worldwide with 66 of them specified as being equipped with:

Air gun arrays	34
Gas/sleeve exploder arrays	18
Maxi-pulse gun	7
Vaporchoc	5
Air gun (single)	2

Two of the reported 14 contractors operated over one half of the vessels; Western Geophysical with 22 and Geophysical Service Inc. with 17 vessels.

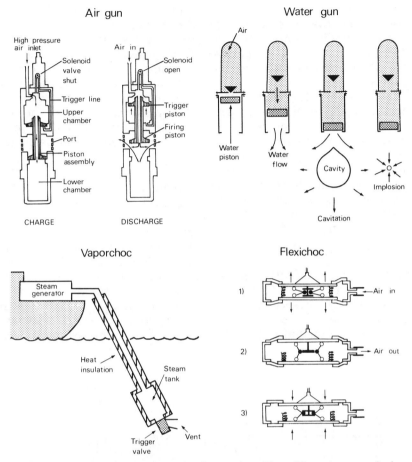

Figure 3/20 Schematic diagram showing the principle of operation of four different non-explosive seismic sources (after McQuillin and Ardus, 1977).

Top left: PAR Air Gun (registered trade mark of Bolt Associates. The air gun is patented under British Patent Specification No. 1090363; 1175853 and 1322927).

Top right: MICA Water Gun. (Courtesy: Société pour le Developpement de la Recherche Appliquée (SODERA).)

Bottom Left: VAPORCHOC seismic source. (Courtesy Compagnie Générale de Géophysique (CGG).)

Bottom right: FLEXICHOC seismic source. (Courtesy: Géomécanique.)

3.5.1 Air guns

An individual air gun generally consists of a system of two high pressure chambers (see figure 3/20), connected and sealed by a double-ended piston. During the charging cycle, air at high pressure (say 2000 psi) is fed into the upper chamber and bleeds through the hollow piston into the lower chamber. The piston assembly is held in the downward position because the area of the trigger piston is larger than that of the firing piston. To fire the gun, an electrical pulse opens the solenoid valve and a slug of high pressure air is delivered to the underside of the trigger piston. The piston shoots upwards under the pressure exerted on the firing piston releasing the air in the lower chamber into the water. Pressure in the upper chamber then drives the piston back to its initial position and the charging cycle recommences. During a period of a few milliseconds, all the high pressure air in the lower chamber is vented into the surrounding water through centrally located ports and it is the explosive release of this air which provides energy for the initial seismic pulse.

Guns range in size from a few cubic inches (lower chamber capacity) to about two thousand cubic inches, and operate at pressures of 2000–4000 psi. As described earlier, multiple arrays of guns are usually employed, towed behind or alongside the survey ship on a suitable frame at depths of about 10m. The differences in pulse shape generated by a single air gun, a wave-shaped air gun and an array of air guns is illustrated in figure 3/21.

3.5.2 Sleeve exploders

The sleeve exploder, also called Aquapulse, is a seismic source developed by Esso Production Research, and it is used under licence by several contractors and has other alternative trade names. In figure 3/22 an Aquapulse system is

shown as used by Western Geophysical. Propane and oxygen are separately fed into the rubber sleeve and ignited by a sparking plug. The high energy pulse is generated as the sleeve expands; after expansion, the sleeve contracts with venting of gases to the surface, decompression being assisted by cooling and pumping. Multiple arrays such as shown in figure 3/22 are generally required to provide sufficient total energy output for deep penetration.

3.5.3 Maxipulse

Maxipulse is one of the simplest seismic sources currently in use. A small cylindrical charge of nitrocarbonitrate explosive is projected by water pressure down a hose to a submerged gun, striking its detonator on a firing wheel as it is ejected (see figure 3/25). A percussion cap is activated and this fires a delay fuse which detonates the main charge after a sufficient period for the gun to have been towed a few metres from the explosion point.

The firing gun assembly has a gun trans-

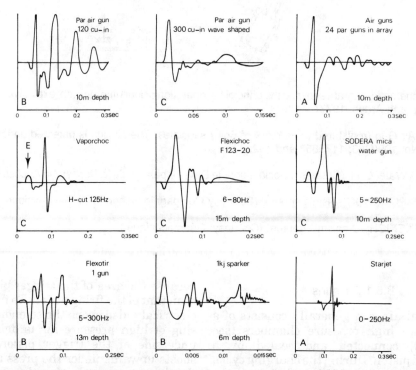

Figure 3/21 Signatures of a range of seismic sources. Letters A, B and C apply to signature type category as assigned by Giles and Johnston in Appendix 3. Note the differing horizontal scales between high and low frequency sources. Vertical scales are not defined but are measured in bar-metres where experimental data have been acquired in deep water using a calibrated hydrophone.

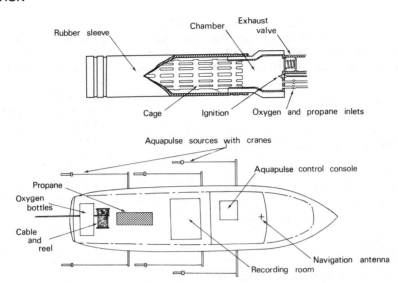

Figure 3/22 Schematic diagram of (above) an Aquapulse seismic source with (below) a ship layout showing a towing arrangement for the deployment of an array of six sources. (Courtesy: Western Geophysical.)

ducer attached to it which is used to record the source signature. As described above, no attempt is made to attenuate the bubble pulse in the acquisition stage; this is adequately performed by specialised pre-processing of the data.

3.5.4 Vaporchoc* and Starjet*

Vaporchoc, used by Compagnie Générale de Géophysique (CGG), is an implosion source fuelled by superheated steam (see figure 3/20). A steam generator is located on board ship and steam is fed into the water through a remotely controlled valve to form a steam bubble. When the valve closes, the steam in the bubble condenses and the bubble implodes under hydrostatic pressure. During implosion, high pressure develops in the water around the wall of the diminishing bubble and acoustic energy is radiated as a primary impulse with negligible secondary oscillation on complete collapse of the bubble (see figure 3/21).

CGG have recently announced a new development of the Vaporchoc principle called Starjet. This system utilises not a single gun but a set of up to four subarrays each of which has four tunable guns. One of the deficiencies of the Vaporchoc source is that acoustic energy is generated during expansion of the steam

bubble; this can be seen on the signature (figure 3/21) marked E. The array is so designed that steam is injected into the water through four differently designed guns. The bubbles so produced collapse at varying intervals after injection, thus it is possible (see figure 3/23) to tune the guns of the subarray to minimise the undesired effects of 'forerunners'. In practice, the implosion peaks are lined up by a 6ms spacing between periods. The emitted signal is not minimum phase; thus it is necessary to

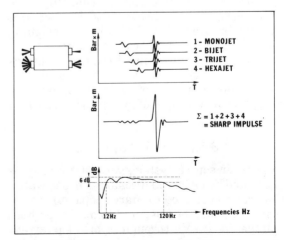

Figure 3/23 A Starjet subarray of four guns is timed so as to minimise the amplitude of forerunners (Courtesy: CGG).

*CGG Trademark

apply a wavelet processing sequence. Near-field signatures are recorded onboard the survey vessel of each subarray along with their sum. The far-field signature of Starjet is shown in figure 3/21.

3.5.5 Flexotir

The Flexotir source was developed by Institut Français du Pétrole. A small explosive charge is flushed down a hose into a submerged perforated steel cage of diameter slightly larger than that attained by the bubble at maximum expansion. The charge is fired electrically. Oscillation of the bubble is inhibited and the bubble pulse attenuated by the flow of fluids in and out of the perforations. As seen in figure 3/21, the signature of a single gun is quite complex, but as with air guns, improvements can be attained by operation of an array of sources.

3.5.6 Water guns

The water gun is a variety of air gun which has been designed by SODERA (Société pour le Développement de la Recherche Appliquée) as an implosive seismic source which generates an intrinsically clean signature (see figure 3/21). On firing, the compressed air in the gun chamber rams a piston forward ejecting a volume of water through a number of ports. When the piston is arrested, cavitation occurs and the main acoustic energy is generated by implosion of the water body through cavity collapse. No bubble pulse is produced; the air in the chamber is vented and the charging cycle recommences. One advantage of this source is that its features allow flexibility in designing source arrays which can be effective in attenuating coherent noise. Because operations can be easily switched between air guns and water guns, a number of contractors now offer the option of both types of source and equip their vessels accordingly.

3.5.7 Sparker systems

In sparker source systems, acoustic energy is generated by electrical discharges in sea water. Generators are used to charge capacitor banks which can then be triggered to discharge high voltage (3–10kV) through spark tip arrays towed in the water. Low energy sparkers, 100J–5kJ, are widely used in single channel seismic profiling as part of shallow geological studies, engineering site surveys, etc. (see McQuillin and Ardus, 1977). High energy sparkers, energy up to 200 kJ, are used as a source for conventional seismic work, one advantage being the relatively low operational cost of using this type of system. In recent years, a more significant use of the sparker as a source for multichannel seismic acquisition is in the field of shallow gas detection, in particular for drill site surveys. Here sparkers are operated in the 3–5 kJ range and data acquisition aims for high resolution data in the 0–1s two-way reflection time range. Sections are processed to give true amplitude recovery (see Chapter 4) display as well as conventional displays.

3.5.8 Comparison of marine source signatures

In the foregoing, it has been emphasised that there are three groupings of sources:

A. 'Tuned' multiple-source arrays, which are designed by number and/or energy-weighting and spacing (in the same manner as detector arrays, described previously) to attenuate each other's bubble pulse oscillations.

B. Single source units which do not attempt to eliminate the bubble pulse (or only partially) but instead use its recorded signature to design cross-correlation or signature deconvolution filters for its subsequent elimination during processing.

C. Single or multiple source arrays in which each source unit is designed either not to generate (implosive) or to attenuate bubble-pulse oscillations.

A summary of energy sources and their type signatures according to the above groupings is given in Appendix 3. Typical actual signatures of several sources are shown in figure 3/21. It will be readily apparent from these signatures that bubble-pulse suppression is not always completely effective and that subsequent signature filtering may still be required in many cases regardless of the type of source.

In assessing the suitability of a particular seismic source for a project and for purposes of comparison between available systems, study of the seismic signature is a useful first approach. The ideal source should emit a pulse with minimal forerunning energy, a narrow main

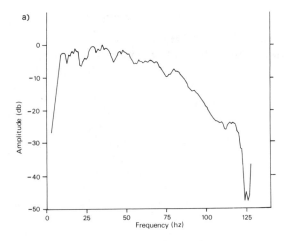

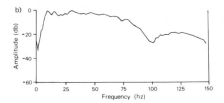

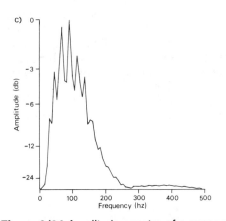

Figure 3/24 Amplitude spectra of a range of seismic sources and source arrays. (a) 905 in³ high pressure airgun array (Western Geophysical). (b) 2594 in³ SLAG subarray (GECO). (c) 48.9 in³ airgun array; 40 with wave shape, 10, 5, 1.9 in³ (Racal-Decca).

peak, and a clean signal after the main pulse. The energy of a source is usually measured from the far-field signature in bar metres (bar-m) peak-to-peak. These measurements are made experimentally using a calibrated hydrophone, and in deep water. For comparative evaluation,

measurements are usually made with no low-cut filter, and a 125 Hz high-cut (see Appendix 3). Measurements with high-cut at higher frequency can give a deceptive value of peak-to-peak amplitude. Source arrays generating peak amplitudes of up to 100 bar-m are currently available, but source energies of 40–50 bar-m are more common.

It is also necessary to ensure that the source emits an essentially flat frequency spectrum. Thus it is necessary to evaluate spectra as well as signatures when evaluating seismic sources. In figure 3/24 the frequency spectra of a variety of sources are compared. When source arrays are used, each individual array configuration will produce a difference in far-field energy spectrum, thus the examples shown must be regarded as typical rather than definitive.

3.6 Geophones and hydrophones

In seismic surveying, two types of acoustic detecting transducers are used, geophones on land and hydrophones in marine conditions and in mud-filled boreholes. We have seen in Chapter 2 that in reflection surveying we are

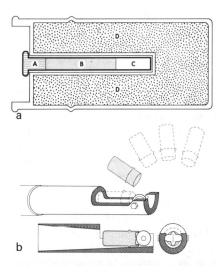

Figure 3/25 (a) Schematic cross-section of Superseis charge used in the Maxipulse system. A — rim-fire percussion cap, B — delay column, C — booster and D — nitrocarbonitrate main charge. (b) Maxipulse gun. Top — perspective view of gun showing charge striking the firing wheel, then successive stages of ejection. Bottom — cross-section and end views of charge in gun at instant of striking the firing wheel. (Courtesy: Western Geophysical.)

attempting to record trains of P-waves, or compressional waves as they pass a specified point. Thus pressure sensitive phones are ideally suited for marine work, recording as they do the pressure changes above or below ambient water pressure. On land, pressure-sensitive phones cannot be used as it is generally impractical to bury the phones in such a way that they would have adequate fluid coupling with surrounding material. The difference between the pressure-sensitive hydrophone and the motion-sensitive geophone has an important implication for their use. We recall that because of the low density and velocity of the air the sea or ground surface is an almost perfect reflector of seismic waves. Signals will thus reach the receiver both directly and after reflection at the surface (figure 3/26). In Chapter 2, we showed that

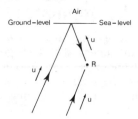

Figure 3/26 Raypaths to a buried receiver. Displacement direction is shown by '*u*'.

where the incident wave is in the medium of higher acoustic impedance, as here, there is a 180° phase change on reflection. Thus an incident compression will be reflected as a rarefaction, and since the reflection coefficient is almost unity, the two signals will have almost exactly the same amplitude. If a pressure-sensitive receiver is placed at a very shallow depth, the direct and reflected signals will arrive at almost the same time, and cancellation will result. Hydrophones must therefore be placed an appreciable fraction of a wavelength below the free surface to receive any signal at all. On the other hand, there is no change in the direction of particle displacement (*u*) on reflection at the free surface, so the direct and reflected signals will add up in phase at a motion-sensitive detector on the surface. Thus particle motion-sensitive land geophones are best coupled to the ground surface, where the signal will be a maximum.

3.6.1 Land geophones

Figure 3/27 illustrates the principle of a geophone designed to measure particle motion by conversion to electrical energy. Arrival of a compressional wave sets the earth's surface in motion; the geophone case with rigidly attached coil is coupled to the ground and moves in sympathy whereas the magnet which is suspended on springs remains effectively stationary because of its inertia. Movement of the coil within the magnetic field induces an electrical voltage across the coil which is proportional to the velocity of the coil with respect to the magnet. An alternative, and more common, design, is to mount the magnet rigidly and suspend the coil around an inertial mass; the principle is identical and such geophones are called moving coil geophones. In both cases the output is independent of frequency at frequencies above the natural frequency of the suspended element; below natural frequency the response is frequency dependent. Modern geophones use a dual coil system; the dual, series-connected coils reduce external interference.

In land surveys, it is important that good coupling is obtained between the ground surface and the geophones. The cases are usually either heavy with a flat base, or light with a coupling spike which is pushed into the soil. Geophones are usually deployed in groups at each detector location, the phones being connected in series. The group is called a string, and each string is attached to a takeout in the

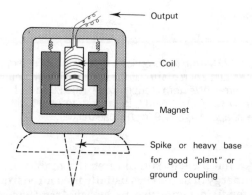

Figure 3/27 Schematic of a velocity, or particle motion, geophone. In this type the coil is fixed and the magnet suspended; alternatively, the coil may be suspended and the magnet rigidly fixed to the geophone case.

main cable which feeds to a specific channel in the recording system. For speed and efficiency during a survey of a seismic line, several cables are laid out so that the number of strings deployed is greater than the number of channels being recorded. After each seismic shot, or a fixed number of shots, a string or a number of strings are picked up from the end of line already surveyed and 'leap-frogged' to the other end. The recording system incorporates a 'roll-along' switch which is used to drop-out and pick-up the appropriate cable take-outs. Geophones of this type will record only vertically vibrating P or SV/P (P-converted shear waves vibrating in the vertical plane) waves. Special horizontal particle motion geophones need to be employed for recording of either SH or SV shear waves generated by the appropriately polarised source. Such geophones are extremely sensitive to proper layout or planting. Either spirit/bubble levels are used for levelling or special gimbal assemblies contain the recording element.

3.6.2 Hydrophones

The principle of the marine hydrophone is very simple. Within the phone a piezoelectric transducer produces voltages in response to pressure changes caused by the passage through surrounding water of seismic pressure waves. Figure 3/28 shows the frequency response of a typical hydrophone. For static cable recording, good responses can be obtained with a simple hydrophone element, but current methods employ continuous profiling using towed streamers and substantial noise is generated through vibration of the cable (strumming) and sudden acceleration/deceleration effects produced by heave acting on the towing vessel, the vessel's movements being transmitted to the streamer. Various methods are employed to reduce these effects:

1. Ship motion is decoupled from the streamer by using an elastic non-active lead-in section; this absorbs the ship's heave motion allowing the cable to be towed at a constant speed through the water.
2. Streamer depth controllers or 'birds' are used to maintain constant streamer depth along the length of the streamer. Each controller clamps to the cable and has fins which are servo-linked to either pre-set pressure sensors, or pressure sensors which can be adjusted by remote control. Typically about 6 to 8 controllers might be attached to a long marine streamer.
3. Lead-in sections to the cable can be faired to reduce noise induced by strumming.
4. Instead of single crystal element hydrophones, dual crystal, acceleration-cancelling phones are used which have a very low sensitivity to horizontal accelerations, one of the main sources of noise problems.

The acceleration cancelling hydrophone is an important recent innovation. A schematic diagram of one type is shown in figure 3/29. Each crystal element consists of an annular piezoelectric ring, metallic-coated on both surfaces and bonded at the open ends by thin convex metallic diaphragms. In the towed streamer configuration, vertical pressure fields produce enhanced responses due to the pressure vectors generated in the annular rings by the diaphragms. Furthermore, axial accelerations generate equal but opposite phase signals in the twin elements in each geophone which when summed effectively cancel each other out. Among the benefits of using streamers incorporating such hydrophones, as well as other noise-reducing design criteria, are firstly that marine surveys can now (as opposed to a few years ago) be conducted at higher tow speeds and in rougher weather conditions, and secondly, improved signal to noise ratio allows the geophysicist to obtain deeper information

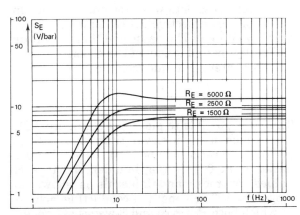

Figure 3/28 Pressure sensitivity versus frequency response curve for a Multidyne hydrophone group in a Prakla-Seismos Piezo Oil Streamer. Pressure sensitivity S_E is related to the termination resistance R_E across the input to a DFS recording system. (Courtesy: Prakla-Seismos.)

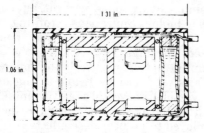

Cross section of complete hydrophone assembly

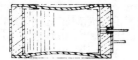

Cross section of hydrophone element

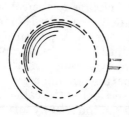

Top view of hydrophone element

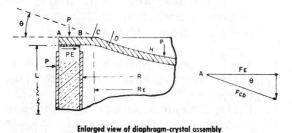

Enlarged view of diaphragm-crystal assembly

Figure 3/29 Schematic of a Multidyne acceleration cancelling hydrophone. (Courtesy: Seismic Engineering Company.)

and better structure delineation after data processing.

3.6.3 Digital telemetry

An increasing requirement for greater vertical (2D) and lateral resolution (3D) from not only seismic interpreters but also geologists and engineers, particularly in production development, has placed greater demands on seismic crews to expand the number of recording groups. The inevitable pressure on the circuitry

design of ever-enlarging analogue spread systems has been circumvented by yet another leap forward in digital applications—digitised telemetry. Marine streamers are available now with up to 480 channels, depending on the noise specification required; the increased channels are permitted by the use of electronic modules within the cable which can perform all the normal functions described in Section 3.8 of this chapter, short of permanent storage which these units ensure by sequentially transmitting digitised samples via coaxial cable to the onboard central control units and tape recorders. One additional advantage of these systems is that leakage and crossfeed are eliminated. In land seismic, the breakthrough in simplifying the circuitry has been by using similar electronics in small portable units which, on command from a central control unit, transmit the digitised and temporarily stored signal samples by VHF FM radio. Up to 1016 channels can be utilised through the simultaneous use of 254 units, each accessing four geophone spreads.

3.7 Shear wave generation and recording

In recent years increased emphasis has been placed on the generation and detection of shear waves (both those which are generated at source and those which result from conversion of compressional waves) (see figure 3/30). Comparison of records of either type with conventional compressional wave records are made in terms of the respective ratios; time (T_P/T_S) or velocity (V_P/V_S). The value of either such ratio is directly related to that of Poisson's ratio σ for the rock interval in question, i.e.

$$\frac{T_P}{T_S} = \gamma_T = \sqrt{\frac{1 - 2\sigma}{2(1 - \sigma)}}$$

Thus it is possible to plot a graph, figure 3/31, from which, using interval–time ratios, values of Poisson's ratio may be read. Poisson's ratio varies from 0 to $\frac{1}{2}$ (for liquids). A range of values for rocks and minerals is shown in Appendix 1. The benefit of conducting surveys in which both P-waves and S-waves are generated and recorded is that when results are displayed, changes in the elastic properties of a rock interval are directly related, and such changes can be related to, in particular, changes in porosity and/or fracturing. For

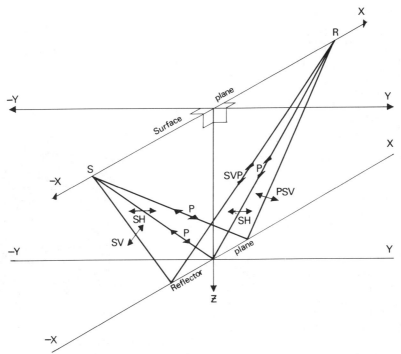

Figure 3/30 Three-dimensional relationship between normal and converted compressional and shear waves. Particle motion for normal shear waves (SH) is both generated and recorded within the *YZ* plane: for all others particle motion is in the *XZ* plane, either in the direction of the ray path (P and SVP) or perpendicular to it (SV and PSV). Polarisation requirements of source and receiver are shown in figure 3/32. It should be noted that because of the change of velocity which occurs when a wave is converted on reflection at a boundary, for converted waves the angle of incidence does not equal the angle of reflection.

example, an increase in porosity reduces the bulk modulus without to such an extent affecting the shear modulus, thus the P-wave velocity is reduced and the value of γ_T becomes anomalously low.

A range of seismic sources can be used, capable of producing a polarised and orientated output waveform. Such sources are designed to produce normal horizontal shear waves (SH), operated in conjunction with a normal isotropic P-wave source. Receiver arrays are similarly designed to receive both normal shear waves (SH) and normal compressional (P) as well as converted compressional waves: PSV corresponding to an emitted P-wave received as an S-wave, and SVP corresponding to an emitted S-wave received as a P-wave. In figure 3/32, wave modes and seismic components are shown.

There are three main types of source:

(i) Dynamite laid as linear charges or in rows of shotholes.
(ii) Directionally alternating horizontal hammer blows against an anchored base plate.
(iii) Anchored horizontal vibrator (Vibroseis* method).

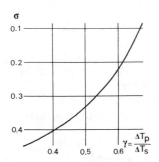

Figure 3/31 Relationship between Poisson's ratio and interval time ratio $\gamma = \Delta T_p/\Delta T_s$.

As the dynamite and hammer sources will generate non-polarised, spherical P-waves, which will be recorded even by the polarised

Polariz-ation Wave type	Horizontal		Vertical	Isotropic
	X	Y	Z	
P			S/R	S/R
PSV	R		S	S
SVP	S		R	R
SH		S/R		
SV	S/R			

Figure 3/32 Wave modes and seismic components. The figure shows the source and receiver design requirements for the wave modes: P, PSV, SVP, SH and SV. For example SH waves are generated by a source which is horizontally polarised in the X-axis (see figure 3/30) and detected by a receiver in the same axis.

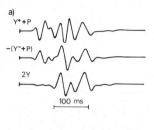

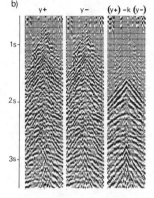

Figure 3/33 Summing of two SH traces with opposite polarity by inversion of one of them to eliminate the P-wave signal and produce residual energy of constant polarity, 2Y. (Courtesy: CGG.)

geophones, a summing procedure is used to eliminate the effect of this received energy. By alternating the polarity of successive transmissions, these may then be subtracted in weighted sums; the shear signals have opposite polarity and sum whilst the compressional wave is eliminated. Figure 3/33 illustrates this process.

Geophones used are horizontally polarised and orientated along the required axis, i.e. Y for SH waves and X for PSV waves.

The principal difficulties to be overcome during acquisition (and processing) of shear wave data include the following:

(i) The method is limited to areas where the use of explosives or anchored surface sources is environmentally acceptable.
(ii) Static corrections are often hard to determine because of uncertain water-table effects.
(iii) NMO corrections are routinely determined for SH waves; however, correction of converted waves involves the determination and application of both P and SV velocity functions.
(iv) To compare P and S-wave sections, time scale adjustments need to be made.
(v) Frequency spectra of S-waves tend to be lower and more narrow-band than for P-waves.

An example of a comparison between P and S seismic sections (figure 3/34) shows an easily identifiable dolomite pod in the P-wave section, the porosity of which can be inferred from the comparison of the two sections.

3.8 Recording equipment

The field equipment required to record reflected seismic signals as detected by arrays of geophones consists basically of four units, each of which performs several functions (figure 3/35). A typical system would be as follows:

Recording amplifier, with
(a) pre-amplifiers,
(b) analogue filters including high and low frequency cut-off and anti-alias,
(c) multiplexer,
(d) gain-ranging amplifier,
(e) analogue to digital convertor.

Control or logic unit, with
(a) time break amplifier,
(b) formatter,
(c) digital gain control,

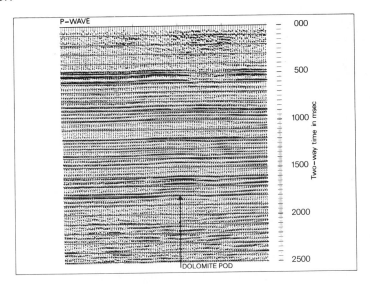

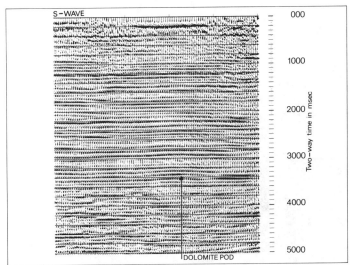

Figure 3/34 Comparison between P- and SH-wave migrated seismic sections from the east shelf of the Anadarko Basin in Oklahoma. (Courtesy: CGG.)

(d) digital to analogue convertor,
(e) demultiplexer,
(f) playback amplifier,
(g) playback filters,
(h) total system controls.

Tape transport unit.
Camera unit.

A total field system would also include power supply, seismic channel and system test circuitry, test signal generator and system fault location units. Present day field equipment is contained in compact, lightweight modules which are capable of being back-packed for operators in difficult terrains, or built into container cabins, small vehicles, etc. for more normal operations. One such system is illustrated in figure 3/36.

Although analogue field systems are still in use in research and academic studies, most seismic exploration is almost exclusively conducted utilising digital recording techniques. Only digital techniques will be described here; information on analogue methods may be

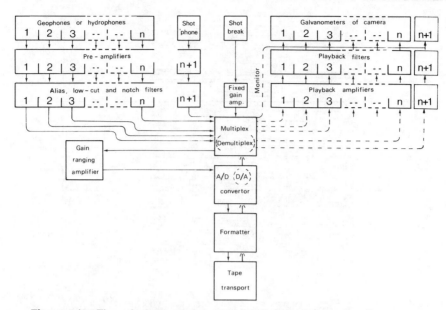

Figure 3/35 The principal components of a digital seismic recording system.

Figure 3/36 Texas Instruments DFS V seismic recording system. (Photo: Techmation.)

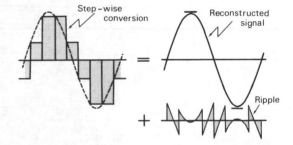

Figure 3/37 Digital sampling and playback (adapted from illustration in: Pictorial Digital Atlas. Courtesy: United Geophysical Co.)

obtained by reference to one of the geophysical text books listed as suggested reading.

Digital field systems are designed to record seismic signals in digital form by sampling amplitudes at appropriate time intervals such that reversal of the process, digital to analogue conversion, will generate an output signal which compares with the original without loss of fidelity. The process is illustrated in figure 3/37. In the example shown a 62.5 Hz wave (dashed line) is sampled every 2ms. On play-back, conversion to an analogue electrical signal produces a signal which may be conceived of as the original signal (smooth line) reduced to 97 per cent amplitude plus a ripple of higher order harmonics. Reproduction of the original signal is achieved by filtering with an analogue high-cut filter, thus removing the unwanted high frequency harmonics.

3.8.1 Recording amplifier

The functions of the various components of this part of the system are outlined as follows:

(a) Pre-amplifiers These provide a means of selecting fixed gains for application to the input signal.

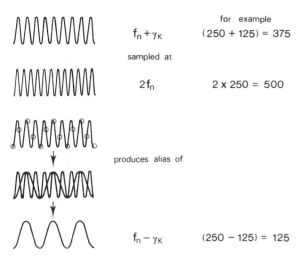

for example

$f_n + \gamma_K$ (250 + 125) = 375

sampled at

$2f_n$ 2 x 250 = 500

produces alias of

$f_n - \gamma_K$ (250 − 125) = 125

Figure 3/38 The aliassing effect of sampling a signal at too low a sampling frequency.

(b) Analogue filters In land work, low-cut filters are tailored to the natural frequency response of the geophones being used. The cut-off is chosen above the natural frequency to avoid harmonic distortion and is generally in the range 7–14 Hz. In offshore surveys, the flat frequency response of hydrophones means that an 'out' or 0 Hz setting could be applied; however, a low-cut filter is usually applied to reduce streamer and other low frequency environmental noise.

High-cut filters must be applied with cut-off being related to the digital sampling interval being used. If the seismic signal is insufficiently sampled, aliassing may occur and on playback artificial low frequencies may appear. The frequency below which aliassing does not occur is termed the Nyquist frequency and this is one half the sampling frequency, i.e. for 2ms sampling the Nyquist frequency would be 250Hz. If f_n is the Nyquist frequency and a signal of higher frequency $(f_n + \gamma_k)$ is sampled at the interval $1/2f_n$, aliassing produces an output signal of frequency $(f_n - \gamma_k)$. This is illustrated in figure 3/38. A signal of 375Hz is sampled at 2ms (500Hz) to produce an aliased signal of 125Hz; thus for 2ms sampling it can be seen that a high-cut anti-alias filter is necessary removing all signals of frequency higher than 250 Hz. Figure 3/39 shows a suite of filter amplitude response curves which might be used for 2ms sampling acquisition.

The third type of analogue filter which is sometimes used is a notch filter. This type of filter is used, for example, to eliminate high-voltage power line interference, which is often a serious problem in land operations. Figure 3/40 shows the amplitude response curve for one such filter.

(c) The multiplexer The previous steps of amplification are performed simultaneously but separately on each seismic channel. It would however be a most inefficient and cumbersome method of digitally recording the data if each channel were to be separately recorded on tape. The function of the multiplexer is to switch through all data channels in sequence, allowing enough time (a few microseconds) for each signal to charge a capacitor to the correct signal voltage level. Thus at each sample time, for example every 2ms, every channel is sampled and the voltage level stored ready for output as a single data channel to further stages of amplification and analogue to digital conversion.

(d) Gain-ranging amplifier This part of the system is at the heart of modern seismic instrumentation and on it depend the high dynamic range and fidelity of the recording process. In analogue recording, gain control is applied directly to boost or attenuate the input signal to keep it within the system's usually stringent amplitude limits. Several methods can be applied, but they all suffer from the same drawback; the signal is irretrievably altered and the process cannot be reversed unless a sophisticated method of recorded programmed gain is employed.

The principle of the digital gain amplifier hinges on its ability to simultaneously apply to the signal, and record separately, a required value of variable gain. Thus, by reversing the procedure, the true amplitude can be recovered on playback, and this is particularly valuable in detailed stratigraphic and hydrocarbon studies. Current limitations on true amplitude recovery are confined to those associated with available display methods. The process by which the amplifier operates is comparative. As described above, after multiplexing, each portion of the sample signal is stored in a capacitor and successively compared with reference voltage outputs. Current amplifiers are of either the binary or quaternary floating point types, i.e. they involve 2^1, 6dB gain steps or 2^2 (=4), 12dB

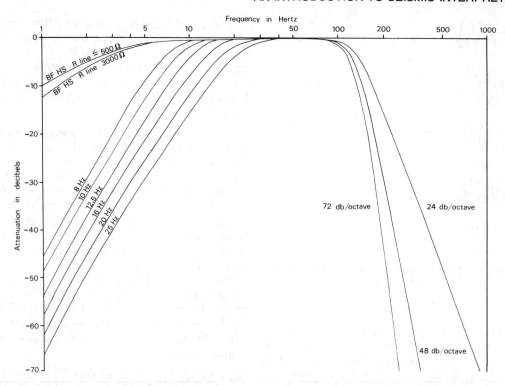

Figure 3/39 Response curves for a typical range of high and low cut-off options available for filtering input signals to a seismic recording system. This set is for use with a 2 ms sampling frequency. (Courtesy: Sercel.)

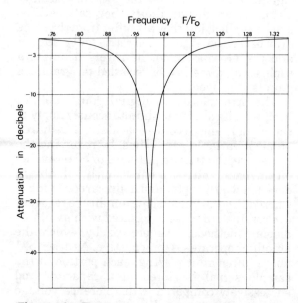

Figure 3/40 Response curve for a notch filter designed to attenuate power-line interference. (Courtesy: Sercel.)

gain steps. As an example the method used in a Sercel SN338B system is illustrated in figure 3/41. At the beginning of the sample processing cycle, the amplifier is set at the intermediate gain of 2^8. The output is fed to two comparators with two references 12dB apart. If the voltage falls within the bracket of the two voltages, the

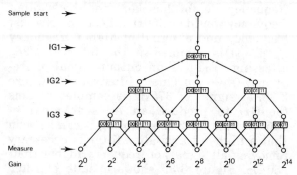

Figure 3/41 Comparator system for applying digitally recorded gain to a seismic signal. (Courtesy: Sercel).

gate is unchanged. If the output voltage falls outside this bracket, the control logic opens another gate so as to increase or decrease the gain. After three such comparisons, the amplifier has reached the proper gain as can be seen in the schematic of figure 3/41. The time required to make these comparisons and decisions is 10 μs. After the third decision is made a conversion order is transmitted to the analogue to digital convertor. The gain is thus established in the range 2^0 to 2^{14}, or 84dB.

(e) Analogue to digital convertor After each sample has had the required gain established and applied, the resultant amplifier output sample is converted into a binary code word, generally of 15 bits, made up of sign plus fourteen 0 or 1 binary digits.

It is appropriate here to consider the intrinsic signal to noise ratio of contemporary instrumentation. The maximum dynamic range of reflection signals as measured at detector output is of the order 100dB above natural background noise. If we include potentially recoverable signals below ambient level, an additional 20dB can be included. Recorder and amplifier design thus aims for noise to be held down 30dB below ambient level, and this is generally attained. The desired 120dB signal range requires, therefore, a 130dB dynamic range of the recorder and this is adequately attained by an 84dB gain range amplifier linked to an analogue to digital signal convertor of the type described above.

3.8.2 Control or logic unit

The functions of this part of the system are as follows:

(a) Time break amplifier This handles the input from the time break (shot initiation) signal with output to a separate channel to the seismic signal channels. In figure 3/35 it is defined as the shot break fixed-gain amplifier.

(b) Formatter This instrument organises the digital word output from the analogue to digital convertor (A/D convertor in figure 3/35) ready for writing on tape. Current industry accepted formats are 9 track SEG B or C, although 21 track formats are also in use. Header information is fed to the tape in advance of the data record.

(c) Digital gain control For playback purposes through the monitor camera, the recorded signals must be adjusted to fall within the dynamic range of the galvanometers of the camera. Automatic gain control is generally used.

(d) Digital to analogue (D/A) convertor This converts back the digital record onto analogue form for display on the monitor camera. Because of the limited dynamic range required, the most significant bits of the digital record are processed.

(e) Demultiplexer Again, to allow display on the monitor camera it is necessary to reverse the multiplexing process through a demultiplexer.

(f) Playback amplifier This adjusts the amplitudes of the output channels simultaneously prior to filtering.

(g) Playback filters These provide a range of analogue filter settings, high- and low-cut, which are adjusted to give the best quality monitor camera record.

(d) Total system controls This unit includes plug-in connections to all other instruments and indirectly to array inputs etc. The geophysical observer has all the necessary switches and controls to operate the complete shot-firing and signal recording process from start to finish.

3.8.3 Tape transport unit

Digital tape decks are used to record the output from the formatter onto either 9-track $\frac{1}{2}$ inch tape or 21-track 1 inch tapes in standard SEG formats. Data are recorded at 1600 bpi at speeds of between 10 and 120in/s. Systems are also available which utilise 6250 bpi tapes, thus resulting in considerable saving in tape costs as well as logistics of transport and storage.

3.8.4 Camera unit

The camera unit contains galvanometers which are driven by the output from playback channels. Wiggle trace records are produced for each channel on light-sensitive dry-write paper. Two types of read-after-write records are used. For monitoring purposes, the direct input signals can be displayed, or alternatively, as a means of obtaining a provisional seismic section, the data can be played back and displayed after application of gain control and filtering.

3.8.5 Demultiplexed field recording

A significant saving in processing time and costs can be achieved if field data are demultiplexed and recorded in a trace-sequential format. By employing such a recording system it is also possible to undertake vertical stacking and other processing as part of the field acquisition procedure. Application of such techniques to land acquisition is relatively straightforward where acquisition rates are comparatively low. Modules which can be integrated into marine recording systems are a relatively new (and costly) innovation.

3.9 Shot-firing control

In marine surveys, seismic acquisition is a continuous automated routine. Most survey vessels depend for navigation and position-fixing on either integrated satellite navigation systems, or on radio-navigation systems, usually linked to a computer and data-logger which provides control of the shot-firing sequence. For CDP stacking it is important (see figure 3/1) that distances between shot-points are constant. The complete interfacing of all systems is well illustrated in figure 3/42.

In land surveys, the firing routine is manually controlled by the observer. For surface sources the observer simply presses a firing button to initiate the shot. On dynamite crews, the observer is connected to the shooter's shot-box both by telephone and electrically, and, for safety, the shot initiation is controlled by the shooter.

3.10 Quality control

Quality control of a seismic survey is principally the responsibility of the seismic crew's party chief and observer. Modern equipment incorporates special fault finding and monitoring facilities which allow faults to be quickly discovered and, hopefully, remedied. On most surveys, the client, who is paying for the survey, will have a representative attached to the project, and it is the responsibility of the client's representative to assist in drawing up the detailed job specification and ensure that this is adhered to at all times except, with his approval, when modification or relaxation can be seen to be advantageous to the client.

3.11 Survey position-fixing and accuracy

3.11.1 Land surveys

Seismic lines on land need to be surveyed to obtain the position of shot points and the geophone spreads as well as their elevations. Prior to shooting, the shot points and the centre point of geophone spreads can be staked out on the assigned uniform spacing. Subsequently, the surveying crew makes detailed position fixes and elevation determinations using conventional land survey techniques. In remote areas, it is important to link the survey to the permanent and accurately located bench marks or beacons in the area, or to obtain accurate geodetic control from astronomical observations or by using a satellite receiver.

Modern technologies are playing an increasing role in improving on the accuracy of position fixing. Vehicle mounted inertial navigational equipment, which utilises 3-dimensionally oriented accelerometers, when properly referenced can provide a high degree of accuracy. Laser beam equipment with accuracies in the order of 1cm over many kilometres is also being introduced. Even with conventional equipment, the surveyor's speed and efficiency in the field is being improved dramatically with the use of minicomputers which with suitable data memory and plotter output peripherals can provide real-time base map generation.

In general, there is little likelihood that for any individual project, the positional inaccuracies would be so large as to cause significant seismic mis-ties (see Chapter 6) at line intersections. Only when the interpreter is handling older data from surveys of different vintages and contractors in remote areas may it be necessary to check for inconsistencies between the different surveys. Readjustment to a single datum or reference point will be necessary before interpretation commences and a comparison of line elevations at intersections can be used as a check on consistency.

3.11.2 Marine surveys

Methods of marine seismic survey positioning out of sight of land start, however, with one's feet firmly on the ground and with one's eyes raised occasionally spacewards; in this environment radio position fixing is necessary, often in conjunction with integrated satellite

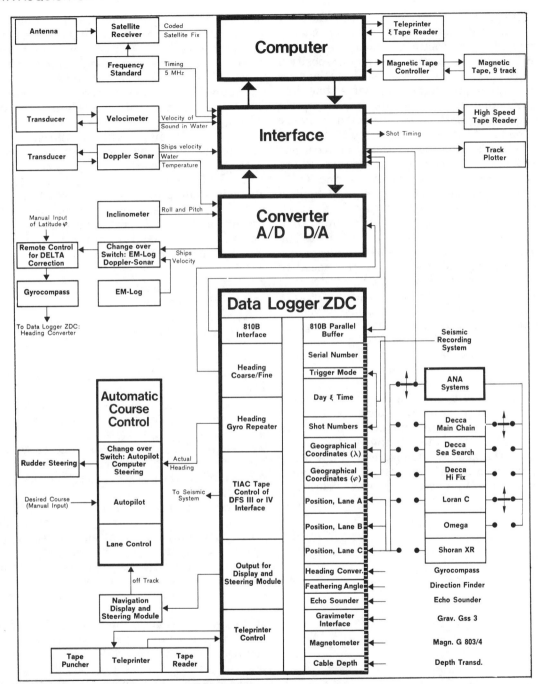

Figure 3/42 Integrated navigation and data acquisition system. (Courtesy: Prakla-Seismos).

navigation methods. Generally, the higher the frequency of the radio system, the higher the accuracy, but the less the range. Thus, the lower-frequency, longer-range (hundreds of kilometres) systems may have inaccuracies as large as hundreds of metres, whereas the high frequency short range systems are accurate to within a few metres, but can only be used in areas close to the coast or in the vicinity of a fixed platform and drilling rigs.

Radio-positioning systems obtain fixes by determining distance values between survey points and shore or platform based transmitter/receiver reference stations. At any fix-position distance values to two or more such stations are obtained at approximately the same time thus defining the location of the fix. A simple position-fixing pattern is illustrated in figure 3/43, and in this case we see the method applied to a high precision (better than 5m accuracy) survey of a limited area, an estuary. Shore stations are positioned at A, B and C. A transmitter on the survey vessel at D sends a coded radio signal to each of the stations in turn and this activates the transmitter which then sends a responding pulse. By measuring the elapsed time between transmission and reception of signal and dividing by the speed of propagation of radio waves a range to each shore-station can be measured. Motorola Mini-Range III and Decca Trisponder systems operate in this fashion. (For further reading, see Chapter 2 of McQuillin and Ardus, 1977.)

In comparing data from different surveys, the interpreter must again be aware of the possible uncertainties regarding position-fixing. The results of a single survey can show good internal consistency in terms of the fit of line intersections, but comparison of surveys using different position-fixing systems may indicate the need for adjustment, there being with all such navigation systems a difference between absolute accuracy and relocation accuracy. The accuracy with which a site may

be relocated using an identical deployment of positioning equipment is higher than the absolute accuracy of any position-fix with reference to a worldwide co-ordinate system. It is sometimes possible to calibrate a positioning chain and define fixed (as opposed to random) errors which exist within the area of cover. Again, in comparing data from different surveys it is important to check which corrections have been applied for such errors. Although radio-positioning systems such as Decca Hi-fix, Pulse 8, Loran-C, Toran etc. are still widely used to give accuracies of 100m or better (see figure 3/44) an ever increasing proportion of seismic surveys are now located using integrated satellite navigation systems. A typical configuration might be:

Satellite navigation receiver,
Four-element doppler sonar,
Gyro compass,
Interface to radio-positioning systems.

Some systems are in use which employ integration of satellite and inertial navigation systems.

Satellite navigation depends on reception of signals on board ship from a group of satellites operated by the US Government; these are the Transit satellites and at the time of writing there are six in orbit all occupying polar circuit orbits circling the globe every 107 minutes (see figure 3/45).

A satellite position is obtained only during periods when a satellite passes above the horizon as viewed from the survey vessel. The frequency of such occurrences varies with latitude, but good fixes every 1–3 hours are common. Signals are received through an omnidirectional aerial. These signals contain precise data on the satellite's orbit and the system also measures the doppler shift of the transmission of two carrier wave frequencies. A computer analyses this data to give the ship's position with reference to a satellite derived geodetic system. For accurate results the ship's speed and heading need to be well defined during the reception period of signals from the satellite, hence the need for integration with a gyro compass as well as either a radio-positioning system, or more commonly, a four-element doppler sonar. This latter device gives an accurate measurement of the ship's speed over seabed by transmitting pulses of

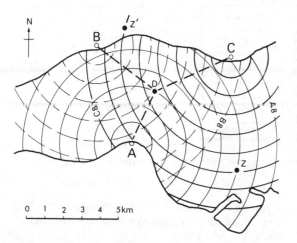

Figure 3/43 Range-range radio position-fixing pattern in an estuary (after McQuillin and Ardus, 1977).

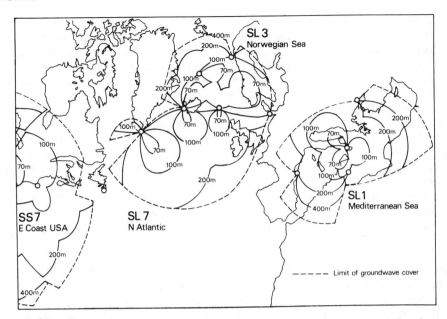

Figure 3/44 Loran-C coverage of North Atlantic and Mediterranean showing calculated positional accuracies; 67% probability.

sound in four narrow beams as illustrated in figure 3/46. Comparison of the frequencies of the fore and aft received signals gives a measure of the doppler shift from which can be computed the vessel's along-course speed. Comparison of the frequencies of the port and starboard received signals gives similarly a measure of the ship's sideward drift. A computer is at the heart of the system and this is often used to control the shooting cycle during seismic surveys to give added precision to the location of shot-points for CDP stacking. The

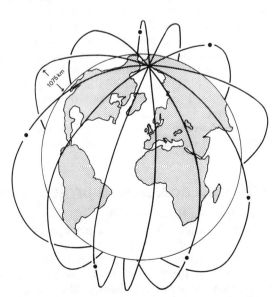

Figure 3/45 Orbit diagram of the Transit navigation satellites (after McQuillin and Ardus, 1977).

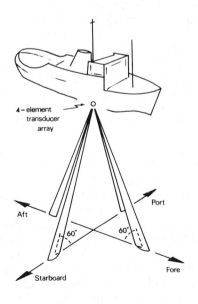

Figure 3/46 A 4-element doppler sonar (after McQuillin and Ardus, 1977).

complexity of the many relationships between the ship's movement, location of each shot, timing and data recording and the simplicity with which modern analogue sensors and digital computers can handle this is vividly summarised in the data acquisition flow chart for a state-of-the-art seismic survey vessel shown in figure 3/42.

Contractors' reports or independent quality control reports usually contain estimates of survey accuracy. However this may vary within an area surveyed, in particular if both deep and shallow water areas are under investigation. Doppler sonar can only be operated with present-day equipment in water depths of up to between 200m and 500m, depending on equipment. In deeper water the signals from seabed are too weak and the sonar system tracks on thermal layers within the body of sea water; in such circumstances positional data are generally of much lower accuracy. Integration with radio-navigation systems or an inertial navigator can obviate this difficulty. For best results, at least two good satellite passes should be recorded on each survey line. In surveys made up of a large number of short lines it can be costly to meet this condition and if positional uncertainty is suspected, lines surveyed in the absence of this control may need adjustment.

In most circumstances there should be no difficulty obtaining good line intersection ties between surveys made on the one hand with good radio navigation position-fixing and on the other with an integrated satellite navigation system. Nevertheless it should be noted that the values of latitude and longitude derived from Transit satellite data are in terms of a particular satellite derived world geodetic system and that it is necessary to apply corrections to such data to obtain positions in terms of other local or global systems such as the British National Grid or UTM. Often these corrections are applied by computer before data are output from the system but this may need to be checked if mis-ties occur between surveys.

Given that the proper precautions have been taken, most modern marine data in water depths within sonar range, or within the range of good radio navigation, should have been located to better than 50–100 m, and any necessary adjustment between different surveys should not greatly exceed this figure. As reviewed earlier, surveys using short range

radio communication may achieve high precision accuracies of approximately 5–10 m. A further advance in marine seismic survey position-fixing is expected with the release for commercial use of US Defence Global Positioning System (IGPS) satellites. By 1988, when the six satellites currently available have been increased to 21, the current 1–3 hour delay between fixes from orbiting transit satellites will be obviated. Instantaneous or real-time fixes will be provided with quoted accuracies of between 15–30m, based on single frequency use, and 10m when a dual frequency code is released for use.

3.11.3 Drilling rig site surveys

.Where an exploration or development drilling location is the end product of the seismic interpretation, the geophysicist has an important role to play in providing precise coordinates. In land work where economically attractive prospects are often of small areal extent, locating the crest of a structural feature or the optimum pay zone thickness of a stratigraphic play may require the relocation of a specific trace on the seismic line and a re-surveying of the same in the field before moving the drilling rig on site.

In the offshore area, because of the high cost, risk and potential rewards of success, a complete site re-survey around a proposed location may be undertaken. This is often done in conjunction with a bottom sampling and shallow sparker reflection survey, which will not only confirm geological and geophysical parameters, but will also aid engineers in identifying any sea bottom and shallow drilling problems such as unconsolidated sediments and gas pockets. For very precise surveys, seabed transponders may be used, to ensure exact location of a rig on a site surveyed previously by geophysical methods (see figure 3/47).

3.12 Map scales and projections

Often the results of more than one survey will be incorporated into a seismic interpretation of an area. One of the interpreter's first tasks in such a situation is to prepare base maps at a suitable scale for the interpretation in hand. A broad reconnaissance study might be mapped at 1:250 000 scale, a regional mapping exercise at 1:100 000, a more detailed mapping investiga-

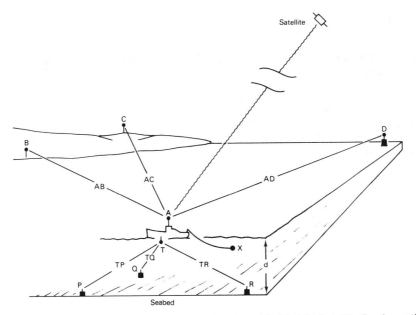

Figure 3/47 Position fixing at sea. Radio-positioning by ranging: AB;AC;AD. Satellite fix. Acoustic transponder ranging: TP;TQ;TR. Correction must be applied for different aerial positions, transponder location T and the layback of the seismic source AX,TX.

tion where a dense seismic grid and drilling data are available at 1:50 000, and a very detailed field study at 1:10 000. In some areas, field survey maps at non-metric scales may be used during data acquisition, but these are usually adjusted to an appropriate metric scale for interpretation work.

Scale adjustment may not be the only problem. It is possible that different surveys within an area, or surveys adjacent to each other, have been mapped using different map projections or co-ordinate systems, based on different central meridians, datums and spheroids. Therefore information on map projection etc. should be checked before mapping data from different surveys.

The most common map projection used in offshore exploration is the Transverse Mercator projection, and there is now wide acceptance of the Universal Transverse Mercator (UTM) system in exploration work. In such a system a relationship is defined mathematically between a grid co-ordinate system and latitudes and longitudes. Position with reference to local topography is then specified by reference to geodetic surveying of the area and a datum point which is assigned an appropriate latitude and longitude value. In Europe, the European Datum is used, and the actual location of the

datum point is the Helmert Tower near Potsdam. The North American datum block which extends from Panama to the Canadian Arctic is referred to a fixed point at a location called Meades Ranch, USA. Other areas of the globe are mapped with reference to other internationally recognised datum points. Thus any map should state not only the projection but also the datum to which topographic positions are related. A further complication is that as mapping systems have developed, different mapping agencies have adopted different spheroids as a means of computing the relationship between co-ordinate system, topographic position and latitude and longitude. The translation of positional data, mapped using one system into another can therefore be a complex problem, though easily handled by modern computers. The interpreter's principal concern is that of checking data to ensure that there is complete consistency in the mapping methods used, and that where adjustments have been necessary, these are accounted for and documented. As an example of the magnitude of errors which can arise if proper adjustments are not made, in the North Sea datum shifts between the British National Grid system, UTM using the European Datum and Satellite data are of the order of 50–100m.

References

B. S. Evenden, D. R. Stone and N. A. Anstey, *Seismic prospecting instruments*. Gebruder Borntraeger, Berlin.
Vol. 1: *Signal characteristics and instrument specifications* by N. A. Anstey (1970).
Vol. 2: *Instrument performance and testing* by B. S. Evenden and D. R. Stone (1971).
These volumes give a full account of seismic instrumentation design and performance.

J. T. Hornabrook, 'Seismic re-interpretation clarifies North Sea structure'. *Petrol. International*, **14** (1974), pp. 45–53.

F. S. Kramer, R. A. Peterson and W. C. Walter, (eds), *Seismic energy sources – 1968 Handbook*. Bendix United Geophysical, Pasadena (1968).

R. McQuillin and D. A. Ardus, *Exploring the geology of shelf seas*. Graham and Trotman Ltd., London (1977).

W. H. Mayne, 'Common reflection point horizontal data stacking techniques', *Geophysics*, **27** (1962), pp. 927–38.

Chapter 4

Data Processing

To convert the field recordings into a usable seismic section requires a good deal of data manipulation. At the very least, we want to correct our CDP gathers for normal moveout and add them together, i.e. stack them. Before doing this, we should correct our data for the effect of near-surface time delays; these adjustments are called static corrections. In addition, both before and after stack, various filtering processes can be applied, with the object of improving signal to noise ratio and increasing vertical resolution. Finally, we shall want to convert our seismic section into a form showing the true spatial disposition of the reflecting surfaces; this is achieved by the process of migration.

There is no unique processing sequence which can be applied to all seismic data; it is always necessary to balance improvement in quality against processing cost. A clear idea of the use to which the section will be put is therefore essential. The seismic interpreter has an important role to play here. If he can define a zone of particular interest, efforts to improve the seismic sections can be concentrated on this zone; it may be necessary to accept degradation of record quality in other parts of the seismic section in order to get the best definition of this prospective zone. In all except virgin exploration areas, the interpreter can also help the processor by supplying information about the area based on previous experience. For exam-

ple, it is often possible to give a generalised curve for variation of seismic velocity with depth to the processor, who can use this as a starting point for his own more detailed velocity analysis. Sometimes it is possible for the interpreter to give warning of particular problems, such as the presence of strong multiples in a particular area.

All modern seismic sections are extensively annotated with details of the processing sequence used. It is part of the interpreter's job to read and digest this information, so that he is aware of any peculiarities that may influence the data. This can be especially important when a prospect is to be mapped using data from different surveys, with different acquisition and processing parameters. After working with a data set for some time, an interpreter is usually acutely aware of its deficiencies, and may be able to suggest ways in which the data could be reprocessed to improve their quality.

4.1 Demultiplexing and amplitude manipulation

As we saw in Chapter 3, the raw data are recorded in multiplexed form. The first step in processing is therefore to re-order the data so as to reconstitute the signal for each receiver-shot combination. This signal consists of a series of amplitude values, spaced at whatever sampling interval was used during recording. This is usually 4ms in marine work, but in land or high resolution work a shorter interval is often used so that higher frequencies can be recorded without aliassing.

Since the recording amplifier gain for each sample is itself recorded on the field tape, the sample values can be adjusted to the amplitudes they would have had at a constant gain setting. In recent years, it has become known that lateral variations in the amplitudes of reflectors carry much useful information, so it has become the practice to avoid artificial tampering with the recorded amplitude values. It is, however, necessary to correct for the gradual decrease of amplitude down the record due to the spherical spreading of the source wave, and absorption along the path.

Spherical divergence would be easily corrected in a homogeneous earth, where amplitude falls off as the reciprocal of distance from the source. It would then be sufficient to apply a compensating gain which increased linearly with TWT. In the real earth, velocity usually increases with depth, leading to bending of the raypaths and a faster decline of amplitude with TWT (figure 4/1). In this case, an estimated velocity–depth relation can be used to calculate the correction required. To correct for absorption effects, a constant correction of a few dB per second of TWT can be applied in addition.

4.2 Muting

The first seismic arrivals at the receivers are not generally the reflections we wish to record, but instead energy which has propagated at shallow depth as refractions or surface waves. These unwanted arrivals can have large amplitudes, and would be a serious source of noise if not suppressed. They are edited out of the data by zeroing each trace before a certain time, which increases with increasing offset. This process is called muting. Some care is needed to make sure that all the unwanted signal is rejected; incorrect muting can cause bands of noise or even spurious events in the shallower part of the section.

4.3 Statics

Near-surface time delays arise from two distinct causes. Firstly, in the case of land data, there may be elevation changes along the line, and so we need to correct our travel-times to what they would have been if all our shots and receivers had been on a horizontal datum plane. Secondly, there may be strong velocity variations in the near-surface. On land, there is generally a surface 'weathered layer' whose seismic velocity is much lower than normal. This layer is usually some tens of feet thick, but may occasionally reach a thickness of several hundred feet. Often, the thickness of this layer is very variable along a line, leading to significant time delays of magnitude dependent on the positions of shot and receiver.

For marine data, the first type of static correction is straightforward. The source and hydrophone streamer are towed at constant depth, so the only correction required is a constant shift of all records to convert the TWTs to what they would have been with shots and receivers at the sea surface. The second type of static problem is found in marine data when there are rapid changes in water depth along a line, such as might for example be found over a submarine channel system.

If not allowed for, static shifts degrade the seismic section in two ways. If there is rapid variation of the shift along a line, with significant change within a spread length, then traces will not line up properly after NMO correction, and the quality of the stack will be degraded. Much more sinister is the effect of more gradual changes in shift along a line,

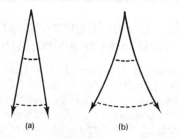

Figure 4/1 Spherical divergence: (a) constant velocity medium, (b) velocity increasing with depth.

which can introduce entirely spurious structures into the section.

4.3.1 Land statics: shots located in bedrock

Most land seismic surveys will encounter variable surface deposits ranging from clays through sands and gravel to glacial boulder clays, ice and permafrost. If dynamite is used, then where it is economically practicable, shot-holes should be drilled into solid bedrock and charges set there. Shots below the weathering zone can be simply corrected to a datum plane (see figure 4/2). Corrections consist of elevation and uphole computations. The time correction T_D to datum E_D is given by:

$$T_D = \left[\frac{2E_D - (E_S - D_S) - (E_G - D_{SG})}{V_B} \right] - T_{GUH}.$$

where E_S is the true elevation of the ground surface at shot-hole location, D_S is the depth of shot beneath ground surface, E_G is the elevation of the geophone station, D_{SG} is depth of a shot-hole located near the geophone station, and T_{GUH} is the uphole time measured at the geophone location: this is measured by placing a shot at depth D_{SG} and a geophone at the surface at location E_G and measuring the time between initiation of shot and the time-break measured at the geophone. If no shot-hole is sited at a particular geophone station, an uphole time T_{GUH} can be interpolated from uphole time measurements at adjacent shot-holes.

It should be noted that in the equation, E_D is the height difference between the local datum plane being used for adjustment of statics and the regional datum to which E_S, E_G and E_D are referenced. Generally the

regional datum is taken as sea-level and sea-level referenced elevations will be used. The velocity of bedrock, V_B, is obtained from a deep uphole survey which is always conducted in each new area. A deep hole is drilled to a point substantially below the base of the interface between weathered layer and bedrock and shots are fired at various levels in the hole to provide a plot of depth against time. The slope of the curve associated with bedrock shots gives the required velocity. An alternative method, where borehole caving might be a problem, is to shoot a series of shots at ground level with a hydrophone located at various levels in the borehole. A field example of an uphole survey plot can be seen in the upper right-hand inset of figure 4/3.

4.3.2 Land statics: shots located in weathered zones

Where it is impracticable to locate all shots in bedrock, due either to the thickness of the weathered zone or the use of non-explosive seismic sources, refraction theory must be utilised. First break times of the records are plotted against distance as shown in the actual field example of figure 4/3. In the simple schematic case, illustrated in figure 4/4, the slope $1/V_W$ is the reciprocal of the weathered layer velocity V_W and the slope $1/V_B$ is the reciprocal of the bedrock velocity V_B. Where the two lines intersect at the crossover distance, there is onset of first arrivals from bedrock. The refracted waves line is projected back to the zero X-axis to obtain an intercept time T_i. If the horizontal source-receiver distance is X, then clearly the travel time T is given by:

$$T = \frac{X - 2D_W \tan\theta_c}{V_B} + \frac{2D_W}{V_W \cos\theta_c}.$$

Thus at $X = 0$,

$$T_i = \frac{2D_W}{\cos\theta_c} \left(\frac{1}{V_W} - \frac{\sin\theta_c}{V_B} \right)$$

and since, by Snell's Law,

$$\theta_c = \arcsin \frac{V_W}{V_B},$$

then $T_i = \dfrac{2D_W \cos\theta_c}{V_W} = \dfrac{2D_W \sqrt{(V_B{}^2 - V_W{}^2)}}{V_B V_W}.$

By plotting all first breaks, and inserting the

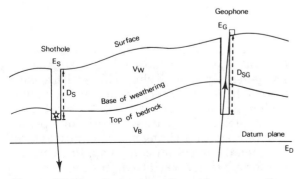

Figure 4/2 Elevation and weathered layer correction.

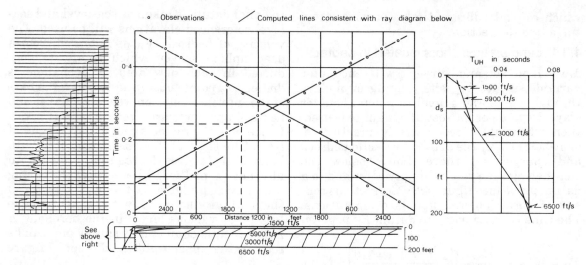

Figure 4/3 Field example of the use of refraction theory to correct for weathered layer (adapted from K. E. Burg, 1952).

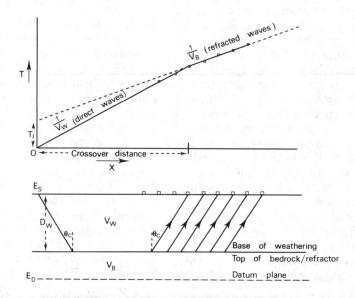

Figure 4/4 Schematic layout of a refraction survey to determine weathered layer thickness and velocity.

appropriate values of T_i, V_B and V_W, depth to bedrock refractor can be computed and a datum correction applied:

$$T_D = \frac{-2D_W}{V_W} + \frac{2(E_D - E_S + D_W)}{V_B}$$

where E_S and E_D are the surface and datum plane elevations. Here it is assumed that both shot and geophone are at ground surface. The T against X plot will vary for each shot location and spread as values of V_W, V_B, T_i, D_W and T_D vary. In multifold shooting, there are generally enough shot-points within each spread to allow interpolation of T_D values giving adequate definition of the weathered layer along the entire length of spread. Where elevation changes are significant, an additional correction can be applied correcting for elevation variation, in this case:

$$T_E = \frac{E_G - E_S}{V_B}.$$

If shots are located in boreholes, an uphole time T_{UH} will have been measured at each shot location and this should be incorporated into the correction; it will be subtracted from the total correction time, and correspondingly the shot depth D_S is subtracted from D_W.

The above refraction corrections all assume idealised and simple earth structure. Severe elevation changes can be particularly troublesome, affecting the refractor plot and therefore the intercept time: it may be necessary to make the elevation and uphole corrections before constructing plots of T agonist X, especially where good data exists on weathered layer velocities.

If the bedrock surface is undulating, the simple $T - X$ plot is apt to be hard to interpret. Hagedoorn (1959) has described a useful approach to these more complex geometries.

4.3.3 Residual statics

By taking advantage of the data redundancy inherent in the CDP method of shooting, one can refine estimates of static corrections. These methods give good control of variations in near-surface delay within a spread length, but give little information on more gradual changes. The methods described in the previous section will therefore have to be applied first, before any attempt at automatic residual static calculations.

4.3.4 Automatic static picking

Various methods of calculating statics corrections from the CDP data have been proposed. We shall outline just one of them, to illustrate the principles. Figure 4/5 shows, as an example, the traces available in a 4-fold cover system. Each trace is labelled by its shot-point and receiver numbers. We shall work with traces that have already been corrected for elevation and/or weathering, and also for NMO, by the methods discussed later in this chapter. Each trace of a CDP gather (a vertical alignment in figure 4/5) is compared with a reference trace, which may be the near-offset trace. The object is to find for each trace a time shift which will line it up as well as possible with the reference trace. Hands-off calculation of the required shift may

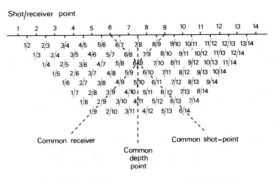

Figure 4/5 Traces available in a 4-fold CDP cover survey. Trace label a/b means shot-point a, receiver b.

be performed by using the cross-correlation function (also described later in this chapter). By these means, the entire table of figure 4/5 may be filled in, with a time-shift value for each trace. We can now average the values along a common-receiver line to obtain a time-shift applicable to this receiver alone (assuming that the shot values average to zero). Similarly, the average of the values along a common shot-point line gives a static shift applicable to that shot-point. Finally, the static correction applicable to each individual trace is the sum of the appropriate receiver and shot-point values.

Such a method is sensitive to the correctness of NMO removal. This can be checked by examining the columns of figure 4/5 for systematic variation of time-shift with increasing offset. If such an effect is present, it suggests incorrect NMO removal. This can be rectified by fitting a low-order polynomial to the column data and retaining only the residuals.

4.3.5 Marine statics

It is usual, though not invariable, to apply a bulk shift to marine data to convert the TWT to what it would have been if source and receivers had been towed at the sea-surface. The correction is simply

$$\frac{d_S + d_R}{v}$$

where d_S is the source depth, d_R the receiver depth, and v the velocity of sound in sea water. It is not usual to allow for tidal variation in the height of the sea-surface, though in some coastal areas this can amount to over 10ms in TWT. Bulk shifts may also be applied to allow for a constant time-delay between the electrical signal that fires the source and the actual detonation.

Severe variations in sea-floor topography, such as are found over submarine channels, particularly near the edges of continental shelves, give rise to significant static corrections. The velocity of sea water is about 1500m/s, and can be accurately calculated if the temperature and salinity are known, from the equation

$$V = 1449 + 4.6T - 0.05T^2 + 0.0003T^3 + (1.39 - 0.012T)(S - 35) + 0.017Z,$$

where V is in m/s, T is in °C, Z is the depth in metres below the sea surface, and S is salinity in parts per thousand. This velocity is much less than that usually found in the underlying bedrock and so sea-floor depressions will cause artificial lows in all succeeding reflectors. Required static corrections can be derived in the same manner as land elevation corrections by replacing the anomalous time intervals with those recalculated using the seabed/bedrock velocity, since the sea floor topography is known from the echo sounder record which is always obtained simultaneously with the seismic data. Such corrections are not made routinely, however, as the seabed is often rather flat over large areas.

A similar problem arises when buried near-surface channels are filled with soft sediments of seismic velocity much lower than the solid rock in which the channels are incised. Correction is more difficult in this case, as the seismic velocity in the soft sediments will not be accurately known. It may be possible to work out the time shifts required by flattening or structure smoothing a horizon passing just below the channel, using a residual static cross-correlation algorithm as described earlier in Section 4.3.4.

4.4 Signal processing concepts

In later sections we shall describe some of the types of filtering that may be applied to the seismic traces. Some aspects of signal processing theory that are essential background to this discussion will be presented in this section.

4.4.1 Linear filters and convolution

In general, a filter is a device which converts an input signal into an output signal in some deterministic way, so that we always get the same output for a given input. The 'device' may be either a physical 'black box' working on

analogue signals, or a computer algorithm working on a digitally sampled signal.

One way to specify the effects of a filter is through its impulse response. This is the output that would be obtained from an input signal consisting of a sharp spike. This can easily be visualised for a digitally sampled signal (figure 4/6).

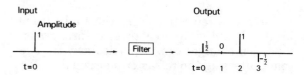

Figure 4/6 Impulse response of a filter. Note that $t = 0$, 1,2,3 . . . refers to successive sample increments so that for example at 4 ms sample interval the actual times would be $t = 0,4,8,12$ . . .ms.

All the filters we shall be concerned with are linear. This means that if we have an input which is the sum of two functions A and B, the output will be the sum of the outputs produced from A and B individually. In this case, we can calculate the output for any input by applying the impulse response to each sample in turn (figure 4/7). Algebraically, if we have an impulse response function f and an input g, then the output h is given by

$$h(t) = g(t).f(0) + g(t-1).f(1) + g(t-2).f(2)$$
$$+ \ldots + g(1).f(t-1) + g(0).f(t)$$
$$= \sum_j g(t-j).f(j).$$

This process is known as *convolution* and is often written symbolically as

$$h = g * f.$$

4.4.2 Fourier analysis

Rather than using the impulse response, it is often convenient to describe the effect of a filter in terms of its response to sine waves of different frequencies. It then becomes important to know if we can describe an input signal as the sum of a series of sine waves, so that we can calculate the output from a filter by adding the output derived from each individual input sine wave component.

The methods of Fourier analysis make possible such decomposition into sine waves. If we have a signal which is defined over a

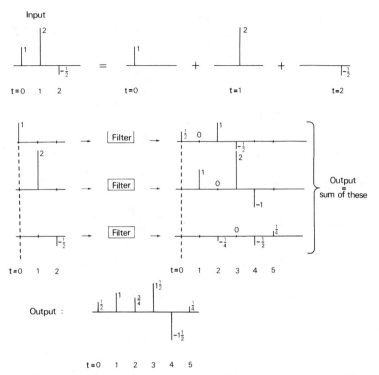

Figure 4/7 Example of application of filter shown in figure 4/6.

length of time T, then within this interval we can equate the function to a sum of sinusoidal signals having frequencies $1/T$, $2/T$, $3/T$ ..., N/T, ..., plus a constant term. This decomposition is possible provided the function is single-valued and has only a finite number of finite discontinuities; these conditions will certainly be satisfied by any signals of interest to us. To express the function exactly, an infinite number of terms is in general required, but the sum of the series approximates more and more closely to the function as N increases.

Thus, if we have a function $f(t)$ defined in the interval $t = 0$ to $t = T$, it may be written as

$$f(t) = a_0 + \sum_{j=1}^{\infty} a_j \cos 2\pi jt/T + \sum_{j=1}^{\infty} b_j \sin 2\pi jt/T.$$

At each frequency there will in general be both a sine and a cosine term present in the series. It is often useful to speak of the amplitude A and phase ϕ at each frequency, defined by:

$$A_j = (a_j^2 + b_j^2)^{\frac{1}{2}}$$
$$\phi_j = \arctan b_j/a_j.$$

Given that the decomposition into such a series is possible, it is easy to find an expression for the coefficients a_j and b_j. If we multiply the above expression by $\cos 2\pi kt/T$ and integrate from 0 to T, we obtain:

$$\int_0^T f(t) \cos 2\pi kt/T \, dt = a_0 \int_0^T \cos 2\pi kt/T \, dt$$

$$+ \sum_{j=1}^{\infty} a_j \int_0^T \cos 2\pi jt/T \cos 2\pi kt/T \, dt$$

$$+ \sum_{j=1}^{\infty} b_j \int_0^T \sin 2\pi jt/T \cos 2\pi kt/T \, dt.$$

The first term on the right-hand side is

$$a_0 \int_0^T \cos 2\pi kt/T \, dt = \frac{Ta_0}{2\pi k} [\sin 2\pi kt/T]_0^T$$

$$= 0 \text{ (for } k \neq 0).$$

A typical term within the summation of the last term is

$$c_j = b_j \int_0^T \sin 2\pi jt/T \cos 2\pi kt/T \, dt =$$

$$\frac{b_j}{2} \int_0^T (\sin 2\pi (j-k)t/T + \sin 2\pi (j+k)t/T) dt$$

$$= -\frac{b_j T}{4\pi}\left[\frac{\cos 2\pi(j-k)t/T}{(j-k)} + \frac{\cos 2\pi(j+k)t/T}{(j+k)}\right]_0^T$$

if $j \neq k = 0$.

If $j = k$, then $c_j = \frac{b_j}{2}\int_0^T \sin 2\pi(j+k)t/T\,dt$

$$=\frac{-b_j T}{4\pi(j+k)}\left[\cos 2\pi(j+k)t/T\right]_0^T = 0 \text{ also.}$$

A typical term within the summation of the middle term is

$$d_j = a_j\int_0^T \cos 2\pi jt/T \cos 2\pi kt/T\,dt$$

$$=\frac{a_j}{2}\int_0^T (\cos 2\pi(j-k)t/T + \cos 2\pi(j+k)t/T)\,dt$$

$$=\frac{a_j T}{4\pi}\left[\frac{\sin 2\pi(j-k)t/T}{(j-k)} + \frac{\sin 2\pi(j+k)t/T}{(j+k)}\right]_0^T$$

$$= 0 \qquad \text{if } j \neq k$$

If $j = k$, then $d_j = \frac{a_j}{2}\int_0^T (1 + \cos 4\pi jt/T)\,dt$

$$= \frac{a_j T}{2}.$$

Therefore, $a_j = \frac{2}{T}\int_0^T f(t)\cos 2\pi jt/T\,dt.$

By a similar argument, it may be shown that

$$b_j = \frac{2}{T}\int_0^T f(t)\sin 2\pi jt/T\,dt.$$

and $\qquad a_0 = \frac{1}{T}\int_0^T f(t)\,dt.$

We shall now work two examples in detail to illustrate the method.

First, consider the function $f(t) = t(T - t)$. The expressions for the coefficients are:

$$a_0 = \frac{1}{T}\int_0^T t(T-t)\,dt = \frac{1}{T}\left[\frac{Tt^2}{2} - \frac{t^3}{3}\right]_0^T = \frac{T^2}{6}$$

$$a_j = \frac{2}{T}\int_0^T t(T-t)\cos 2\pi jt/T\,dt$$

$$= \frac{2}{T}\int_0^T \frac{T}{2\pi j}Tt\,d(\sin 2\pi jt/T)$$

$$-\frac{2}{T}\int_0^T \frac{t^2 T}{2\pi j}d(\sin 2\pi jt/T)$$

$$= \frac{T}{\pi j}[t\sin 2\pi jt/T]_0^T - \frac{T}{\pi j}\int_0^T \sin 2\pi jt/T\,dt$$

$$-\frac{T}{\pi j}[t^2\sin 2\pi jt/T]_0^T$$

$$+\frac{2}{\pi j}\int_0^T t\sin 2\pi jt/T\,dt \text{ (integrating by parts)}$$

$$= \frac{T}{\pi j}\cdot\frac{T}{2\pi j}[\cos 2\pi jt/T]_0^T$$

$$-\frac{2}{\pi j}\cdot\frac{T}{2\pi j}\int_0^T t\,d(\cos 2\pi jt/T)$$

$$= \frac{-T}{(\pi j)^2}[t\cos 2\pi jt/T]_0^T + \frac{T}{(\pi j)^2}\int_0^T \cos 2\pi jt/T\,dt$$

$$= \frac{-T^2}{(\pi j)^2} + \frac{T}{(\pi j)^2}\cdot\frac{T}{2\pi j}[\sin 2\pi jt/T]_0^T$$

$$= \frac{-T^2}{\pi^2 j^2}.$$

$$b_j = \frac{2}{T}\int_0^T (Tt - t^2)\sin 2\pi jt/T\,dt$$

$$= \frac{-2}{T}\cdot\frac{T}{2\pi j}\int_0^T (Tt - t^2)\,d(\cos 2\pi jt/T)$$

$$= \frac{-1}{\pi j}[(Tt - t^2)\cos 2\pi jt/T]_0^T$$

$$+\frac{1}{\pi j}\int_0^T (T - 2t)\cos 2\pi jt/T\,dt$$

$$= \frac{T}{\pi j}\cdot\frac{T}{2\pi j}[\sin 2\pi jt/T]_0^T$$

$$-\frac{2}{\pi j}\cdot\frac{T}{2\pi j}\int_0^T t\,d(\sin 2\pi jt/T)$$

$$= \frac{-T}{(\pi j)^2}[t\sin 2\pi jt/T]_0^T$$

$$+\frac{T}{(\pi j)^2}\int_0^T \sin 2\pi jt/T\,dt$$

$$= \frac{-T}{(\pi j)^2}\cdot\frac{T}{2\pi j}[\cos 2\pi jt/T]_0^T = 0.$$

So $t(T - t) = \dfrac{T^2}{6} - \dfrac{T^2}{\pi^2}\cos 2\pi t/T$

$$-\frac{T^2}{4\pi^2}\cos 4\pi t/T - \frac{T^2}{9\pi^2}\cos 6\pi t/T \ldots$$

Figure 4/8 shows how the function is approximated by a number of terms. For this smoothly varying function, a few terms of the series

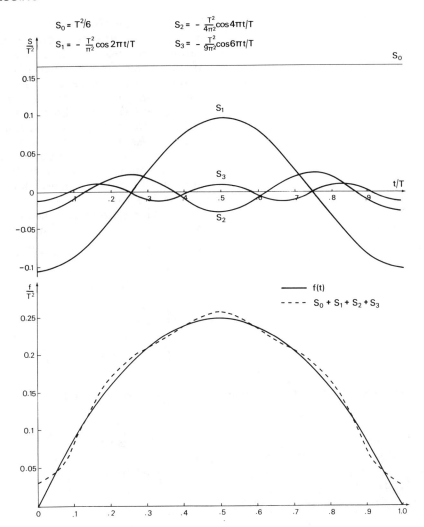

$S_0 = T^2/6$ $\qquad S_2 = -\frac{T^2}{4\pi^2}\cos 4\pi t/T$

$S_1 = -\frac{T^2}{\pi^2}\cos 2\pi t/T$ $\qquad S_3 = -\frac{T^2}{9\pi^2}\cos 6\pi t/T$

Figure 4/8 Fourier decomposition of $f(t) = t(T-t)$.

are enough to give quite a good approximation. The amplitude and phase spectra are easily found, for

$$A_j = T^2/\pi^2 j^2 \text{ for } j > 0$$

$$A_0 = T^2/6$$

and $\phi_j = \arctan\left(\dfrac{0}{-T^2/\pi^2 j^2}\right) = 180°$

The phase is therefore the same for all terms; as figure 4/8 shows, each term has a negative maximum aligned at $t = 0$. The amplitude spectrum is shown in figure 4/9. As one might intuitively expect, this smoothly varying function is rich in low frequencies and poor in higher frequencies.

Another example is provided by the 'box-car' function,

$f(t) = 1$ for $t = T/4$ to $3T/4$
$f(t) = 0$ otherwise.

In this case we shall have

$$a_0 = \frac{1}{T}\int_0^T f(t)\,dt = \frac{1}{T}\,[t]_{T/4}^{3T/4} = \tfrac{1}{2}$$

$$a_j = \frac{2}{T}\int_0^T f(t)\cos 2\pi jt/T\,dt$$

$$= \frac{2}{T}\cdot\frac{T}{2\pi j}[\sin 2\pi jt/T]_{T/4}^{3T/4}$$

$$= 0 \qquad \text{for } j = 4n$$

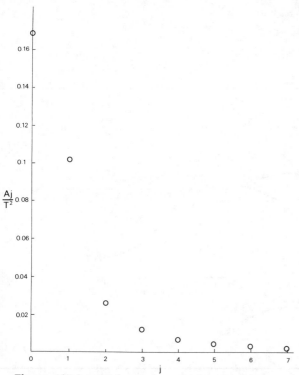

Figure 4/9 Amplitude spectrum for function of figure 4/8.

So, with $a_j = 0$
for $j = 4n$

$-2/\pi j$ for $j = 4n + 1$
$0 \qquad$ for $j = 4n + 2$
$2/\pi j \quad$ for $j = 4n + 3$, where $n = $ any integer.

$$b_j = \frac{2}{T} \int_0^T f(t) \sin 2\pi jt/T \, dt$$

$$= \frac{-2}{T} \cdot \frac{T}{2\pi j} \left[\cos 2\pi jt/T \right]_{T/4}^{3T/4}$$

$$= 0.$$

Thus $f(t) = \tfrac{1}{2} - \dfrac{2}{\pi} \cos 2\pi t/T + \dfrac{2}{3\pi} \cos 6\pi t/T$

$- \dfrac{2}{5\pi} \cos 10\pi t/T \dots$

Figure 4/10 shows how the function is approximated by an increasing number of terms. The function is poorly approximated by the first few terms of the series, owing to the discontinuities at $t = T/4$ and $3T/4$. In this case, the phase shows an alternation between 0 and 180° for successive non-zero

terms. The amplitude spectrum is shown in figure 4/11. Only the odd harmonics (odd j values) are present, but their amplitudes fall off rather slowly with increasing j.

As we take more and more terms, the approximation to the function gradually improves, but the overshoot peaks adjacent to the discontinuities are found to remain present. As more terms are included, the overshoots become narrower, but do not reduce in amplitude; even with a very large number of terms, an overshoot remains.

Altering the width of the 'box-car' alters the Fourier representation in an interesting way. Consider the function

$f(t) = 1$ for $t = T/2 - \tau$ to $t = T/2 + \tau$
$f(t) = 0$ otherwise.

Then we shall have

$$a_0 = \frac{1}{T} \left[t \right]_{T/2 - \tau}^{T/2 + \tau} = 2\tau/T$$

$$a_j = \frac{2}{T} \cdot \frac{T}{2\pi j} \left[\sin 2\pi jt/T \right]_{T/2 - \tau}^{T/2 + \tau}$$

$$= \frac{1}{\pi j} \cdot 2 \cos \pi j . \sin 2\pi j\tau/T$$

$$b_j = -\frac{2}{T} \cdot \frac{T}{2\pi j} \left[\cos 2\pi jt/T \right]_{T/2 - \tau}^{T/2 + \tau}$$

$$= \frac{1}{\pi j} \cdot 2 \sin \pi j . \sin 2\pi j\tau/T$$

$$= 0.$$

As $f(t)$ comes to represent a sharper pulse, i.e. $\tau \to 0$, then

$$\sin 2\pi j\tau/T \to 2\pi j\tau/T$$

so $a_j \to 4 \dfrac{\tau}{T} \cdot \cos \pi j$.

Therefore, all the a_j will come to have the same size, independent of j. The amplitude spectrum is therefore flat; all frequencies are present with equal amplitude. This result can be proved more generally; an ideally sharp pulse is a signal whose amplitude spectrum is flat.

4.4.3 Filters in the frequency domain

Fourier analysis allows us to think about the filtering of time series from a standpoint

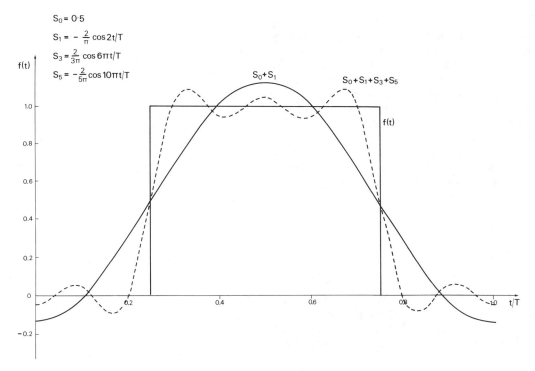

$$S_0 = 0.5$$

$$S_1 = -\frac{2}{\pi}\cos 2t/T$$

$$S_3 = \frac{2}{3\pi}\cos 6\pi t/T$$

$$S_5 = -\frac{2}{5\pi}\cos 10\pi t/T$$

Figure 4/10 Approximation of a box-car function by Fourier series.

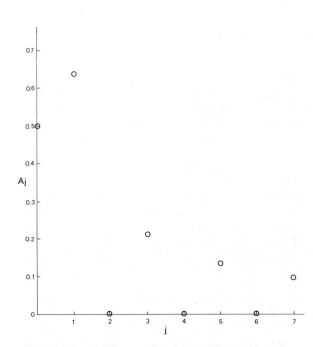

Figure 4/11 Amplitude spectrum for function of figure 4/10.

different from that of Section 4.4.1. We can express the effect of a filter in terms of the changes caused in the amplitude and phase spectra of the signal. Thus, we can specify two functions $A(j)$ and $\phi(j)$. The effect of the filter is then to be calculated as follows. First, we find the amplitude and phase spectra of the input signal by the method described above. The amplitude and phase at each frequency is then modified by the filter responses A and ϕ: for the jth term, the amplitude is multiplied by $A(j)$, and $\phi(j)$ is added to the phase. The output from the filter is then the time series corresponding to the modified spectra.

An example is a simple low-pass filter, which leaves phase unchanged ($\phi(j) = 0$ for all j), but removes all signal at frequencies higher than some particular value:

$$A(j) = 1 \text{ for } j \leqslant j_0$$
$$A(j) = 0 \text{ for } j > j_0$$

Suppose that $j_0 = 3$ and we apply the filter to the function of figure 4/8. The effect of the filter will be to pass unchanged the terms S_0, S_1, S_2 and S_3, but to remove all the higher-frequency terms completely. The function we should recreate in the time-domain from $S_0 + S_1 + S_2 + S_3$ is shown in figure 4/8. In this case, removal of the higher-frequency components would have little effect on the signal. A similar filter applied to the box-car function (figure 4/10) would have a drastic effect, however.

It can be shown that there is a simple equivalence between the time and frequency domain expressions of a filter. The quantities $A(j)$ and $\phi(j)$, which define the filter response in the frequency domain, are themselves the Fourier representation of the impulse response function, which defines the filter response in the time domain.

4.4.4 Cross-correlation and auto-correlation

At various processing stages, we can be faced with the problem of comparing two different traces which are similar but not identical, one of which is delayed in time relative to the other. We wish to find the time shift to apply to one of the traces to align it as well as possible with the other. One example of such a requirement has been mentioned in Section 4.3.4. Cross-correlation is a tool that may accomplish this object.

The principle is shown in figure 4/12. The idea is to slide one trace past the other; at each time delay (sometimes called a lag), the series are multiplied together term by term and the results added. The cross-correlation has a maximum value when the two traces are in best alignment with one another, which in this case occurs at a time-shift of 2 units. Formally, the cross-correlation $\phi_{gf}(t)$ between two functions g and f can be defined as

$$\phi_{gf}(t) = \sum_j g(t + j)\, f(j).$$

Some caution is needed in using cross-correlation to find time-shifts for trace alignment. If the traces are strongly affected by multiples, alignment of the multiple on one trace with the corresponding primary on the other may give a high value of the cross-correlation at an incorrect time-shift, as in figure 4/12 at a lag of 5 units. Also, cross-correlation performs poorly when two traces are similar in shape but one is stretched a little relative to the other.

A useful concept is that of correlation of a

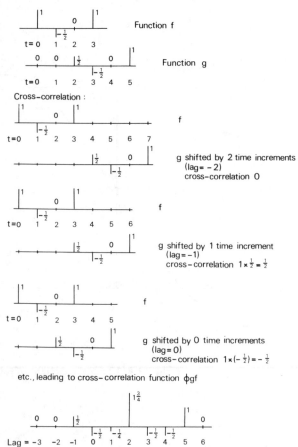

Figure 4/12 Example of cross-correlation.

signal with itself. This is called an auto-correlation function (ACF). Two examples are shown in figure 4/13; the autocorrelation (unlike the cross-correlation of two different signals) is obviously symmetrical about zero delay, and so we need to show the function only for positive lags. In figure 4/13(a) we have a slowly varying waveform. The ACF is clearly giving us some information about this signal, but it is not obvious how much. It can, in fact, be shown that the amplitude spectrum of the signal may be derived from the ACF by Fourier analysis, but that there is no information on the phase spectrum of the signal contained in the ACF. Thus, to reconstruct a signal from its ACF, we would have to make some assumption about the phase spectrum.

The example of figure 4/13(b) is a crude representation of a waveform with an attendant multiple after four time units. The ACF shows a

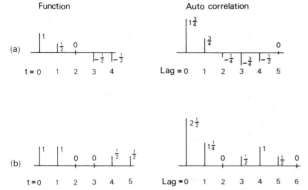

Figure 4/13 Example of the auto-correlation function.

subsidiary maximum at a lag of four units. This suggests that the ACF can be used to recognise the presence of multiples, and estimate their reverberation period.

4.5 Deconvolution

An ideal seismic trace would contain only a series of spikes at TWTs corresponding to reflectors in the earth. A real seismic trace departs considerably from this ideal. The source does not emit an ideally sharp pulse, but rather a wavelet which may have a complex waveform, as we have seen in Chapter 3. On its journey through the earth, this waveform is changed gradually, by absorption and the effect of multiples. The recording instruments will distort the seismic trace, because the ideal

recorder with a flat amplitude and phase response is not physically attainable. The instrument response is fairly easy to measure and correct for, but dealing with the other effects is much harder.

Neglecting the effect of noise, and assuming instrumental response has been corrected for, we can regard the seismic trace as a convolution of the reflection coefficient series of the earth with a wavelet. The aim of deconvolution processes is to undo this convolution to arrive back at the reflection coefficient series. If this is not done, interference between wavelets reflected from closely-spaced horizons will result in complex waveforms which will be very sensitive to the exact reflector spacing. It will then be difficult to identify individual closely-spaced reflectors, or interpret lateral changes in waveforms (figure 4/14).

The decomposition into wavelet and reflection coefficient series is not unique, for we may choose to include multiples either in the wavelet or in the reflection coefficient series. Often it is useful to include the shorter-period multiples, including ghosts and reverberations, in the wavelet, but not the long-period multiples (with time delays greater than a few hundred ms).

4.5.1 Spiking deconvolution

The observed seismic trace S is the convolution of the earth's impulse response R (the set of reflection coefficients we actually want) with the wavelet W:

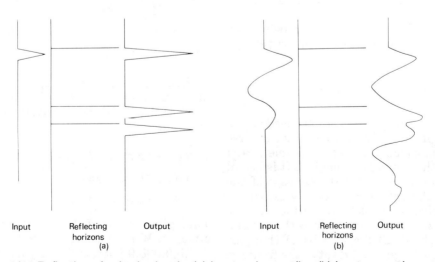

Figure 4/14 Reflection of seismic signals; (a) input a sharp spike, (b) input a complex waveform.

$$S = R * W.$$

Suppose we know an operator D which, when convolved with the wavelet, produces a signal δ consisting of a single unit-amplitude spike at $t = 0$:

$$D * W = \delta.$$

Then applying D to the seismic trace will give

$$D * S = D * R * W = D * W * R = \delta * R = R,$$

since the convolution operation is associative and commutative, and convolution with the spike δ leaves a function unchanged.

Let us for the moment suppose that we know the wavelet W, and wish to design a spiking operator D, often called the inverse of the wavelet W. A full treatment of this problem has been given, e.g. by Robinson (1980). The simplest case is for a 2-element wavelet $(1, k)$. It can be shown that the operator D is then $(1, -k, k^2, -k^3, k^4 \ldots)$. Provided that $|k| < 1$, this operator is stable in the sense that successive terms decrease in size, so truncation after a finite number of terms yields a useful approximate operator. Thus, for the wavelet $(1, -1/2)$, the spiking operator D will be $(1, 1/2, 1/4, 1/8, 1/16 \ldots)$. If we truncate this after three terms we shall have an approximate operator $(1, 1/2, 1/4)$. When convolved with the wavelet, this operator gives $(1, 0, 0, -1/8)$, with fairly good spiking. However, if $|k| > 1$, the inverse is unstable with terms successively increasing in size, so that truncation does not give a useful approximate operator.

Now, a wavelet of any length can be written as a convolution of such 2-element wavelets:

$$W = (W_0, W_1, W_2 \ldots W_n)$$
$$= (a_0, a_1) * (b_0, b_1) * \ldots * (p_0, p_1),$$

where n 2-element wavelets will be needed. We can find the inverse of W by finding the inverse of each component 2-length wavelet, and convolving these inverses together. It is useful at this stage to introduce the concept of the minimum delay (often equivalently called the minimum phase) wavelet. Note first that the wavelet (a_0, a_1) has the same ACF as the wavelet (a_1, a_0). We shall call the wavelet of this pair which has the larger coefficient at the front the minimum delay wavelet, and that with the

larger coefficient at the end, the maximum delay wavelet. A whole suite of wavelets, all with the same ACF as W, can be generated by switching the order of the coefficients within some or all of the component 2-element wavelets. There will be one wavelet of the suite which is formed by convolving a series of minimum delay 2-element wavelets. This is called the minimum delay wavelet of the suite. Similarly, one can build up a maximum delay wavelet by convolving all the maximum delay component wavelets, or a mixed delay wavelet by taking some minimum and some maximum delay components. It is clear that the above process will generate a stable inverse for W provided it is a minimum delay wavelet.

It is not generally obvious from inspection of a waveform whether or not it is minimum delay. It is true that the minimum delay wavelet is front-end loaded, with the largest amplitudes at or near the start of the signal, but this is not a reliable intuitive test; it is, however, possible to devise a mathematical test for the minimum delay property, but this is beyond the scope of this chapter.

For practical use, we need inverses of finite length. It is possible to produce a better approximate inverse than that given by simply truncating the exact inverse. The least error (in an rms sense) due to finite inverse length is caused when we calculate D, of length $(m + 1)$, from the equations:

$$d_0\phi_0 + d_1\phi_1 + \ldots + d_m\phi_m = W_0$$
$$d_0\phi_1 + d_1\phi_0 \ldots + d_m\phi_{m-1} = 0$$
$$\vdots$$
$$d_0\phi_m + d_1\phi_{m-1} + \ldots + d_m\phi_0 = 0,$$

where ϕ is the ACF of W.

Thus, for example, suppose we wish to calculate an inverse of length 3 for the wavelet $(7, -3, 1)$. The ACF of this wavelet is $(59, -24, 7)$, so we have:

$$59\, d_0 - 24\, d_1 + 7\, d_2 = 7$$
$$-24\, d_0 + 59\, d_1 - 24\, d_2 = 0$$
$$7\, d_0 - 24\, d_1 + 59\, d_2 = 0$$

This gives $D = (0.145, 0.062, 0.008)$ whose effect on W is to convert it to $(1.015, -0.001, 0.015, 0.038, 0.008)$, quite a good approximation to the desired spike.

More generally, it is possible to design shaping filters which will convert W to any desired waveform X. The best operator for this purpose is found from a set of equations similar to the above but with the right-hand side replaced by ϕ_{xw}, the cross-correlation of the desired and input wavelets: $\phi_{xw}\ (t) = \sum_j x(j + t) . w(j)$. Thus, suppose we wish to convert $(7, -3, 1)$ to $(3, 1)$. Then we should have:

$$59\, d_0 - 24\, d_1 + 7\, d_2 = 18$$

$$-24\, d_0 + 59\, d_1 - 24\, d_2 = 7$$

$$7\, d_0 - 24\, d_1 + 59\, d_2 = 0$$

These equations give $D = (0.428, 0.326, 0.082)$ whose effect on W is to convert it to $(2.996, 0.998, 0.024, -0.080, 0.082)$ which is rather close to the desired wavelet. For a shaping filter of given length, the best results (least error in the output) are obtained when both input and output have the same delay properties, e.g. both minimum delay as in the example here.

If the input wavelet is not minimum delay, better spiking may be achieved if the design output is a spike at a later time than $t = 0$; the error in the result is critically dependent on the choice of spike location.

To summarise, if we know the shape of the wavelet we can convert it to a different, more desirable shape, though whether an operator of realistic length (say a few tens of points) will perform adequately depends on whether the input and output have the same delay properties. The problem in applying this technique to the seismic trace is that the wavelet is not, *a priori*, known. It can, however, be calculated directly from the seismic trace under certain assumptions. If the earth reflectivity coefficients form a random series, and if the effect of noise can be neglected, then it can be shown that the ACF of the seismic trace is the same as the ACF of the wavelet. As we saw in Section 4.4.4, this not enough to define the wavelet, because the ACF gives us no information on the wavelet phase spectrum. However, if we are prepared to assume that the wavelet is minimum phase, then it is reasonable to try to convert it to a spike at zero time. In this case, as in the example above, the right-hand side of our equations for D will have the form $(W_0, 0, 0, \ldots 0)$. We can thus calculate D, using the autocorrelation of the seismic trace in the left-hand side of the equations. The unknown W_0 is a scale factor which applies to the entire solution for D.

Several points may be made about this simple spiking deconvolution. Firstly, it is usually unwise to aim for complete collapse of the waveform to a spike. If we think of our process in the frequency domain, we are trying to convert our wavelet, which may show appreciable energy in the frequency range (say) 5–60 Hz, into a spike, in which all frequencies are present with equal amplitude. This means that very low and very high frequencies will be considerably amplified. Since there will be little or no signal present at these frequencies, we shall simply amplify noise on the trace. Aggressive spiking therefore degrades signal to noise ratio, and a balance between wavelet contraction and noise amplification has to be struck experimentally for a given data set. A technique for partial spiking will be explained in the next section.

Secondly, the process aims to collapse the wavelet, but the resulting waveform is not under firm control. Rather than aim for the unattainable ideal spike, it may be better to try for a more modest reshaping of the wavelet, which may be carried out to rather higher accuracy. Otherwise, it may be difficult to draw inferences from detailed changes of waveform across a section.

Thirdly, the assumptions of a random earth reflection coefficient series and of a minimum phase wavelet may not be true in practice, though they are usually a good approximation.

Finally, the wavelet changes on its passage through the earth, partly because of the low-pass filter effect discussed in Chapter 2, but also because the wavelet includes all reverberations with periods short enough to be included in its length. It may therefore be necessary to design different operators for use over different TWT ranges.

4.5.2 Dereverberation

The key to removal of reverberations from the seismic record is that they are essentially predictable, and therefore may in principle be subtracted out. A simple example will illustrate the point (Backus, 1959). Reverberation in the water-layer is a common problem in marine seismic data (figure 4/15). The reflection coefficient at the water surface is nearly -1; suppose the reflection coefficient at the seabed is k (often large compared with reflection

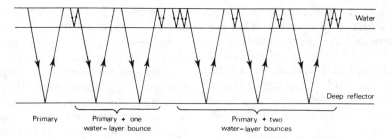

Figure 4/15 Water layer reverberant system.

coefficients at deeper interfaces). If the TWT in the water-layer is T, a primary of amplitude A will be followed by a series of multiples, as follows. There are two different possible paths for the multiple undergoing one water-layer bounce; therefore, the primary will be followed after a time T by a multiple of strength $-2kA$. There are three possible paths for the two-bounce multiple, giving an amplitude $3k^2A$ at time $2T$ after the primary. Thus, the received signal will be $(A, -2kA, 3k^2A, -4k^3A, \ldots)$, where for simplicity we have departed from our usual practice of showing signal amplitudes at the recording sample increment (e.g. 4ms), in favour of an increment time T. If we now convolve our signal with the operator $(1, 2k, k^2)$, we arrive at the result $(A, 0, 0, \ldots)$, as shown in figure 4/16. A very simple operator therefore suffices to remove the entire reverberant train.

In practice, the reverberant energy to be removed will have a more complex time-structure than in this simple example. It may be removed by the method of predictive deconvolution (Peacock and Treitel, 1969). The basic idea is that one can design a prediction operator P, with prediction distance α, such that its effect on a signal S is to produce an estimate of the signal at a future time, i.e.

$$y(t) = \sum_{\tau} S(\tau).P(t - \tau),$$

where $y(t)$ is an estimate of $S(t + \alpha)$. The difference between the actual value of $S(t + \alpha)$ and the predicted value $y(t)$ defines an error series:

$$e(t + \alpha) = S(t + \alpha) - y(t).$$

The error series $e(t)$ thus represents that part of

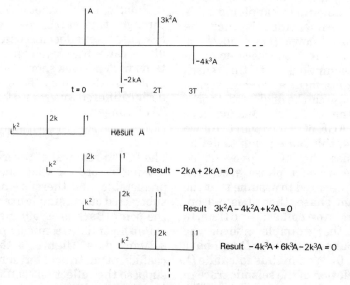

Figure 4/16 Removal of reverberant train.

S which is not predictable from the previous behaviour of the signal. Neglecting the presence of noise, we may hope that the seismic trace is the convolution of an uncorrelated reflection coefficient series with a reverberating wavelet. The above procedure removes the predictable portion of the trace, which will be largely due to the repetitive reverberations, and leaves the signal $e(t)$ which represents the desired reflection coefficient series.

We can compute the error series in a single step by means of a prediction error filter. If the prediction operator is $(p_0, p_1, p_2, \ldots p_n)$, then the corresponding prediction error operator is $(1, 0, 0, \ldots, 0, -p_0, -p_1, -p_2, \ldots, -p_n)$.

$$\underbrace{}_{(\alpha - 1) \text{ zeros}}$$

It can be shown that the best estimate of P is obtained from the equations:

$$p_0\phi_0 + p_1\phi_1 + \ldots + p_n\phi_n = \phi_\alpha$$

$$p_0\phi_1 + p_1\phi_0 + \ldots + p_n\phi_{n-1} = \phi_{\alpha+1}$$

$$\vdots$$

$$p_0\phi_n + p_1\phi_{n-1} \ldots + p_n\phi_0 = \phi_{\alpha+n},$$

where ϕ is the autocorrelation function of the seismic trace. The spiking filter discussed in the previous section is a special case of the prediction error filter, with prediction distance equal to 1.

Various properties of the prediction error operator are demonstrated by Peacock and Treitel. It removes the predictable energy having periods between α and $(\alpha + n)$ time units, so that the ACF of the output will tend to vanish for lags between α and $(\alpha + n)$. By varying α, we can control the amount by which we are contracting the wavelet. As we saw in the last section, too small a value of α may introduce excessive noise. Also, the trace is not scaled by an arbitrary factor, as we found to be the case in spiking deconvolution, so trace-to-trace amplitude variations are preserved. It can also be shown that, for the simple reverberative system with which we began this section, the prediction error operator is the same as the simple operator we used to remove the entire reverberant train.

This approach to deconvolution is superior to the simple spiking deconvolution of the last section, because it is much more under our control. No real attempt is being made to shape the wavelet, however; it is simply passed intact for the prediction distance, and has its reverberant tail removed thereafter. Ideally, we should like to convert the wavelet into some optimum shape, which may be attempted by methods discussed in the next section.

4.5.3 Wavelet processing

From the interpreter's point of view, the ideal wavelet would be a single spike at $t = 0$. As we have seen, this ideal is not attainable. However, not all wavelets of a given frequency content are equally desirable. The minimum phase wavelet has the disadvantage that the arrival time of the reflection is marked by the onset of the wavelet, not by its peak amplitude, which follows at some later time. A better shape is provided by the zero-phase wavelet, which is a symmetrical wavelet with a single peak at the arrival time, and the smallest possible side-lobes. This will make it as easy as possible to pick reflection times accurately, and draw inferences from changes in waveform across a section; it will also make comparison of the seismic section with a well log as simple as possible.

If we knew the actual wavelet, we could convert it to a zero-phase form by using a shaping filter as discussed in Section 4.5.1. Unfortunately, it is not easy to establish the actual wavelet. It may be possible to measure directly the wavelet produced by the source, though this is not straightforward; it is easy enough to measure the signature of a point source, at least in the marine case, but the signature of a source array (such as is most often used in practice) can be recorded only at a hydrophone deep beneath the source, owing to interference effects between the individual elements of the array. Thus marine source signatures can be recorded only in water deep enough so that reflections do not interfere. Various methods have been used to derive the wavelet from the seismic data. These include use of the seabed reflection waveform in marine data, and the derivation from averaged trace auto-correlations on the usual minimum phase assumptions.

4.6 NMO correction and stacking

As we saw in Section 2.6, we wish to correct the seismic traces of the CDP gather for normal moveout (NMO), and then add them together.

We assume that the travel time t at offset x is related to the zero-offset travel time t_0 by

$$t^2 = t_0{}^2 + x^2/v_s{}^2$$

where v_s is a stacking velocity approximately equal to the rms average velocity to the reflector concerned. Determination of the NMO correction is therefore equivalent to analysing the seismic velocity structure of the section. In principle, every CDP gather contains velocity information. To cut down the work involved, it is usual to assume that velocities do not vary rapidly along the section unless there are extreme lateral structural or stratigraphic changes. It is then sufficient to carry out a velocity analysis every few km along a section, interpolating between them as required. The locations of the velocity analyses should be chosen with regard to the geological structure. There may, for example, be a marked change in

the velocities on crossing a major fault or stratigraphic feature. It would then be sensible to make an analysis on either side of the fault, with the locations as close to the fault as possible. However, a fault zone itself should be avoided, because in regions of complex structure some of the seismic events will be diffractions, or reflections from out of the plane of the section, either of which would generate meaningless apparent velocities.

Obviously, it would be possible to measure t by hand at various offsets of a CDP gather, plot a graph of t^2 against x^2, fit a best straight line and thus deduce v_s. A complete analysis would require this process to be carried out for a number of reflectors at different TWTs. In practice, the large volumes of data involved dictate the use of more automated methods.

One possible technique is the constant velocity gather (CVG). If we assume that the

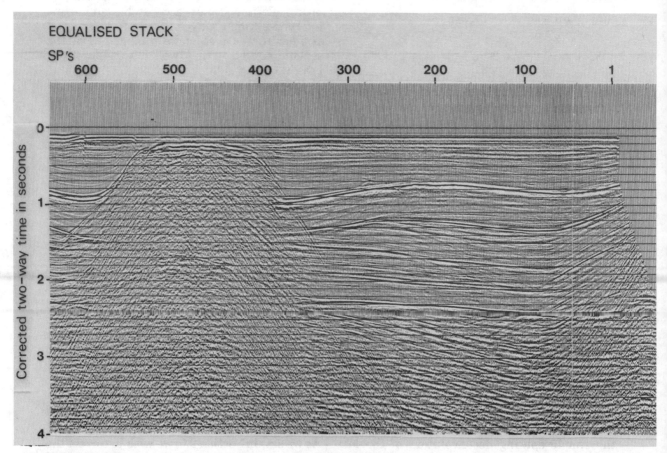

Figure 4/17 Equalised stack over a salt dome structure to which velocity test data in figures 4/18 and 4/19 refer (Courtesy: SSL).

seismic velocity has a constant specified value throughout the raypath, we can calculate the NMO for each trace as a function of TWT. The traces of the gather are corrected for NMO using these calculated values. Where NMO has been correctly removed, a reflector will line up horizontally across the gather. If the true stacking velocity to a reflector is less than the velocity assumed in constructing the CVG, the event will be under-corrected and will bend down toward the larger offset traces; if the true velocity is greater than that assumed, the event will be over-corrected and bend up at the larger offsets. Thus, we can estimate the TWT where our assumed velocity is equal to the desired stacking velocity (figures 4/17 and 4/18). If we repeat this procedure for a series of different assumed velocities, we can build up a picture of stacking velocity as a function of TWT. A disadvantage of this method is that alignments across the gather are difficult to see if signal to noise ratio is poor.

A method that should work better for noisy data is the constant velocity stack (CVS). The idea is to calculate NMO using a constant velocity, just as in the CVG; these NMO corrections are then applied, and the data are stacked. A few (e.g. 10) adjacent gathers are corrected and stacked in this way, and the results displayed (figure 4/19). The best stack (strongest event) is produced where NMO has been correctly removed. It is not easy to pick velocities very precisely by this method, however, as quantitative interpolation between the various velocity panels is difficult.

Another method is to analyse a CDP gather by finding automatically the 'best-fit' value of velocity, as follows. At a given TWT, choice of a stacking velocity defines a hyperbolic time-offset relation across the gather. We calculate these hyperbolae for a wide range of velocities. Along each hyperbola, we then measure the correlation between traces, using, for example, a 50ms window (figure 4/20). This calculation is

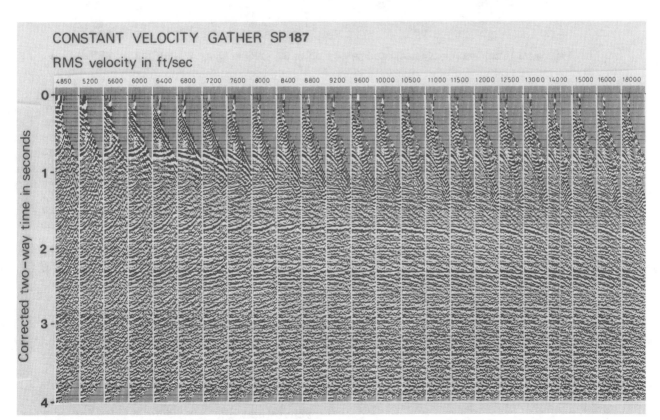

Figure 4/18 Constant velocity gather. Note for example events at approximately 1.7 s TWT with a velocity of about 10 000 ft/s. (Courtesy: SSL.)

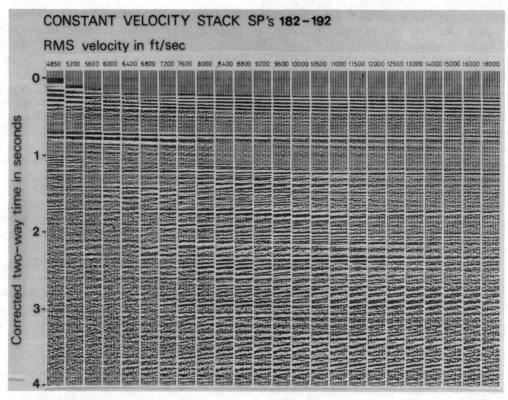

Figure 4/19 Constant velocity stack at same location as the CVG in figure 4/18. (Courtesy: SSL.)

repeated for closely spaced TWTs down the record. The results can be presented in various ways, of which one is shown in figure 4/21. Here, the degree of correlation has been contoured as a function of stacking velocity and TWT. Such a display can be interpreted by hand, joining together the main peaks with straight lines. Note that the strong peaks around 4800ft/s have been ignored. They are due to reverberation in the water layer, and therefore have an NMO appropriate to the velocity of sound in water. Towards the bottom of the record, there are no significant peaks because there are no reflections above the noise level. The velocity curve can be continued by assuming a plausible interval velocity below the last valid pick, and calculating the corresponding rms velocities from the Dix formula (see below). Beyond a TWT of about 4s, the NMO is generally small, and the quality of the stack is not very sensitive to the exact stacking velocity used.

It is useful to check the picked velocities for consistency along the seismic line. They may be contoured as shown in figure 4/22. Sudden changes from one velocity analysis to the next may be related to geological structure (e.g. a major fault), but in the absence of a plausible geological explanation any anomaly should be re-investigated.

Both as a check on the plausibility of a velocity analysis, and also in the hope of

Figure 4/20 NMO hyperbolae on a CDP gather.

Increasing offset

Two-way time

Velocity too high

Correct stacking velocity

Velocity too low

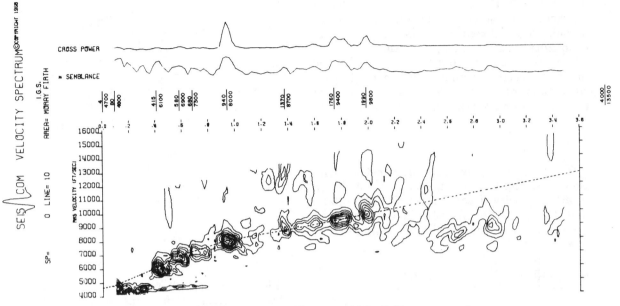

Figure 4/21 Velocity spectrum. (Courtesy: *IGS*; *Seiscom survey*.)

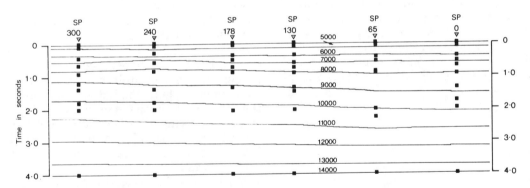

Figure 4/22 Velocity profile along a seismic line (Courtesy: *IGS*; *Seiscom survey data*).

making geological (e.g. lithological) inferences from the stacking velocities, it is often useful to work out the interval velocity between two reflectors. This is the average velocity between two depths, and can be calculated from the Dix formula (figure 4/23):

$$v_{\text{interval}} = \left(\frac{T_2 v_{\text{rms2}}^2 - T_1 v_{\text{rms1}}^2}{T_2 - T_1} \right)^{\frac{1}{2}}.$$

In using this formula, one generally assumes that the stacking velocities are true rms velocities, though it may be necessary to correct them for dip (Section 2.6). The interval velocity calculated from the formula is, strictly speak-

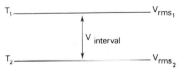

Figure 4/23 Calculation of Dix formula for interval velocity.

ing, the rms average velocity between the two reflectors, but the formula is usually applied over a fairly thin interval in which the velocity is reasonably constant.

The accuracy attainable in velocity determination has been reviewed by Al-Chalabi (1979).

It is not difficult to obtain adequate velocity information for stacking, but it is much harder to make detailed velocity studies, e.g. to detect lithological changes along a section. It can be shown that the standard error σ_{int} in the interval velocity obtained from rms velocities with standard error σ, when the interval is of thickness h and average depth D, is given by

$$\sigma_{int} \doteq \frac{1.4\sigma D}{h}.$$

Thus, for study of sand-shale ratio variations, we might want an accuracy of 5% in interval velocity of, say, a layer between 1000m and 1200m depth. We should then need an accuracy in the rms velocity to top and base of the layer given by

$$\sigma = \frac{h\sigma_{int}}{1.4\,D} = \frac{200 \times 5}{1.4 \times 1100} = 0.6\%.$$

Such an accuracy is not generally attainable.

Stacking velocities can be very sensitive to lateral velocity irregularities. An example of a near-surface feature producing a constant delay of magnitude δ is shown in figure 4/24. The effect is to distort the reflection curvature in the CDP gather by amounts depending on the location of the CDP (figure 4/25). The pro-

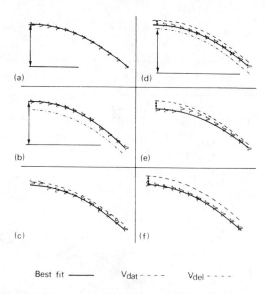

Best fit —— V_{dat} - - - - V_{del} - - - - -

Figure 4/25 Diagrammatic illustration of the change in the best fit hyperbola as the CDP gather moves across a vertical step. (After M. Al-Chalabi, 1979.)

cesses used to determine velocities from a CDP gather all, in one way or another, fit a hyperbola to the $t - x$ data. In these gathers, some traces suffer different time delays from others. The best-fit hyperbola across the whole gather can be very different from the hyperbolae that would fit the individual segments of the gather with the same time-delay. The derived stacking velocities are correct in the sense that they will give the best cancellation of NMO and hence the best stack, but they are very different from v_{rms}.

Once the velocity structure is known, the gathers can be corrected for NMO. As TWT increases, NMO will decrease. Therefore, the upper part of a trace is shifted further by NMO correction than the lower part. This results in a stretching of the trace, causing a shift to lower frequencies. The effect can be very severe on the large offset traces at small TWT, and is a further reason for muting these portions of the trace.

After NMO correction, the traces of the CDP gather can be added together to form the stacked trace. This improves the ratio of signal to random noise by $\sqrt{n}$, where n is the number of traces in the gather. Thus, for a 48-fold cover, an improvement by a factor of almost 7 is achieved. The stack may also be effective in suppressing multiples, because their NMO is different from that of primaries at the same

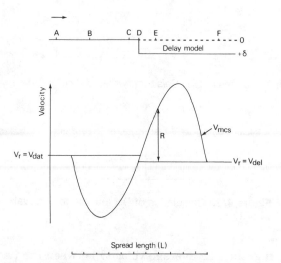

Figure 4/24 The response across a vertical step. V_{dat} and V_{del} are rms velocities in the zones AD and DF, respectively. V_{mcs} is the stacking velocity calculated for different CDPs along the profile. (After M. Al-Chalabi, 1979.)

TWT. Usually, modern marine data have a high fold of stack (e.g. 48 or 96), and the multiples, undercorrected for NMO, will be smeared out and at least partially attenuated. In land surveys, the fold of stack is usually less (e.g. 12), and the effectiveness of multiple removal is critically dependent on the relationship between the signal frequency and the NMO difference between primary and multiple. Thus, on land, it may be necessary to tailor source and receiver patterns specifically to remove particularly damaging multiples. This has already been discussed in Chapter 3 (p. 31). Even in the marine case, stacking will not be of much help in removing, for example, a single peg-leg bounce in a thin water layer, as there will be little difference in stacking velocity between such a multiple and a primary at the same TWT. Such multiples, however, can be removed by the dereverberation techniques discussed in Section 4.5.2.

4.7 Frequency filtering

As we saw in Section 2.5, the frequency of a reflected seismic signal tends to decrease as the path length increases. We can improve the section by filtering it so as to select the range of frequencies where the signal to noise ratio is highest. This will usually mean a time-variant bandpass filter, with the pass-band frequencies becoming lower as TWT increases. Where there is appreciable structure along a profile, it can also be useful to vary the filter along the section so as to keep the frequency content, and therefore character, of a given event roughly constant. Frequency filters are usually chosen by inspection of a suite of displays of a test section of record. This incorporates both narrow pass-band filters and a set of standard broader-band settings (figure 4/26). Choice of filter settings for optimum signal to noise ratio is a subjective process. For best resolution, as much high-frequency content should be retained as possible. Changes in the pass-band frequencies should be made gradually down the record, for sudden changes may cause confusing changes in the character of a dipping event as it passes from one filter to another.

4.8 Display

Unhappily, all final displays have a very limited dynamic range between the smallest and largest signals that can be reproduced. For example, in a wiggle trace display, the range is restricted between the smallest visible wiggle and the largest that does not create confusing overlap between adjacent traces. Generally therefore, seismic traces are 'equalised' before display. This means that amplitudes are adjusted so as to make the average amplitude constant over a window of, say, several hundred ms, which slides down the trace. The process destroys a good deal of the amplitude information present on the unequalised trace. This is a pity, because lateral amplitude variation along a reflector can be very important, as in the 'bright spot' method of gas detection. It may be necessary to produce two displays, one of them suitable for general structural interpretation with fairly severe equalisation, the other unequalised. The latter will preserve amplitude data at the expense of blank zones in low-amplitude parts of the section. Alternatively, amplitude information may be shown as a colour overlay to the section processed for structural interpretation. Sometimes, it is possible to find a compromise by which both structural and amplitude information are to some extent preserved, by using a long window (e.g. 1s) in the equalisation stage.

Several methods of final display are in use. They are all methods of treating a wiggle trace so as to make the correlation between adjacent traces more obvious. A common display is the 'variable area' section (figure 4/27a), in which the troughs of the wiggle are deleted and the peaks are shaded in solid black. The wiggle trace in the trough can be added to the display, thus giving more information about the waveform (figure 4/27b). The proportion of the wiggle to be shaded black can be specified so as to give the best effect. Figure 4/28 shows a variety of display modes. The appearance of the section is quite sensitive to parameters such as the display gain; a subjective choice of display parameters can be made from a suite of test displays of a piece of line.

Display scales are chosen to suit the purpose in hand. For general structural interpretation, vertical scales of 5 or 10cm per second of TWT are usual. Horizontal scales of around 1:50 000 and 1:25 000 are often chosen for these vertical scales respectively. This will usually result in sections without severe vertical exaggeration. For detailed studies of a particular reflector, much larger scales may be used.

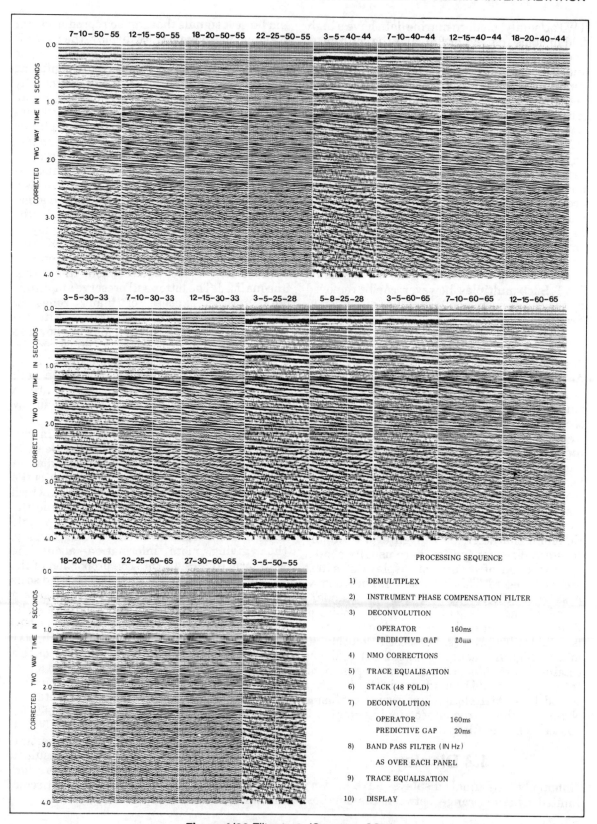

Figure 4/26 Filter test. (Courtesy: SSL.)

Figure 4/27 (a) Variable area seismic section display, (b) variable area and wiggle trace section display.

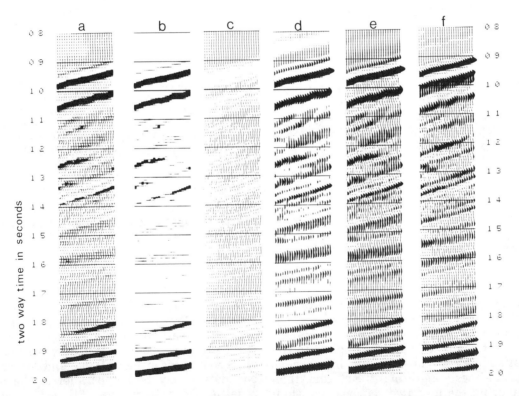

Figure 4/28 Range of seismic display options: Wiggle trace (c) is the normal output of a processing camera galvanometer. Variable area display (d) shades all peaks with uniform intensity. Variable density display (b) prints shading which varies with amplitude, but in this case there is no shape variation. Variable density display can be superimposed on the wiggle trace, such as in (a); similarly variable area display can be combined with the wiggle trace such as in (e). Note that (a) and (e) have had their overall intensity doubled using a two-pass system. For comparison purposes in detailed character analysis or 'bright spot' studies, a reverse polarity section (f) — here in wiggle variable area display — is often specified. (Courtesy: Digitech.)

4.9 Migration

In areas of steep dip, reflector segments appear on the time section considerably removed from their true position (figure 4/29). In the zero-offset case, the reflection received at X actually comes from point A, but is plotted vertically below X at C, where XC = XA in two-way time. The result is that, on the seismic section, reflectors are moved downdip, and show a smaller dip than is actually the case. More complex geometries arise when the reflector curvature exceeds that of the incident wavefront (figure 4/30). Reflections received at certain surface points can then originate from more than one point of the subsurface. The resulting TWT section is not easy to interpret. The presence of a syncline can be recognised, but the exact shape of its sides will remain uncertain. Migration is the process which attempts to reposition reflector segments so as to give the TWT section the geometry of a depth display.

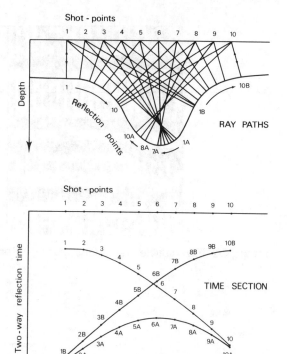

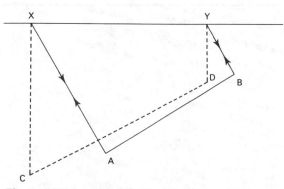

Figure 4/29 Actual event AB produces apparent event CD.

Figure 4/30 Reflections from a concave reflector: curvature of the reflector exceeds that of the wavefront. Rays emanating from each of the source locations 1–10 are reflected at up to three different points (all at normal incidence) on the concave reflector. The resulting time section shows a complex pattern of three reflector curves.

To illustrate some of the problems involved, we shall first discuss a simple manual migration method. This can be used to migrate picked sections. It can also be applied to the migration of a TWT map, by drawing sections along dip-lines, migrating them, and then recontouring the migrated sections. We shall suppose that we are dealing with ideal dip sections, with all seismic events coming from reflectors within the plane of the section. Figure 4/31 shows the geometry. Since we are considering the stacked trace, we need consider only zero-offset reflections. If the velocity is a function of depth alone, then the time difference between traces is

$$\Delta t = \frac{2 \sin\theta . \Delta x}{v_1}$$

where v_1 is the seismic velocity just above the reflector. However, it is unlikely that the velocity is really a function of depth alone. It will be to some extent affected by lithological variation between formations, and formation boundaries may be sub-parallel to the deep reflector. The extreme case of this effect would have the equal-velocity surfaces parallel to the reflector. In this case,

$$\Delta t = \frac{2 \sin\theta . \Delta x}{v_0},$$

where v_0 is the near-surface velocity.

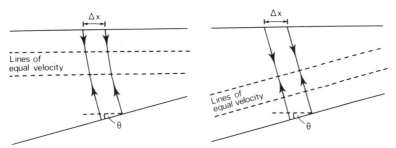

Figure 4/31 Reflection from a dipping surface.

Thus, we can find the true dip, θ, of the interface from the observed Δt and Δx by means of the expression

$$\sin \theta = \frac{\Delta t . v}{2\Delta x},$$

but we do not know what velocity v to use; depending on the pattern of lateral velocity change, the correct velocity may vary from the near-surface value to that found just above the reflector. In the absence of detailed knowledge, perhaps the best we can do is to use an average velocity to the reflector, which will usually be intermediate between these extremes.

Once a velocity has been chosen, the reflector can be swung updip by the proper amount (figure 4/32), and a reflector segment drawn on the cross-section. Repetition of the process for other CDPs allows us to draw the complete migrated reflector. This procedure is called the dip-bar method of migration.

We shall be in difficulties if we try to migrate a section which is not in the dip direction. Suppose that figure 4/32 is a dip section of a two-dimensional structure, and consider a strike seismic section perpendicular to figure 4/32. On this section, the reflector will appear horizontal, with no indication that it requires to be migrated. If, however, we leave it alone, the strike section will not tie with the migrated dip section, even though the unmigrated sections tie correctly. French (1975) has shown that

correct migration of a line not in the true dip direction requires adjustment to the migration velocity. Specifically, if we have a single dipping reflector, with uniform overburden velocity v, then for a profile at azimuth angle ϕ to the true dip direction, the correct migration velocity is $v/\cos\phi$. In the real earth, strike direction may vary with depth, so there is no such thing as a dip line on all structural levels. It may then be necessary to use elaborate methods, involving the tracing of raypaths through a model of the structure. A simpler approach is to choose the nearest we can to a dip line, and obtain a suite of migrations using different plausible migration velocities; the most pleasing result can then be selected.

Manual migration is of no help if the seismic section is very confused by diffractions and the effects of sharply curved interfaces; in this case, it may not be possible to pick out the reflector segments belonging to a single horizon. Various methods exist to migrate the section directly, by computer application of filters operating across many traces. To illustrate the technique, we shall describe the Kirchhoff sum method. By Huygens' principle, we can consider each point of the subsurface to be a scatterer of seismic energy. Each scattering point will produce a curve on the seismic section. If velocity were constant, these would be hyperbolae (figure 2/19). The exact shape of the curve can in practice be calculated from the known velocity-depth relationship. To find a value for the amplitude at a particular TWT on a particular trace (A in figure 4/33) in the migration section, we proceed as follows. The quasi-hyperbolic curve corresponding to a point scatterer at A is superimposed on the section. Wherever it crosses another trace, we note the amplitude value at that point. These values are then summed, and this sum gives the new amplitude at A. It is, of course, inevitable that

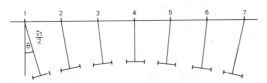

Figure 4/32 The dip-bar method of migration. $\bar{v}$ is the average velocity of the reflector.

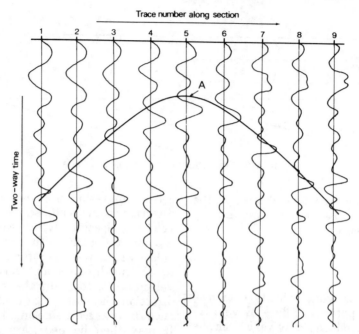

Figure 4/33 Kirchhoff migration. Amplitude A is found by summing amplitudes along the hyperbola.

not all the amplitude values in the summation relate to the scatterer at A; the hope is that energy received from other scatterers will tend to cancel out, but energy truly radiated from A will add up in phase along the curve. Large noise spikes, however, will cause trouble as they will be smeared out across the section, since they will appear in the summation for any scatterer whose hyperbola intersects the spike. To remove the resulting 'smiles' it is necessary to edit the data to remove such spikes before migration. Since amplitude is reduced by divergence of the wavefront with increasing TWT, it is desirable to weight the contribution of each trace so as to give preference to traces near the vertex of the curve. Repetition of the process for every sample of every trace will allow us to build up a complete migrated section.

Another commonly used method is wave equation migration. Conceptually, the method has some similarities with the downward continuation techniques of gravity and magnetic prospecting. It is helpful to approach it via the 'exploding reflector' model. Suppose that, instead of the seismic sources being on the earth's surface, they were actually spread out along each reflector. All the sources are set off

at zero time, and the resulting seismic signals are recorded at the surface. Consider a single ray to the surface from each source point, starting off normal to the reflecting surface. The ray geometry is the same as that of a zero offset reflection section, except that the 'exploding reflector' travel times must be doubled. Equivalently, the seismic velocities in the medium may be halved.

In the exploding reflector model, we could image the reflectors by the following procedure. The seismic signals recorded at the surface give us the pressure amplitude P as a function of distance x along the section and of time t at a depth $z = 0$. In practice, of course, we have only samples of the entire $P(x, z = 0, t)$ function, with a spacing Δx in x equal to the interval between zero-offset traces, and Δt in t equal to the recording time-sampling increment (e.g. 4ms). We must have data in which both Δx and Δt are small enough to allow us to construct $P(x, z = 0, t)$ accurately, without aliassing. Given P on $z = 0$, it is possible to obtain P on a plane at a depth Δz below the surface. The calculation of $P(x, \Delta z, t)$ requires solution of the equation of seismic wave propagation in the material. Various approximations are required, and the way in which these are implemented can be critical for the correctness of the result. Knowledge of the seismic velocity variation with x between $z = 0$ and $z = \Delta z$ will be required. When the calculation has been carried out, we can consider the data that have moved to zero time: $P(x, \Delta z, 0)$. Recollecting that this is the time of source initiation in the exploding reflector model, we can see that these data have been fully migrated. The rest of the $P(x, \Delta z, t)$ field may be carried downward a further depth increment Δz by the same process to give $P(x, 2\Delta z, t)$. By repetition of this process, all the pressure field can ultimately be correctly placed in the (x, z) plane at $t = 0$, the instant when the reflectors explode.

To apply this method to the migration of a zero-offset section (to which we hope our CDP stack approximates), we can use the same procedure except that, as discussed above, the seismic velocities must be halved. Figure 4/34 shows the process in operation on a synthetic seismic model. At each step, the pressure field has been downward continued to the level indicated by the arrow. We see that events above this level have been fully migrated, whereas events below the arrow have not. Successive downward steps carry the migration deeper into the section. Figure 4/35 shows an example of the application of the method to real data. The section is from the thrust belt of the foothills of the Canadian Mountains. Thrusted Cretaceous and Mississippian formations overlie an undisturbed Cambrian economic basement seen at approximately 2.5s two-way time. The improvement in data quality in the migrated section is clearly apparent despite the complexity of the geological structure in the shallow rock layers; a small fold emerges (upper

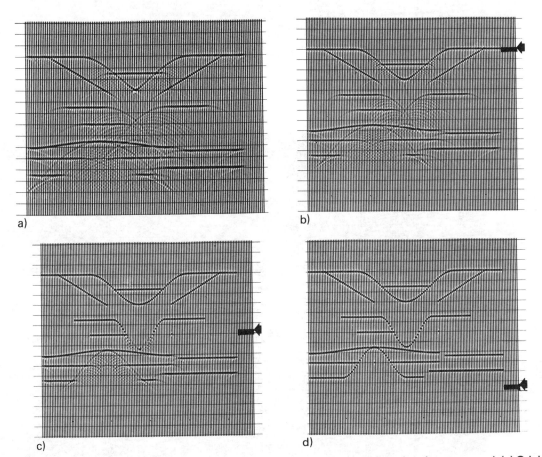

a)   b)   c)   d)

Figure 4/34 Migration: synthetic seismic models which have been wave equation migration processed. (a) Original synthetic section, (b), (c) and (d) migration sections with arrows indicating progressive correction down through section. (Courtesy: SSL.)

CDP stack

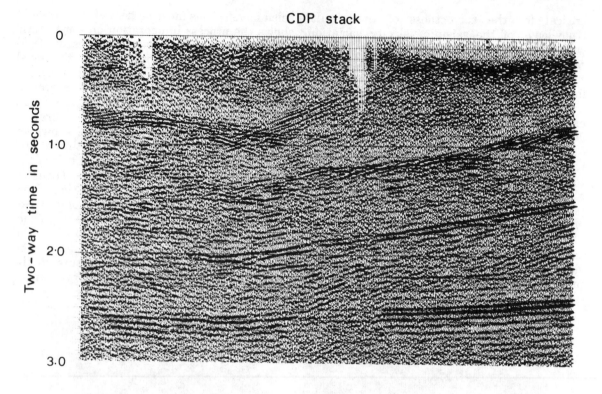

Wave equation migration

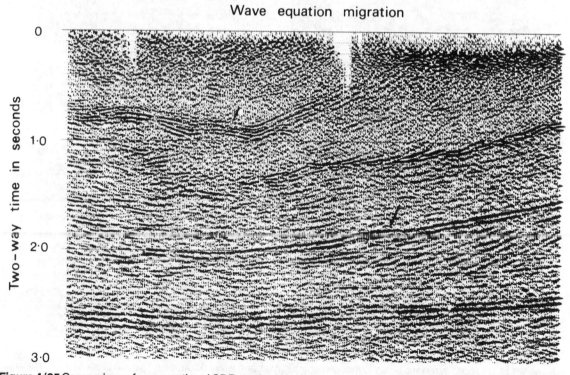

Figure 4/35 Comparison of a conventional CDP stack with a wave equation migrated section from the foothills area of the Canadian Rocky Mountains (Courtesy: Digitech).

arrow) within the synclinal axis, and deeper in the section an upward flexure (lower arrow) of a strong reflector is both amplified and migrated to the right. Throughout, reflection continuity is much enhanced.

In general, the effect of migration is to make anticlines steeper and of smaller area, while synclines tend to be enlarged. Therefore, unmigrated sections tend to give a more optimistic assessment of the volumes of potential hydrocarbon reservoirs when located in structural traps.

4.10 Special displays

The steps described above are those commonly carried out in producing sections suitable for general structural interpretation. For special studies, various other displays are available.

4.10.1 Interval velocity display

By use of the Dix equation, stacking velocities can be converted to interval velocities. Overlays can then be produced showing contours of equal interval velocity. Alternatively, colour-coded interval velocity information can be added to a conventional section (figure 4/36, colour section).

4.10.2 Acoustic impedance sections: inversion

We have seen in Chapter 2 that reflections are generated from interfaces between layers of different acoustic impedance, and that the reflection coefficient is related to the acoustic impedance contrast across the interface. If we have succeeded in producing a seismic trace which approximates the reflection coefficient series in the earth, we can in principle convert this to a trace which shows the variation of acoustic impedance with TWT. This is easier to compare with a well log than the original trace, and opens up the possibility of using each trace as a pseudo-log. For example, we may be able to follow a particular horizon (e.g. a sand layer) and make geological inferences from a gradual change in acoustic impedance along the section (e.g. changes in sand–shale ratio within the layer). Such techniques can be very useful when we have good well control, and wish, for example, to establish exactly where a good reservoir sand encountered in one well grades laterally into a tight equivalent sand found in another well.

The principle is shown in figure 4/37. If we

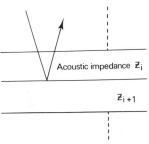

Figure 4/37 Acoustic impedance at an interface between layers i and $i + 1$.

divide the earth into a series of layers, then the reflection coefficient at the ith interface is given (from Chapter 2) by:

$$R_i = \frac{Z_{i+1} - Z_i}{Z_{i+1} + Z_i}$$

Thus, $$Z_{i+1} = \frac{Z_i (1 + R_i)}{1 - R_i}$$

If the acoustic impedance at the surface is known, the acoustic impedance can be calculated down the record if the reflection coefficients are known. The ideal seismic trace is just the series of reflection coefficients we need. For the method to work with a real seismic trace, it should approach the ideal as closely as possible. Specifically, effects of spherical divergence, attenuation, multiples, source wavelet, and random noise should all have been removed as well as possible.

Even so, the resulting acoustic impedance trace will be limited in bandwidth by the lack of very high and very low frequencies in the seismic trace from which it is derived. One can try to replace the low frequency information by deriving the slow variation of impedance down the record from the seismic stacking velocities. Smoothed stacking velocities can be converted to interval velocities, which in turn can be converted to impedance if a relationship between density and seismic velocity is assumed (Section 2.2). The Seislog (TM) display is constructed in this way. Figure 4/38 shows how a sonic log can be expressed as the sum of a gross velocity framework contained in the very low frequencies, and a modulating series of all higher frequencies. A Seislog trace derived from seismic data is compared with an actual sonic log from a nearby well in figure 4/38b. An example of the successful application of Seislog interpretation, extending the boundaries of a gas field, is described in Chapter 9.

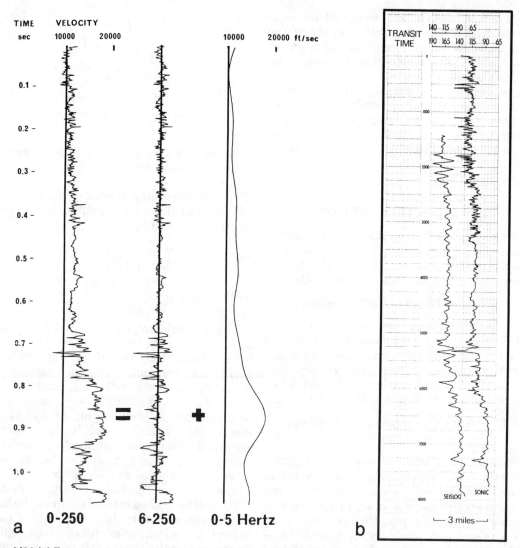

Figure 4/38 (a) Frequency analysis of a sonic log. (b) Comparison of a Seislog and a sonic log from adjacent sites. (Courtesy: Technika Resources Ltd.)

4.10.3 Other seismic attributes

We saw in Chapter 2 that acoustic impedance is sensitive to the presence of quite small amounts of gas in the pore space. This can give rise to 'bright spots' in the seismic section. If, for example, we have a sand encased in shale, where the sand is water-filled the acoustic impedance contrast between sand and shale may be small, giving rise to a weak reflection. However, if the sand has some part of the pore space gas-filled, then the fall in acoustic impedance will give rise to a strong negative reflection from the top of the sand, and a strong positive reflection from the gas-water contact. Therefore, study of lateral variations in seismic amplitude may give a clue to the presence of gas (see the Hewett Field Case History). Some caution is needed, however, for two reasons: amplitude changes may also be caused by lateral changes in lithology, and a bright spot can be caused by very small (and certainly non-commercial offshore) gas saturations.

In order to see lateral amplitude variations clearly, special true amplitude displays can be produced. Essentially, amplitude manipulation is avoided during processing, except where it has a clear physical basis (e.g. spherical divergence correction). In particular, no equalisation is applied before final display. The resulting section (figure 4/39) shows the high amplitude reflectors clearly, but the loss of the lower amplitude reflectors, due to the very limited dynamic range of the display, makes the section unsuitable for structural interpretation. An alternative approach is to use colour-coded overlays to convey attribute information, which can be superimposed on a section normally processed to show structure. Figure 4/40 is a conventional seismic section from the North Sea which can be compared with colour processed displays (figure 4/41, 4/42 and 4/43, colour section).

A gas-induced bright spot should show a negative polarity reflection at the top of the gas column. It is not always easy to see the polarity of a reflection on a normal seismic section; again, colour displays may help. In the colour display shown in figure 4/41 (colour section) the gas reservoir is clearly defined as is the polarity change (see also the Hewett Field Case History).

An example of a combined amplitude and polarity display is shown in figure 4/42 (colour section).

The frequency of the seismic signal may also be an indicator of the presence of gas. In some areas, low-frequency anomalies are found immediately below gas reservoirs. The cause of this effect is not clear. It may be due to selective absorption of the higher frequencies in the gas reservoir; alternatively, the lower velocities in the gas column may cause localised misstacking, with smearing and loss of the high-frequency information. Neither of these explanations, however, is adequate to explain the magnitude of the effect sometimes observed. An instantaneous frequency display of the same North Sea structure previously illustrated is shown in figure 4/43 (colour section).

4.10.4 3D survey processing

It is possible to carry out a seismic survey to give a very dense coverage of an area, e.g. with a line spacing of 50 m. In this case, CDPs form a square grid across the area and therefore a

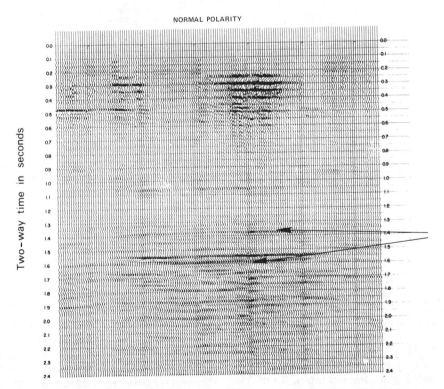

Figure 4/39 Example of a true amplitude display. Arrow points to high amplitude events which occur at both top and bottom of a hydrocarbon contact. (Courtesy: Phoenix/SSC.)

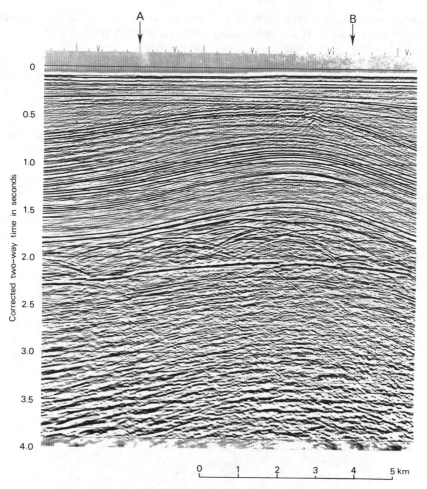

Figure 4/40 Conventional seismic section across a gas reservoir.

three-dimensional time/distance matrix rather than merely a series of lines (see Chapter 3). This opens up the possibility of migration in three dimensions, with potentially a great improvement in data quality in complex areas. The cost is high, however, and such surveys are more suited to the production than to the exploration phase of development.

By suitable selection of CDPs, it is possible to produce from the data set a series of sections across the area in any desired azimuth. A set of closely-spaced sections will allow the unambiguous interpretation of fault patterns. Interpretation can be carried out without much of the tedious tying of loops characteristic of interpretation of a normal seismic grid, since with a line spacing of, for example, 50 m, the change in structure from one line to the next will be small.

With 3D processed data, as well as being able to produce vertical sections along any desired azimuth, it is also possible to produce horizontal slices. Seiscrop* sections are an example of this type of display. Such sections may be used for two purposes.

First, by following a particular event through a series of time slices, and tracing its position, it is possible to build up a contour map directly from the time slice sections. In figure 4/44 (colour section) this method is applied to a 3D prospect in Peru (Brown, 1982). The Seiscrop sections, 4ms apart, are displayed in dual-polarity, black for peaks and red for troughs. The red event at the centre of the 2632ms level is first circumscribed, then the same event in each successively deeper section is similarly traced

* Trademark of Geophysical Services Inc.

until the 4ms contour interval map is complete. This example is over a simple geological structure; interpretation of more complex faulted structures would require reference to vertical as well as horizontal sections during the mapping process.

A second application is in the study of stratigraphic features such as channels. An excellent example of this type of application is shown in figure 4/45 (colour section) from the Gulf of Thailand (Brown, Graebner and Dahm, 1982). The section shown covers a much larger area than in the previous example. Here the parameter expressed in colour is that of reflection amplitude, negative red and positive blue. At this shallow depth, 196ms TWT, beds are flat lying and it is in the amplitude anomalies that the deltaic features are well expressed. Brown *et al.* interpret this section as a meandering channel; part of a prograding delta, progradation southwest to northeast across the survey area.

References

M. Al-Chalabi, Velocity determination from seismic reflection data, in: *Developments in Geophysical Exploration Methods I*, ed. A. A. Fitch, Applied Science Publishers, London (1979).

M. M. Backus, Water reverberations, their nature and elimination. *Geophysics*, **24** (1959), pp. 233–261.

Alistair R. Brown, Structural interpretation from horizontal seismic sections. *Geophysics*, **48** (1983), pp. 1179–1194.

Alistair R. Brown, R. J. Graebner and C. G. Dahm, Use of horizontal seismic sections to identify subtle traps. *AAPG Mem.*, **32** (1982), pp. 47–56.

K. E. Burg, Exploration problems of the Williston Basin. *Geophysics*, **17** (1952), pp. 465–480.

W. S. French, Computer migration of oblique seismic reflection profiles. *Geophysics*, **40** (1975), pp. 961–980.

J. G. Hagedoorn, The plus-minus method of interpreting seismic refraction sections. *Geophysical Prospecting*, **7** (1959), pp. 158–182.

K. L. Peacock and S. Treitel, Predictive deconvolution theory and practice. *Geophysics*, **34** (1969), pp. 155–169.

E. A. Robinson, *Physical applications of stationary time series*. Charles Griffin, London, (1980) pp. 302.

Chapter 5

Seismic Ties to Well Data

When the interpreter comes to establish a tie between his seismic sections and a borehole section he faces the problem of making a direct correlation between patterns of reflectors which are scaled vertically in terms of two-way reflection time and the realities of sub-surface geology as determined by lithological logging of rock chippings and cores obtained from a borehole. The geologist's lithological log is of prime importance in that it provides the basis for identification of reflectors in terms of boundaries between rocks of different type. Other geological work on the cores and chippings aims to establish the age and stratigraphy of the geological section and the presented results of exploration drilling normally include a lithostratigraphic log (rocks described in terms of lithology) as well as a chrono-stratigraphic scale (the rock units subdivided according to age).

It is standard industry practice that at various stages during the drilling of a well and upon reaching total depth (referred to on logs as TD) geophysical logging tests are made with a variety of instruments. These are lowered to the bottom of the well, as drilled at the time of logging, on a wire line which is usually a multi-core electrical cable on which the logging tools can be suspended. The logging tools are then drawn upwards through the borehole, measurements of various parameters being made either continuously or by tests at selected

horizons. The processed results of these geophysical tests provide data which allow identification of the interrelation between the seismic section time scale and the borehole section depth scale and thereby direct correlation between reflector pattern and stratigraphy. These measurements also provide data on the physical properties of the rocks penetrated by the borehole and such data are important to a geological understanding of the variation in reflector pattern which can be seen in seismic sections throughout an exploration province. As far as the seismic interpreter is concerned, the geophysical logging methods of most value to him are gamma-ray logging, compensated formation density logging, compensated sonic logging and well velocity surveys. The results of these are most usefully combined to provide a synthetic seismogram which is a process which aims to produce from the borehole physical data a computed seismic section display which should be comparable with an actual seismic section surveyed through the well site.

5.1 Logging tools

Wherever possible, during logging, a number of individual tools are assembled together so that each logging run will allow concurrent acquisition of data for a group of parameters. Thus the results of most logging runs are displayed as a multi-trace vertical profile showing the measurements pertaining to the tool assemblage used (see for example figure 5/2 which shows a display of borehole diameter as measured by a caliper log, a gamma-ray log and a bulk density log). For a fuller description of logging techniques, see Schlumberger, 1972.

5.1.1 Caliper log

Because the variations in borehole diameter must be accounted for in processing the results of most geophysical tests in open boreholes, a device is attached to the logging tool assemblage which measures variation in the width of the hole throughout the interval logged. Results of a caliper logging run can often provide a useful indication of lithological and porosity variations and these are used in the geological interpretation of any suite of logging tests. In figures 5/2 and 5/5 the results of two separate logging runs are recorded and on each occasion a caliper log was obtained. As is often the case when obtained at different stages of drilling a well, caliper data can be compared to give an indication of lithology. The logs in figure 5/5 were recorded on the earlier run, and it is interesting to note how some intervals in the section have appreciably increased in diameter between runs. Drilled hole diameter is 8 in (the heavy line on the caliper log) and measured diameter ranges from 8 in to, in the log (figure 5/2), in excess of 15 in. Two well-defined zones show a very marked increase in diameter between runs, a broad zone between 3330 and 3420 ft and a bifurcated zone between 3510 and 3590 ft. Such changes in diameter can indicate either solution in the case of salt or

caving in the case of poorly consolidated rocks. In both zones, velocities are low, thus solution of salt is not the likely cause. It should be noted however, that apparent velocities and densities are lowered by caving and/or solution regardless of whether the caved rock or salt is of higher or lower velocity than the bounding media.

5.1.2 Gamma-ray log

The gamma-ray logging tool measures the natural radioactivity of the formations in the borehole and it is used as a means of identification of shales, some evaporites and of igneous rocks and igneously derived sediments. An important feature of the log is that it can be used in a cased borehole and can substitute for electrical logs as a means of correlation between boreholes. Separate runs in the same borehole can be seen in figures 5/2 and 5/5. This log is usually only used as an aid to seismic correlation if sonic and density logging data are not available. For example, if in the same prospect area as that of the well (figures 5/2 and 5/5), another location was drilled and only logged with gamma-ray (and, say, a suite of electrical tools), correlation of gamma logs between wells would probably identify the zone approximately between 3400 and 3520ft depth which is marked by low gamma-ray activity, high velocity and high density. We shall see later (figure 5/13) that this interval generates strong seismic reflections. Thus it may be possible to obtain a well-to-well seismic correlation, even though one well lacks the more significant logs.

5.1.3 Compensated formation density log

The formation density logging tool is constructed as shown in figure 5/1. The instrument is an arrangement of a gamma-ray source and two shielded detectors which measure the intensity of gamma-ray back-scatter from the rock formation (Tittman and Wahl, 1965; Tittman and Johnstone, 1964). The amount of back scatter is approximately proportional to the electron density in the rock, which is proportional to the rock's bulk density. Because a layer of mud-cake commonly forms on the wall of the borehole it is necessary to compensate for its effect, whence the use of two detectors as shown in figure 5/1. Natural radioactivity of rocks does not affect the results obtained but unlike the gamma logging tool this can only be successfully used in uncased holes.

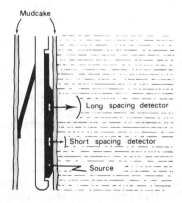

Figure 5/1 Formation density logging tool (Courtesy: Schlumberger).

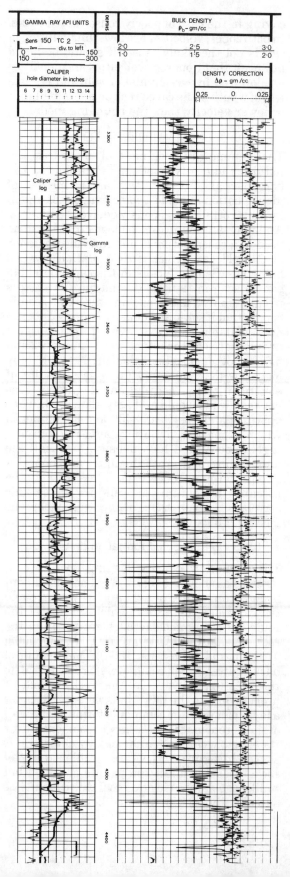

Figure 5/2 Formation density log, Western Canada. Bulk density and density correction are displayed to the right of the depth scale; gamma and caliper logs to the left. Note the extensive caving as shown on the caliper log. In acoustic studies, corrections for caving would be necessary.

In figure 5/2 an example is shown of a log from Western Canada with a range of rock types from Lower Cretaceous coals, density 1.35g/cc, to pre-Cretaceous carbonates with a density of 2.8g/cc. Like the sonic logs (see below) the density log is invaluable to the reservoir geologist in determining porosity variations. This is due to the fact that there is a large density contrast between rock matrix and pore-space fluids. Furthermore, when combined with a neutron density log (not described here) it provides a means of detecting gas accumulations.

A recently developed alternative method of measuring density variations in rocks by use of a logging tool is through use of a borehole gravity meter (Howell, Heintz and Barry, 1966). Use of this tool requires the determination of gravity to a very high accuracy at selected locations in the hole and at present tests are usually undertaken as a means of studying prospective intervals rather than as a means of obtaining a continuous profile of density variation throughout a deep well. However, the results are reliable, accurate, and the zone tested is not restricted to the immediate vicinity of the borehole wall. It seems likely therefore that this tool will have wide application in the future.

5.1.4 Compensated sonic log

The sonic logging tool measures the reciprocal of velocity in rocks of the walls of a borehole (Guyod and Shane, 1969; Kennet and Ireson, 1971; Kokesh and Blizand, 1959; Kokesh, Schwartz, Wall and Morris, 1965). As shown in figure 5/3 it consists of a pair of units each with a sound generator and four receivers. Pulses of sound are refracted along the wall of the borehole and the time differences are measured between receipt of these pulses at each of the groups of two receivers. The time differences are divided by the distance between the two receivers to give a transit time in microseconds per foot. The two units pulse separately and the average of the values observed is used to compensate for errors caused by irregular

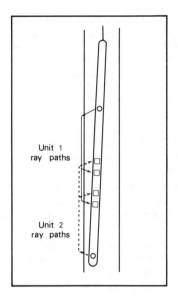

Figure 5/3 Sonic logging tool (Courtesy: Schlumberger).

borehole configuration or angular deviation of the instrument. The log is normally integrated to give total travel time on a scale of milliseconds and interval times may be obtained by counting scaled interval units (see figure 5/4). Interval velocities are obtained by determining the reciprocal of the transit time.

It should be noted that the instrument occasionally skips a cycle due to non-arrival at a far receiver or a late arrival at a near receiver, the transit times being, respectively, too long or too short. This can be readily recognised by the sudden deflection on the curve and compensated for manually. This logging tool can only be run in an open, uncased hole.

An example from western Canada is shown in figure 5/5. Rock types range from coals (130μs/ft or 2.35km/s) through shales and sands to carbonates (50μs/ft or 6.1km/s). Like the density log this log can be used by the reservoir geologist as a porosity tool. This depends on the fact that there is a large velocity difference between rock matrix materials and pore fluids.

5.1.5 Well velocity surveys

A well velocity survey is the most direct method of identifying the relationship between sub-surface geology and seismic reflection data (Seismograph Service Ltd, 1976). The technique involves detecting sound from a near surface source with a pressure geophone (hydrophone) at selected levels within the fluid-filled borehole. These levels are usually chosen with reference to major changes in formations in the geological section. In offshore wells the source is usually an air gun whereas on land, dynamite in shot holes may be used or

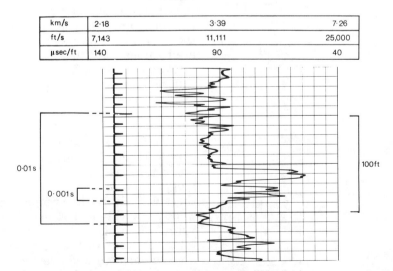

km/s	2·18	3·39	7·26
ft/s	7,143	11,111	25,000
μsec/ft	140	90	40

Figure 5/4 Derivation of integrated total travel time from continuous velocity log (CVL). On the right is shown part of a normal CVL log and on the left is displayed the integrated total travel time for this part of the section. By totalling from the surface downwards the two-way travel time to a marker horizon can be calculated. Interval velocities can also be derived. In contrast with seismic sections, travel times are one-way.

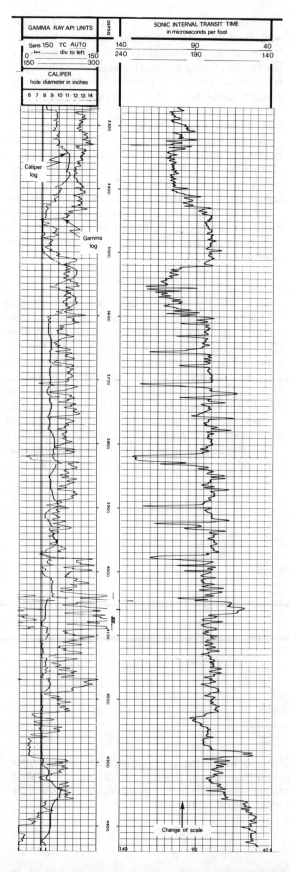

Figure 5/5 Borehole compensated sonic log, Western Canada. This log covers the same interval as in figure 5/2. Note the change in borehole diameter between logging runs as indicated by caliper log. CVL values range from 130 to 50 μs/ft (2.35 to 6.1 km/s) reflecting a range of rock types from coal, through shales and sandstones to carbonates.

an air gun submerged in a water-filled hole. A complete set of equipment is illustrated in figure 5/6.

When surveys are being carried out it is important that certain requirements are met of which the most important are as follows:

1. The source must be a sufficient distance away from the hole, proportional to the expected test level depths, to avoid refracted first arrivals via the casing.
2. If it is desired to make a test close to an important stratigraphic boundary, the hydrophone should be located at approximately 5 m below that boundary to avoid adverse effects of poor hole conditions such as caving or loose fill in the vicinity of the test.
3. For each measurement the logging tool must be locked and the wire line slackened off to avoid cable noise.

An annotated monitor record is shown in figure 5/7.

Data reduction includes the correction of the recorded times to true vertical and these can then be used to calibrate a sonic log. Corrections are by linear or differential shift whichever is most appropriate. The final presentation is as a calibrated acoustic log which can be directly overlain on a seismic section which intersects the borehole. An example of such a calibrated log is shown in figure 5/8.

In most circumstances the well velocity survey is the last logging run made before production casing is fitted or before the well is plugged and abandoned. In such circumstances there will always be pressure to complete the work as quickly as possible. Nevertheless, the value of a good well survey to future interpretation work in a province cannot be over-emphasised and every precaution should be taken to ensure that an adequate number of levels has been tested and that where necessary several shots have been recorded at each level, depending on data quality.

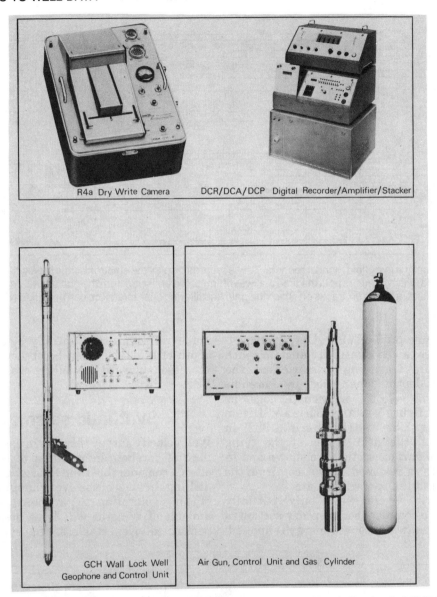

R4a Dry Write Camera   DCR/DCA/DCP Digital Recorder/Amplifier/Stacker

GCH Wall Lock Well
Geophone and Control Unit

Air Gun, Control Unit and Gas Cylinder

Figure 5/6 Well velocity survey equipment (Courtesy: Seismograph Service Ltd (SSL)).

5.1.6 Vertical seismic profiles

Through use of digital acquisition equipment it is possible to derive additional data than those required to produce a calibrated sonic log. If a sufficient time interval is sampled, the data from each test level provide a record which is equivalent to a reflection seismic trace with a deeply buried detector; because the hydrophone is buried, both upward and downward travelling waveforms (first 'breaks' or arrivals) will be recorded from reflecting horizons above and below the detector's location (figure 5/9(a)), as well as multiples generated in the time progression. Beyond the first arrivals of the monitor record (figure 5/9(b)) such secondary events can be clearly seen. Adaptions of standard processing techniques which remove unwanted reverberant events are applied and followed by separate velocity filtering of the two types of waveforms. These are recombined to provide a product displayed in a form similar to that of a variable area seismic section as a

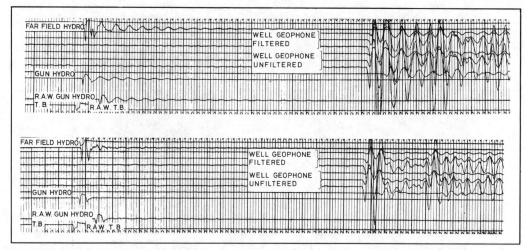

Figure 5/7 Velocity survey field monitor records. These records display the electronic time break, the signal from a gun hydrophone which monitors both the time and waveform of the emission from the gun as well as the downhole detector signal displayed on six traces on which the gain and filtering characteristics can be varied (Courtesy: SSL).

vertical seismic profile (VSP). Upward travelling primaries can be identified coinciding with the downward travelling primaries at the appropriate depths when tied by borehole lithologic and velocity or acoustic logs, as illustrated in figure 5/10(a) which is a VSP from a northwestern USA well (Balch *et al.*, 1982). In figure 5/10(a) the near traces of the tying conventional surface section are shown and in figure 5/10(b) a reversed format confirms the excellent fit of the two sets of data.

Particularly where wells only partially penetrate the complete sedimentary section of an area and sufficient source energy is applied,

major reflectors substantially below the total depth of a borehole can be observed from the processed upward travelling waveforms of a VSP.

5.2 Synthetic seismograms

Well velocity survey results are usually the last item of borehole information to be delivered, often a considerable time after abandonment of drilling. In view of the ever-changing priorities of an exploration department, a sufficient amount of tying-in will have been, by then, performed. Nevertheless, the synthetic seis-

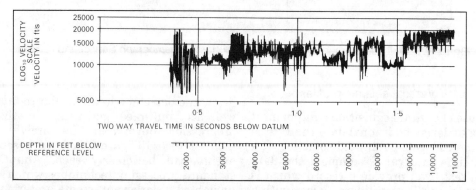

Figure 5/8 A calibrated sonic log. Measured time intervals from the velocity survey are used to correct and calibrate integrated travel times from the sonic log. The sonic log is then redisplayed against a vertical linear two-way travel time scale to match seismic sections through or adjacent to the well. A non-linear depth scale is also displayed. When using such logs, interpreters are advised always to check to which datum depths are referred, e.g. derrick floor or mean sea-level. (Courtesy: SSL.)

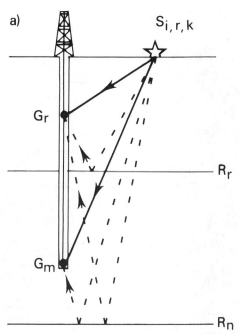

Figure 5/9(a) Principle of well velocity survey and vertical seismic profile (VSP). Source energy originating at S generates downward travelling waveforms directly, and upward travelling waveforms via reflectors R_1 to R_n, which are recorded at detector locations G_1 to G_m; multiples are not illustrated. Note that for each detector location the sum of the upward and downward travel times equals the normal surface to surface two-way reflected time. Processing of all primaries provides a VSP.

b)

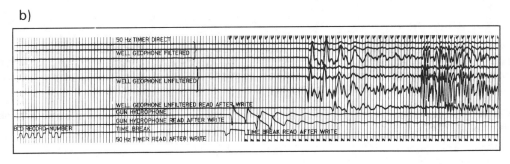

Figure 5/9(b) Well velocity monitor record with filtered and unfiltered traces at three gain settings. Note secondary event 0.385 s after the first arrival; its polarity is reversed suggesting a reflected upward travelling wave. Other events may be multiples of downward travelling waves or additional primaries and multiples of upward travelling waves. The primaries can be enhanced to form a vertical seismic profile (Courtesy: SSL).

mogram can be of great value to the interpreter and it is best presented by splicing it to an interpreted seismic section through the well location (Peterson, 1955; Baranov and Kunetz, 1960).

The process of production of a synthetic seismogram is illustrated in figure 5/11. Acoustic impedance is calculated by multiplying seismic velocity and density, and reflection coefficients are calculated from impedance changes, as seen in Chapter 2. The calculation of reflection coefficients is rather sensitive to imperfections in the well logs. It is common to observe intervals of spurious readings on both sonic and density logs. These are sometimes caused by washouts leading to incorrect readings over the interval of an enlarged borehole. Other causes include incorrect splic-

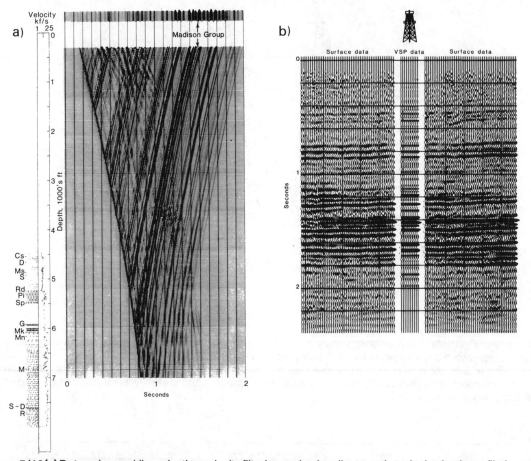

Figure 5/10(a) Data enhanced (by selective velocity filtering and gain adjustment) vertical seismic profile from the Sam Gary Madison no. 1 well, Bell Creek oil field, Montana. Downward events (primary and multiples) increase in time with depth, upward events (primaries and multiples) decrease in time with depth. Lithologic and acoustic logs are depth-tied and surface seismic data are time-tied.

Figure 5/10(b) Surface seismic data tied by upper traces of processed vertical seismic profiles from (a) above (after Balch *et al*., 1982).

ing of data from different runs of the logging tool, and instrumental noise. If they are not corrected, these errors will often generate large spurious reflections. Careful editing of the logs is therefore needed for good results. It is easy to delete spurious spikes, but the replacement of lost data over washed-out zones requires considerable judgement.

For comparison with the seismic trace, the reflection coefficient series must be convolved with a suitable wavelet. Choice of the wavelet is critical for the appearance of the final synthetic seismogram. If we have a near-ideal seismic section, free from multiples and converted to zero phase, it may be sufficient merely to filter

the reflection coefficient series to have the same bandwidth as the seismic section. Otherwise, choice of a wavelet is difficult. Indeed, it may be more useful to work the problem the other way round, and use the comparison between synthetic seismogram and seismic section to deduce the wavelet; this can then be used as input to wavelet processing of the section (Chapter 4). It is not generally obvious to what extent multiples should be included in the synthetic seismogram. This depends on how far they have been removed from the seismic section, by deconvolution and by stacking. It may be helpful to produce synthetic seismograms both with and without multiples, to

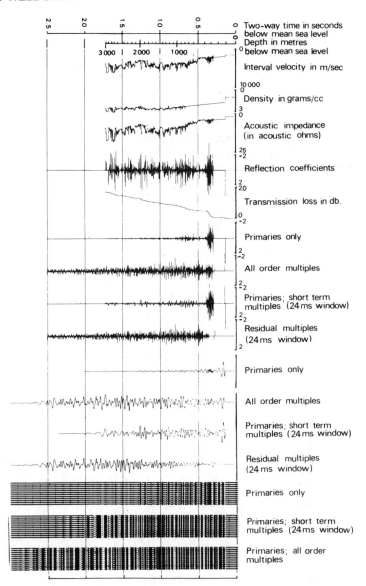

Figure 5/11 Range of synthetic seismogram displays derived from density and interval velocity logs (Courtesy: SSL).

check their influence on the section.

Finally, in comparing synthetics with seismic data in areas of complex geology, one should bear in mind that a single trace of the section is an average response over a lateral extent of perhaps several hundred metres, whereas the borehole logging tools respond to formation properties over a distance of less than a metre around the borehole.

If a computer is not available, it is possible to construct by hand a crude synthetic seismogram by dividing the sonic and density logs into fairly homogeneous blocks over the region of interest. This can be a quick check on possible thin-bed tuning and hydrocarbon indicators.

5.3 The composite log

The interpreter who utilises the results of well logs and surveys is heavily dependent on

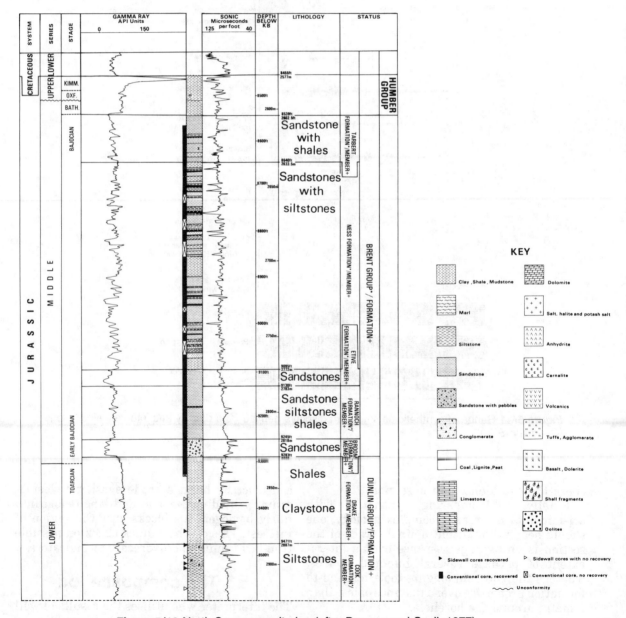

Figure 5/12 North Sea composite log (after Deegan and Scull, 1977).

geological identification as well as age dating. Before a final composite log is produced, geological studies are necessary and in particular by way of undertaking lithological and stratigraphic correlations from drill cuttings and cores by comparison with other wells. Also, proper geological identification will require a range of palaeontological, micropalaeontological and/or palynological studies of the core and drill cuttings. Geophysical logs will be used to estimate formation correlation between wells by comparison of, for example, sonic and gamma-ray logs. The final geological analysis of a borehole is detailed in a composite log which usually consists of:

1. A lithological description including palaeontological information.
2. A borehole compensated sonic log.
3. A gamma-ray log.
4. A caliper log.
5. Formation identification.
6. Information on drilling rate (optional).
7. Information on drilling mud weight (optional).

Where the well has penetrated and/or detected hydrocarbons, pertinent data will be listed which may be utilised by the geophysicist in seismic hydrocarbon indicator studies. A typical composite log from the North Sea is shown in figure 5/12.

5.4 Tying well data

If a seismic section can be tied directly to well information, the synthetic seismogram or VSP (with the correct polarity and frequency band-width) should be overlain or spliced in at the appropriate location. As discussed earlier in the section on seismic pulses and the earth as a filter, it is unreasonable to expect a perfect match in amplitude, frequency and phase. Where there is a good fit, often a time mis-tie will be found due to errors in any of the applicable corrections for NMO, weathering, elevations, depths of guns, cables etc. Other mis-tie effects may be induced by phase distortion in the recording or playback instrumentation. Correlation should therefore be made on an interval best-fit basis as has to be done with the synthetic seismogram which has been produced from an uncalibrated sonic log. Static errors should always be investigated but may be unresolved.

In the absence of, or in conjunction with a synthetic seismogram, the logged velocity trace from a well velocity survey or sonic can be

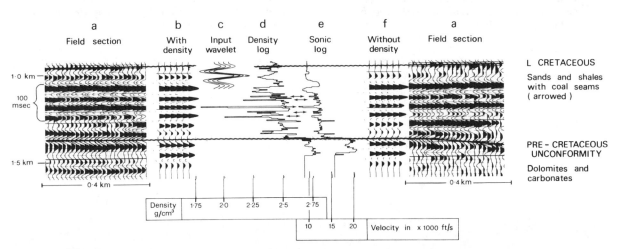

Figure 5/13 Example of borehole tie from Western Canada. The density and sonic logs (d) and (e) are rescaled versions of the logs shown in figures 5/2 and 5/5. Synthetic seismograms (b) and (f) are prepared by convolving the input wavelet (c) with reflections derived from the geophysical logs. Displays (b) to (f) are spliced into the field section (a). The correspondence between field section and synthetic sections is good. A group of reflections can be seen to be associated with the coal seams and it should be noted that although in some cases reflectivities are doubled when calculated using density information (coal has a very low comparative density), the synthetics show only minor amplitude variation because of the thinness (5–17 ft) of the seams (Courtesy; Digitech Ltd.).

overlain on or spliced into a seismic section and velocity contrasts aligned with appropriate peaks and troughs (see figure 5/13). Where the above are not available, interval velocities can be plotted on a suitable time scale from a continuous velocity log and compared as above. If neither synthetics nor well velocity surveys are available, the integration on a sonic log (see figure 5/4) can be plotted manually as an interval velocity curve and used similarly. If no sonic logs are available, a formation density log can be used to give a very qualitative indication of the relationship between the geological section penetrated by a well and the equivalent seismic section. If there is some confidence in the velocities derived regionally from seismic data, the gamma-ray or formation density trace can be converted, depth to time, and empirical correlations made.

When seismic lines do not tie directly to boreholes, various approaches can be adopted for reflection identification but mainly in this situation there is a considerable reliance on intuition and there can be no substitute for a good borehole tie.

References

A. H. Balch, M. W. Lee, J. J. Miller and Robert T. Ryder, The use of vertical seismic profiles in seismic investigations of the earth. *Geophysics*, **47** (1982), pp. 906–918.

V. Baranov and G. Kunetz, Film synthetic avec reflexions multiples theorie et calcul pratique. *Geophys. Prospect*, **8** (1960), pp. 315–25.

C. E. Deegan and B. J. Scull (compilers), A proposed standard lithostratigraphic nomenclature for the central and northern North Sea. Rep. Inst. Geol. Sci., no. 77/25: *Bull. Norw. Petrol. Direct.*, no. 1 (1977).

H. Guyod and L. E. Shane, *Geophysical well logging*, Vol. 1, *Introduction to acoustical logging*. H. Guyod, Houston (1969).

L. G. Howell, K. O. Heintz and A. Barry, The development and use of a high precision downhole gravity meter. *Geophysics*, **31** (1966), pp. 764–72.

P. Kennett and R. L. Ireson, Recent developments in well velocity surveys and the use of calibrated acoustic logs. *Geophys. Prospect.*, **19** (1971), pp. 395–411.

F. P. Kokesh and R. B. Blizard, Geometric factors in sonic logging. *Geophysics*, **24** (1959), pp. 64–76.

F. P. Kokesh, R. J. Schwartz, W. B. Wall and R. L. Morris, A new approach to sonic logging and other acoustic measurements. *Journ. Pet. Tech.*, **17** (1965), No. 3 (March).

R. A. Peterson, Synthesis of seismograms from well log data. *Geophysics*, **20** (1955), pp. 516–38.

Schlumberger Ltd. *Log interpretation principles* Schlumberger, New York (1972).

Seismograph Service Ltd. *Air-gun land and marine well velocity surveys*. SSL, Holwood, UK (1976).

J. Tittman and J. S. Wahl, The physical foundations of formation density logging (Gamma-gamma). *Geophysics*, **30** (1965), pp. 284–294.

J. S. Wahl, J. Tittman and C. W. Johnstone, The dual-spacing formation density log. *Journ. Pet. Tech.*, **16** (1964), pp. 1411–1416.

Chapter 6

Geophysical Interpretation

The primary object of geophysical interpretation is usually to prepare contour maps showing the depth to a series of reflectors which have been picked on the seismic sections. The work falls into several parts which are described below. Often the geophysicist will be involved in a seismic investigation from the earliest stage of survey planning through to interpretation of the data. It is therefore appropriate to consider here the input that the interpreter can make before the seismic sections arrive on his desk.

6.1 Quality control of survey and processing

In planning a survey grid, it is important to think carefully about the purpose for which the data is required. In the case of a regional survey, a rectangular grid is often best, preferably orientated parallel to regional dip and strike, with dip lines being the more closely spaced (Hornabook, 1974). Line spacing may be dictated by constraints on overall cost, but should be at least as close as the distance between major features of the area (for example, major faults or salt structures). Lines through boreholes are especially desirable, since they will often give a firm indication of reflector identity. Surveys intended to solve some more specific geological problems are generally made at a stage when a reasonable amount of data is already available, so that selection of a survey pattern will depend on the problem and the pre-existing data. Other parameters to be specified include the total record length, which depends obviously on the time to the deepest expected event, and the sampling interval, generally taken as 4ms but occasionally as small as 1ms if a particular geological situation requires high resolution, for example, for reservoir delineation. Specifications for maximum permissible ambient noise, source mis-fire rate, number of geophones malfunctioning, feathering angle of cable (at sea), and deviation from intended survey line are fairly standardised, but it is worth remembering that tighter limits may have to be applied if true-amplitude processing is contemplated. During the survey, the geophysicist has a role as client representative either on board the seismic ship, or with the land crew, monitoring progress and ensuring that specifications are met.

Liaison between the interpreter and the processor of seismic data is of great value. Some aspects of the interpretation of the data are implicitly carried out at the processing stage (Fitch, 1976). This is especially true of the velocity analysis stage where events can be enhanced or made to disappear completely in the subsequent stack. Geological knowledge of the area is therefore of great value, both in checking suggested velocity distributions for plausibility (especially in the presence of anomalies such as velocity inversions) and in indicating which events are of most importance to the understanding of the particular area, for it is to these that particular attention should be given. Comments on filter settings, deconvolution tests, and final display mode can all be very valuable. Specification of special processing required, such as migration, can also often be made at this stage. Quality control consists not only of checking for consistency of reflectors at line intersections, and of velocities along lines, but generally making sure that any anomalous features of the sections are not artefacts of processing, such as, for example, a sudden change of processing parameters without geological justification.

6.2 Picking a survey: reflection identification

It is usually best to start picking a survey by inspecting lines through boreholes. Not only do the well logs give a useful geological picture, but they also show where strong reflections might be expected. It is sufficient for a first look to use the sonic log, anticipating an event at each major seismic velocity change, provided that the bed giving rise to it is at least one wavelength (say 100 m) thick. Sometimes, however, density varies in the opposite sense to velocity, and changes in acoustic impedance, and therefore reflection strengths are low despite large velocity changes; this is often the case with salt layers. If there is any doubt at all over reflection identity, it is necessary to use a time-depth log incorporating the results of well check shots, if one exists, to relate the seismic events to the geological horizons. If there is no such log various approximate methods can be used (see p. 111). For an example of such a borehole tie, see the Moray Firth Case Study (p. 203).

The next stage is to examine briefly the entire survey. Often it is best to start with the dip lines which are usually easier to interpret, and lay them out in sequence. In this way it is possible to follow the major structures across the area, and at this stage it is worth marking the main faults, treating apparently major structures visible on only one line with some suspicion (Garotta, 1971). Although reflector discontinuity indicates the presence of a fault, diffraction patterns are often useful in locating

the fault plane precisely; the plane should pass through the apex of the hyperbola. There is often a tendency for reflectors to continue for a short distance across a fault plane on unmigrated sections; the resulting appearance resembles a thrust fault (figure 6/1). Caution should therefore be exercised in marking reverse-throw faults on seismic sections unless independent evidence for their existence is known. It is now possible to mark on the section those reflectors whose identity is known from boreholes. Fine-pointed coloured pencils are a convenient method, since a uniform colour code can be established throughout an area. For each reflector, the first peak should be marked.

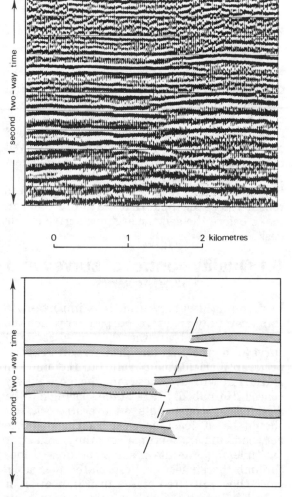

Figure 6/1 Interpreted unmigrated seismic section showing typical normal faulting (Courtesy: IGS; *Seiscom survey*).

Usually, faulting or other structure makes the following of reflectors difficult within a short range of a borehole, since it will commonly have been drilled on a structural high within a basin, and therefore in an area of tectonic disturbance. The problem of correlation across such structures thus arises. It is often possible to correlate across small faults by using strike lines to carry the interpretation around the ends of them; in the case of major structures it is usual, in the absence of any other information, to use the reflection character as a guide. It can be helpful to fold a section so as to bring into juxtaposition the undisturbed area on the two sides of the fault, when the correct correlation is sometimes obvious; doubtful cases should however be treated with suspicion since the character of a reflector is highly dependent on processing parameters which may change greatly over a major fault. Geological plausibility is also a useful guide; faults whose throws diminish markedly downwards should cause reconsideration, although such effects are geologically possible and are not uncommon on seismic sections where velocities generally increase with depth (Tucker and Yorston, 1973).

Finally, the picking of the entire survey should be tied together, making sure that all line intersections are consistent. This is a very powerful check on the correctness of the interpretation, but some problems often occur. It can be useful to make a rough map of the main structural features at this stage. Figure 6/2 shows a fictitious example of such a map. In figure 6/3 the intersection ties from a closed loop are illustrated.

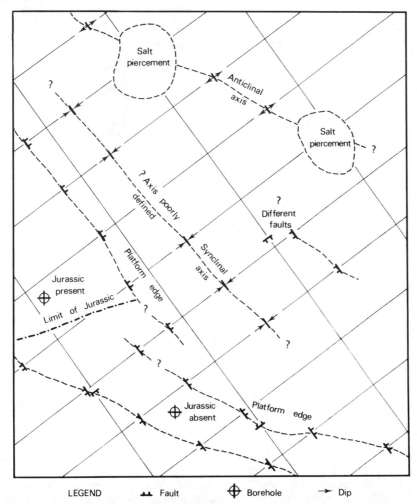

LEGEND   ⊥⊥ Fault   ⊕ Borehole   → Dip

Figure 6/2 A schematic structural map of a seismic interpretation project area.

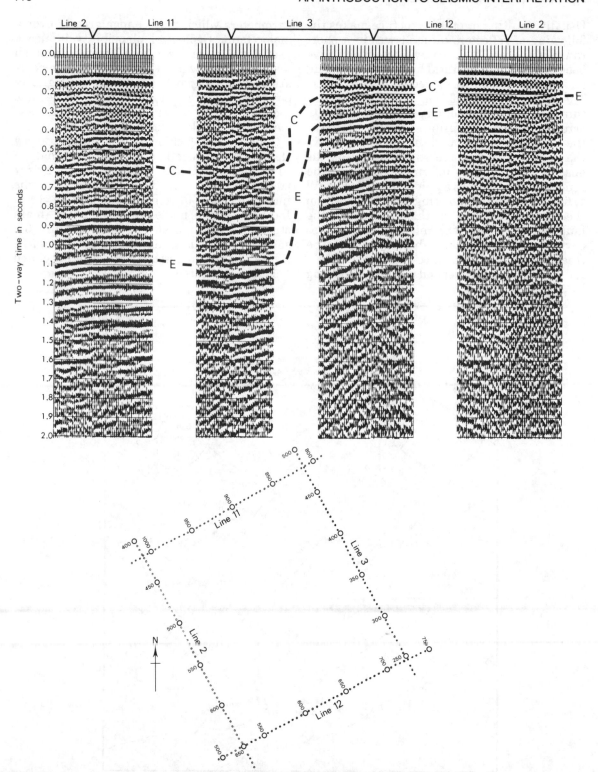

Figure 6/3 Seismic ties around a closed loop. Horizons C and E in the Moray Firth (see Chapter 10) are shown at the intercepts of lines 2 and 11, 11 and 3, 3 and 12, 12 and 2 as shown on the shot point map. A complete interpretation of lines 3 and 11 is shown in Chapter 10. Horizon E can be traced continuously around the loop whereas horizon C is absent at the south end of line 2 and the west end of line 12. Note the change of reflector quality at the various intercept locations (IGS data; *Seiscom survey*).

6.3 Mis-ties and their causes

Closing of loops can be made much more difficult by mis-ties between sections where they intersect. Usually the data from any one survey will tie together more or less perfectly, but lines from different surveys will not. Quite often a bulk shift of one survey relative to the other by a few tens of msec will be sufficient to give a good tie, but sometimes the shift needed will vary with TWT. Let us examine some of the common causes of mis-ties.

Especially with older marine data, poor position-fixing can cause gross mis-ties. Water depths are usually written at intervals along the section. These are usually accurate, and so can be used to demonstrate the presence of a navigational error and to suggest how the line positions should be adjusted. If a line crossing a survey grid needs the same adjustment in a particular direction at every intersection, the interpreter can have some confidence that faulty positioning is the cause of the mis-ties.

Differences in processing can make the intersecting sections look very different from one another. It may then be difficult to know how to adjust them, especially if one section has a much higher vertical resolution than the other. Differences in deconvolution parameters, filter settings, and stacking velocities can all cause problems. Another cause of difficulty is reversal of polarity between sections. On older data, the polarity may be unknown or even wrongly stated. It can be hard to decide whether a polarity reversal is present if it is combined with shifts due to other causes. Differing statics corrections can also cause mis-ties, but are easier to deal with because they produce only a bulk shift between data sets.

Working with migrated sections can also cause problems. We saw in Chapter 4 that a strike line will not be corrrectly migrated on the usual 2D assumptions, and will therefore not tie a migrated dip line. It may sometimes be necessary to use the unmigrated sections to establish the correct tie, which can then be transferred to the migrated sections.

6.4 Digitisation

The first step in making a map from the picked seismic sections is the measurement of two-way time to the picked events along each section. These data will then be transferred to the base-map in order to produce a contour map of structure in two-way time.

The simplest method of reading the times off the sections is to perform the operation directly by hand, using a scale appropriate to the section. The horizontal intervals between readings depend on the complexity of structure visible on the section; ideally, the frequency of reading should be such that interpolation between adjacent points by straight lines is adequately precise. Much depends on the scale of the final map; for example, in regional mapping at a scale of 1:100 000, the sections might be measured every 1km except in areas of greater complexity. If, however, the data will be mapped at a larger scale at a later date, it is probably worth taking this future requirement into account from the beginning. Positions of faults should be recorded and horizons timed on both the up and down thrown sides.

Reduction of data by hand in this way can become tedious if large amounts of data are involved. Systems exist for the direct digitisation of seismic records into computer-readable form, so that maps can be automatically prepared by machine. Typical digitiser hardware is illustrated in figure 6/4. A seismic section is fixed to the table and header information giving the line name, names of horizons to be digitised, etc. is typed in from a keyboard. A cursor is then moved along each horizon in turn and the (x, y) co-ordinates of each point are recorded by the machine. Finally the (x, y) co-ordinates of the corners of the section are recorded so that the software may later calculate the scale factors in x and y directions and apply a correction for the skewness of the seismic record with respect to the (x, y) axes of the table. Output from the machine can be magnetic tape, paper tape, or cards; a preferable system is to connect the digitiser on-line to a computer, in which case data reduction can proceed simultaneously with digitisation, and errors can be seen and corrected before any incorrect information has been filed.

Various software packages exist for reduction of the digitised data to the equivalent of the manually-read values. Typically, data is scaled and converted to shot point versus two-way-time values, which are then interpolated to a uniform specified shot point increment. Static corrections, if specified in the header information, may be made automatically.

Figure 6/4 A digitising table with operator digitising a seismic section (IGS photo).

In general, data reduction by machine does not offer large savings in time unless a number of horizons are to be digitised on each section. This is because of the care required to enter correct header information, without which the subsequent reduction will be erroneous. Checking and correcting of the automatically prepared data files can also be time-consuming. The main advantages of having the raw data in machine-readable form are that updating of maps consequent on the receipt of new data is much simpler, and that depth conversion (see p. 124) is facilitated. A flow chart for a typical system is shown in figure 6/5 (see also Paulson and Merdler, 1968).

6.5 Map construction

Once seismic sections have been interpreted the next object is to produce a contour map of two-way-time to each horizon. Even where depth conversion is intended, this stage should not be omitted as the two-way-time map is directly related to the seismic sections and will need much less radical revision as new data are acquired than will the depth map, construction of which involves the use of (often very uncertain) velocity functions. If the two-way-times have been read off the section by hand, it is necessary next to post these on to a shot-point map. Before contouring the values it is desirable to mark all the faults on the map and decide how to join them together making sure

that faults on maps of different horizons are coincident. Often it is necessary to refer back to the original sections to determine which faults are of similar appearance. Knowledge of dominant geological trends is helpful. Contouring is then carried out between the faults. Contour interval depends on the depth resolution required of the map, but a useful rule of thumb is that contours should have an average spacing of about 1cm if the resulting map is to be both legible and in reasonable detail.

If the data have been digitised by machine, the next step is to digitise the shot-point map so that a map of two-way-times posted in their correct positions can be produced automatically. It is possible to proceed from here by computerised contouring, but few software packages available at present perform well in the presence of faults, and hand contouring is often preferable.

It is difficult to give general advice on hand contouring, as most problems can be solved by common sense. Many people find it useful to begin contouring the data roughly so as to identify the main trends, followed by a detailed revision; it can be useful to draw synclinal and anticlinal axes on the map so as to ensure that all contours intersecting them turn along the same axis. Minor mis-ties become apparent at this stage and can be resolved by reference to the original sections. Particularly in regional-scale work, it is desirable to ignore an isolated value where it disturbs the trend.

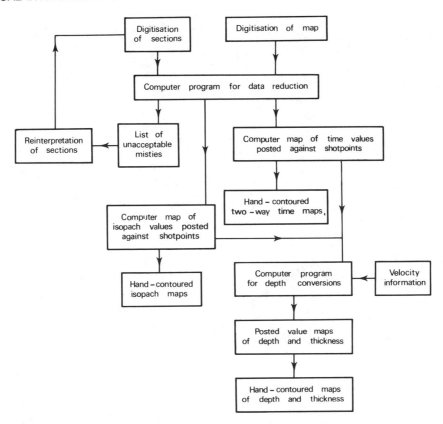

Figure 6/5 Flow chart of a scheme for digitising and depth converting the results of a seismic interpretation project.

An awkward problem presents itself when migrated sections have been used. In structurally complex areas, this is often essential as the unmigrated data may be almost uninterpretable. Usually, migration velocities will not have been corrected for the effect of a traverse not being exactly in the dip direction (Chapter 4). As a result, migrated dip and strike lines will not tie at their intersections. Contouring then should be based mainly on the dip lines, with much less weight attached to the strike lines. In areas of less structural complexity, it may be preferable to construct a map from the unmigrated sections, and then migrate the map by methods such as those discussed in Chapter 4. Migrated maps are significantly different from unmigrated ones when dips exceed about 5°.

When contouring has been completed, it should be checked against the original sections, especially in regions of complex structure (for example, in areas much disturbed by faulting) and in the vicinity of closed highs if they may be

of hydrocarbon significance. The final map is then prepared; usually it shows faults of different throws by different widths of line (increasing in width as the throw becomes greater), and usually also some contours (say every fifth) are bolder than the rest to facilitate a quick appreciation of structure (see also Hintze, 1971). A typical two-way time map is shown in figure 6/6. In figure 6/7 a typical mapping scheme is illustrated; this defines the symbols to be used, contour and fault symbols, line widths and style of presentation.

6.6 Velocity maps

In order to convert the two-way time map to a depth map, we need to know the velocity distribution over the area. The sources of information on velocities are well sonic logs on the one hand and seismic stacking velocities on the other. We shall consider each in turn.

In an area where drilling has been extensive but seismic data are sparse or of poor quality,

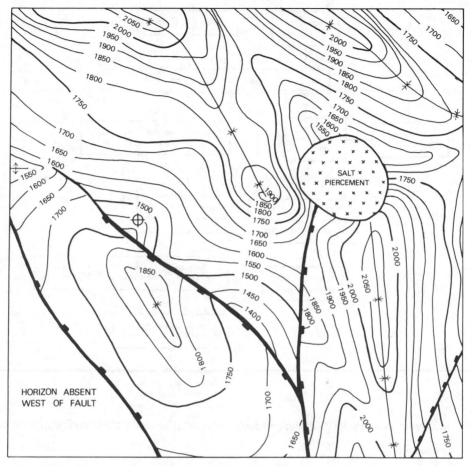

Figure 6/6 A typical two-way-time map.

velocities derived from seismics will be unreliable and it may be necessary to rely primarily on well velocity data. From the sonic log, the average velocity in each formation can be deduced. At first sight, it seems that all that is required is to contour this information for each formation of interest. Unfortunately, however, wells are not normally scattered uniformly over an area; they are concentrated on the high areas within a basin. The problem then arises as to how to extrapolate the well velocities into the deeper parts of the basin. One method is to measure the variation of velocity with depth of burial; in some cases a good correlation is obtained, but if an uplift subsequent to burial has occurred at some time the degree of recovery of compaction on release of overburden pressure will depend on the lithology. Therefore, a substantial scatter on a velocity-depth graph is common. If no other information is available, however, the best depth gradient of velocity may have to be used to extrapolate away from the wells.

Often, the situation is the reverse of the above, and well data are sparse whereas stacking velocities are available along the seismic lines. It is usual to assume that these stacking velocities are rms velocities to the reflectors concerned. We saw in Chapter 4, however, that lateral velocity changes over the length of a CDP gather may result in stacking velocities that are considerably different from the rms values. As a result, stacking velocities usually show considerable scatter over a map. It is therefore necessary to plot the stacking velocity data on a map and contour it by hand. A good deal of subjective smoothing and data rejection is usually needed.

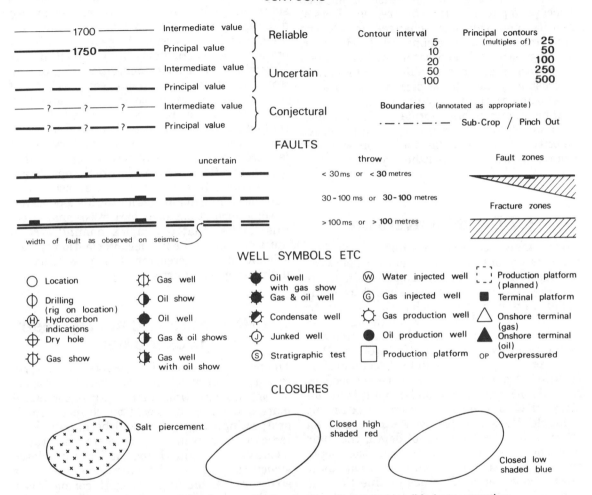

Figure 6/7 A mapping scheme with standard symbols used in oil industry mapping.

The velocity that we need to convert TWT into depth is not the rms velocity, but the average velocity. This is defined for a 'layer-cake' earth of n plane horizontal layers of which the kth has velocity v_k and TWT thickness t_k by

$$v_{av} = \frac{\sum\limits_{k=1}^{n} v_k t_k}{\sum\limits_{k=1}^{n} t_k}$$

The average velocity is usually a few percent less than the rms velocity. A conversion factor may be obtained by calculating v_{av} and v_{rms} from the sonic log at a well. Sometimes it suffices to assume a constant conversion factor over an area, but it may be necessary to contour the conversion factors obtained from several wells, and guess how to extrapolate them across the area.

A better approach may be to calculate the interval velocity from the Dix formula for each lithological unit above the reflector concerned. Contour maps of these interval velocities should reflect the controlling variables of lithological variation and depth of burial. It should be easy to introduce geological intuition into the contouring of such maps; it is much harder to make geological sense of an rms velocity map to a horizon, where the effects of velocity and thickness changes in all the overlying units are compounded. However, treating each unit separately involves much more work, since several reflectors of no direct interest may have to be picked, just so as to depth convert an underlying horizon.

If no other reliable information is available, it is sometimes necessary to use an average velocity for a particular formation, deduced from a knowledge of its lithology.

6.7 Depth conversion

Given maps of two-way-time to various horizons and of the average velocities between the formations, it is in principle simple to multiply times by velocities to arrive at depth, building up the depth to each horizon in turn by a 'layer-cake' method. These depth values can then be posted and contoured in the same way as the two-way time values (Paturet, 1971).

In practice, the amount of computation required if a large area is to be mapped on several horizons is formidable, and it is at this stage that a computer becomes extremely valuable. Various approaches can be used for the automated calculation of depth; the main problem is to make sure that the depth conversion is carried out in such a way as to ensure consistency of values at line ties and near wells. One method is to represent the velocity data by a series of values on a square grid, the grid size being small enough to represent adequately the complexity of the data. Such a grid can be constructed by reading off values by hand from the velocity contour maps, or can be machine-generated from the contour map after the contours have been digitised. At each shot point where depth conversion is required, the computer then calculates an average velocity value by interpolation from several neighbouring grid points, which is used in the depth calculation for that shot point. An advantage of this method is that well data can be included together with the grid points, and can be heavily weighted so that lines in the vicinity of wells will tie into them exactly.

6.8 Isopachs

Isopachs may readily be constructed by subtracting the time or depth values to two different horizons at each shot point. They are often useful in assessing the geological history of an area, particularly in unravelling the history of sedimentary deposition. Again, the work involved in manual computation of isopach values can be considerable, and such maps are more readily generated by the computer from the digitised seismic information. For a fully automatic system, however, it is necessary to have a rather sophisticated system of reflector nomenclature if isopachs are to be correctly calculated in the case where outcrop or faulting makes for the disappearance of some formations over part of the map area; this implies that a complex naming system must be used in the header information of each seismic line, increasing the possibility of digitiser operator error.

6.9 Reporting and management presentation

The primary product of a geophysical interpretation of a set of seismic sections is a series of maps; two-way time maps, depth contoured maps and, in some cases, isopach maps. In oil prospecting, these maps will be annotated to indicate the presence of prospective structures, usually by highlighting in colour closed structural highs, or other potential trap structures. In mining and engineering investigations, important features such as faults, steep dips, structural axes, pinch-outs or interval thickness variation are likely to be the important features which need highlighting on the finally draughted maps. In either case it will be necessary to present these maps, along with an interpretation report to either the client who has commissioned the interpretation, or to company management, with recommendations for further investigation or development of the prospect under investigation. At this stage close collaboration between geologist and geophysicist is essential.

A typical report will include most of the following sections:

1. An introduction describing the aims of the interpretation.

2. A description of the data used with notes on acquisition and processing parameters if relevant, and comments on the quality of the data. If static adjustments have been necessary to allow fit of data from different surveys, these should be noted.

3. A description of the geology and structural framework of the area, illustrated with relevant diagrams.

4. A section describing ties between wells and the seismic sections including a list for each well of seismic times to the tops of important horizons, in particular those which have been mapped.

5. A section on the selection of marker horizons which have been mapped giving evidence for stratigraphic identification of these horizons. It is usual to present sample interpreted sections to illustrate typical structural features and reflector quality.

Horizons may have been selected because of prospectivity significance rather than reflector quality.

6. A detailed description of the interpretation of each horizon mapped – this usually describes the main features of the map or set of maps associated with the horizon in question.

7. A description of the main prospective structures (in oil exploration) or mining or engineering hazards etc. depending on the aim of the investigation.

8. Recommendations for further work; more seismics, other geophysical investigations, drilling. If drilling is recommended a prognosis will be necessary for each suggested site giving a best estimate of the expected geological section. Estimates of depth are necessary to those intervals where cores will be required, as well as to

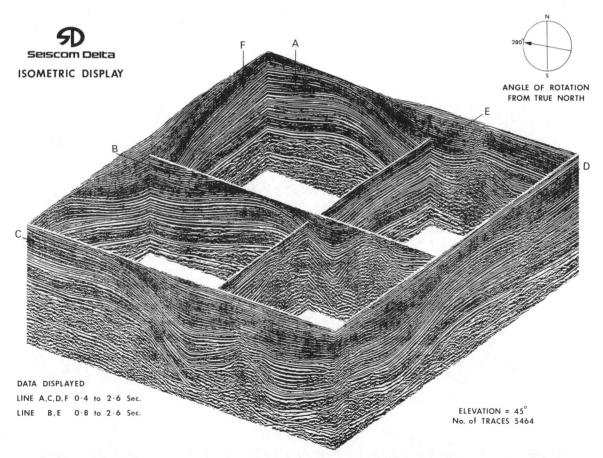

Seiscom Delta

ISOMETRIC DISPLAY

ANGLE OF ROTATION
FROM TRUE NORTH

DATA DISPLAYED
LINE A,C,D,F 0·4 to 2·6 Sec.
LINE   B,E   0·8 to 2·6 Sec.

ELEVATION = 45°
No. of TRACES 5464

Figure 6/8 Isometric section projection of a group of seismic 'loops' prepared to aid three-dimensional comprehension of a structurally complex area (Courtesy: Seiscom Delta).

those of principal economic interest.

9. The aim of the study may be to provide a more general prospectivity assessment of the area to give guidance to management on future development policy, in which case a classification and grading of structures will be required. Detailed drilling recommendations would then depend on follow-up surveys and these would only be undertaken if the company concerned was able to acquire the necessary exploitation licence.

10. A summary of work undertaken, list of maps and technical appendices.

Other sections of a specifically geological nature may be necessary to supplement the geophysical interpreter's input to such a report. Visual aids are extremely important in conveying an interpreter's concepts and, based on the principle that a picture is worth 1000 words, any illustrations that reduce the complexity of the seismic data to a simple, understandable form should be included; in particular, any specialised data processing such as described in Chapter 4 or successful modelling as described in Chapter 9 should be highlighted. The three-dimensional aspects of an interpretation, often difficult to grasp by the casual or busy observer, can be enhanced by the use of isometric presentations of tying seismic sections such as shown in figure 6/8, or by presentation of computer produced isometric views of key horizons such as seen in figure 6/9.

A seismic interpretation report, particularly of a large offshore area, can be quite voluminous. Presentation of the main conclusions of a report to management, partners, clients or government licensors etc. can be accomplished in a visually effective manner by the preparation of a montage. Using, for example, a one metre high reproducible base, designed to go through a normal dyeline machine, the key isochron or velocity converted depth maps, seismic lines, and other supporting geophysical information (such as gravity or

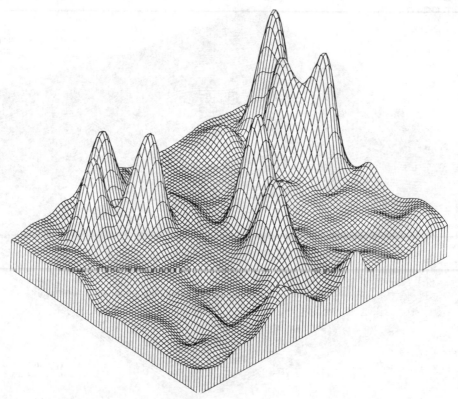

Figure 6/9 Isometric projection of a Devonian reef surface in Western Canada. Vertical scale is greatly exaggerated, the maximum relief being approximately 180 m; the grid square size is approximately 100 m (Courtesy: Mobil Oil Canada Ltd).

magnetics) or seismic modelling can all be brought together in one visual summary.

6.10 Seismic interpretation exercise

The bulky nature of full-scale seismic sections makes it difficult to present within any text book format an adequate data set for exercise purposes. A number of oil companies have released data for such training which can normally be obtained by universities and other training establishments, by arrangement. The Institute of Geological Sciences (IGS) in the UK has also made up a training package, details of which are given in Appendix 7 (McQuillin and Donato, 1981). The present exercise is based on the IGS package. Data are derived from a seismic survey undertaken by IGS in the Moray Firth area, offshore Scotland, the subject of the case history in Chapter 10. Before commencing this exercise, it is advised that Chapter 10 be read.

6.10.1 Description of the data

In this chapter, uninterpreted versions are provided of IGS lines 5, 6 and 7 as well as parts of 10, 11 and 12 from the Moray Firth survey (for locations see figure 10/6). IGS lines 5 and 6 provide good dip sections across the basin and most of the principal features of the Mesozoic geology are to be seen on them, especially if studied in conjunction with lines 3 and 4 displayed in Chapter 10 (colour section). In addition to uninterpreted lines displayed in this chapter (figure 6/10), coloured up interpreted versions are printed (colour section) as part of Chapter 10. It is advised that the interpreted versions should not be studied in too great detail before undertaking this exercise. It must be noted that the sections printed here are at a very reduced scale compared with the size normally used in interpretation work (usually 10cm per second TWT). However, structures are not complex and it should be possible to achieve the main object of the exercise which is to identify, using a borehole correlation shown in figure 10/7, two marker horizons on line 7, the Lower Cretaceous Horizon B and the (near) base Jurassic Horizon E, then pick these horizons around the loops formed by lines 7–11–6–12–7; 6–12–5–11–6; 11–6–10–5–11. For completeness, the exercise should include identification of principal reflectors using

actual well logs, not figure 6/10. A set of the appropriate well logs, along with full scale sections, maps, etc. is included in the IGS package, see Appendix 7.

6.10.2 Interpretation of the seismic lines

Reflecting horizons are usually 'picked' on the seismic sections using coloured pencils. To conform with the colour code used on the interpreted sections, it is suggested that a green pencil be used to pick Horizon B and a red pencil for Horizon E. Starting from well 12/26–1, which is located on the upthrown side of a small fault at Horizon E level, the two reflectors can be followed along line 7 to the intersection ties with lines 11 and 12. It is apparent from the sections that the intersection 7–12 is on a high structure, which can also be seen on lines 6 and 5, whereas the intersection 7–11 is a simpler structural area. It is thus better to follow the reflectors through the deeper part of the basin first along line 11 through intersections 11–6 and 11–5, then around the loop 11–6–10–5–11. Having checked that this loop closes without mis-tie, with reference to the interpreted sections (see colour section) if necessary, lines 5 and 6 should be interpreted through to the intersections 5–12 and 6–12. The final stage is then interpretation of lines along the structural high where Horizon B is shallow and not easily identified in the poorly resolved data close to seabed.

6.10.3 Features revealed by the seismic data

Although the structure of this area is more comprehensively treated in Chapter 10, it is useful here to briefly summarise the principal geological features revealed by the data. It can be seen that the sedimentary basin is divided into two sub-basins by a prominent horst which trends east-north-east across the area. The Great Glen Fault can be seen on the northwestern ends of lines 5 and 6 as a complex zone with thick Jurassic sediments to the southeast. A possible antithetic fault is seen on line 6 associated with the Great Glen Fault. To the northwest and southeast of the horst, the base-Cretaceous is an easily recognised strong reflector relatively unaffected by faulting. The deeper base-Jurassic reflector is a strong event but more heavily faulted. Major faults, however, such as the Great Glen Fault and those bounding the horst continue to be active through the Jurassic into the Cretaceous.

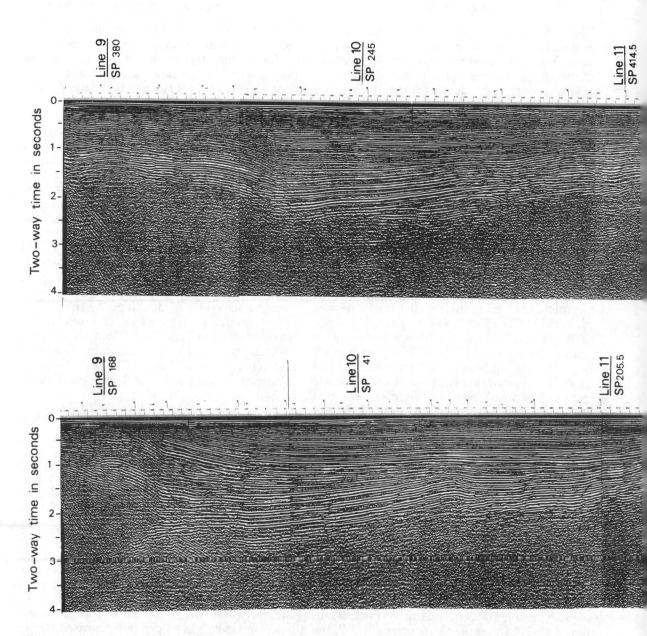

Figure 6/10 Uninterpreted seismic lines from the Moray Firth area; lines 5, 6, 7, 10, 11 and 12. (Courtesy: IGS).

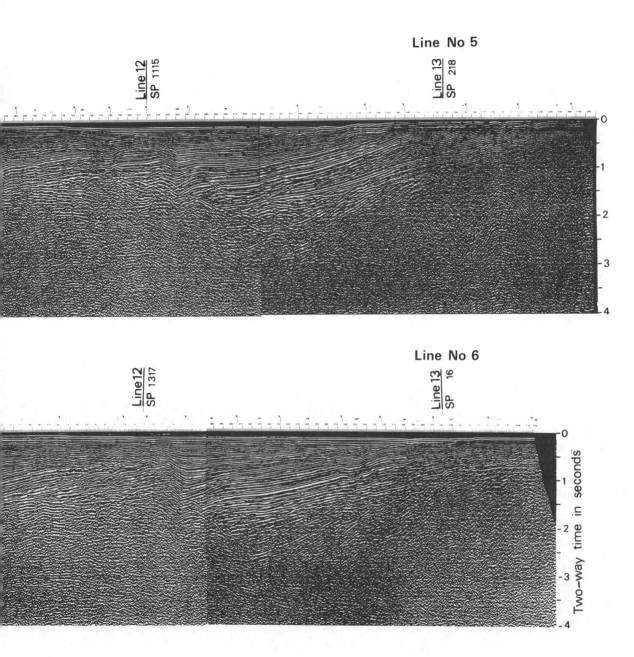

Line No 5

Line No 6

Two-way time in seconds

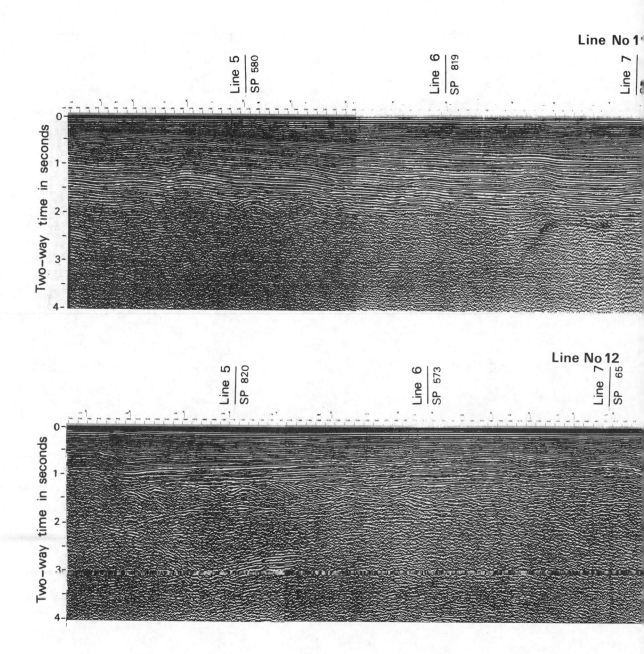

Line No 10

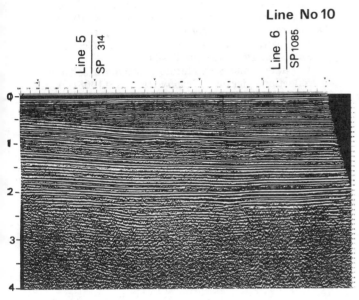

Well
12/26−1

Line No 7

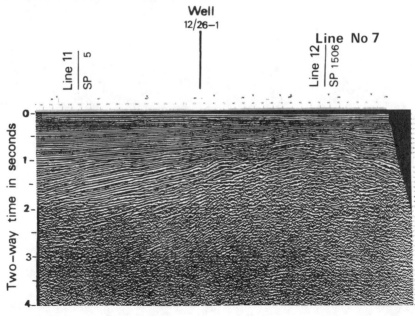

Sediments thin over the horst and thicken on the downthrown sides of the faults indicating fault controlled sedimentation as well as possible intermittent erosion of sediments from areas which were structurally high throughout development of the basin.

References

A. A. Fitch, Seismic reflection interpretation. Gebruder Borntraeger, Berlin (1976).

A. Garotta, Selection of seismic picking based upon the dip, move-out and amplitude of each event. *Geophys. Prospect.*, **19** (1971), pp. 357–70.

W. H. Hintze, Depiction of faults on stratigraphic isopach maps. *Bull. AAPG.*, **55** (1971), p. 871.

J. T. Hornabrook, Seismic re-interpretation clarifies North Sea structure. *Petrol. International*, **14** (1974) pp. 45–53.

R. McQuillin and J. A. Donato, Moray Firth seismic interpretation training package. *Inst. Geol. Sci. (Marine Geophysics Unit)* report no. 113 (1981).

D. Paturet, Different methods of time-depth conversion with and without migration. *Geophys. Prospect.*, **19** (1971), pp. 27–41.

K. V. Paulson and S. C. Merdler, Automatic seismic reflection picking. *Geophysics*, **33** (1968) pp. 431–49.

P. Tucker and H. Yorston, Pitfalls in seismic interpretation. *SEG Monograph No. 2* (1973).

Chapter 7

Geological Interpretation

Interpreting seismic sections, producing time, depth and isopach maps is a task which depends on the interpreter's ability to pick and follow reflecting horizons (reflectors) across an area of study. He has to be able to correlate across faults and across zones where reflectors are absent because of geological discontinuity, for example, between basins either side of a structural high or between fault blocks beneath an unconformity. Reflectors usually correspond with horizons marking the boundary between rocks of markedly different lithology. Such a boundary does not always occur exactly at a geological horizon of major chronostratigraphic importance, such as the base or top of a System or Series, but may be simply a seismic marker horizon which occurs close to that boundary. This problem can be resolved by correlation of seismic and borehole data, and synthetic seismograms have particular relevance to such studies. In this chapter we shall discuss the elements of structural geology and lithological variation in sedimentary rocks as discernible on seismic sections and in those respects which are important to the problems of seismic interpretation.

7.1 Lithology of common sedimentary rocks

As stated above, we are principally concerned in seismic interpretation with geological struc-

ture in thick accumulations of sedimentary rocks. However, firstly we will consider one term in common use in seismic interpretation, that of economic (or seismic) basement. The term basement is also widely used in geology to signify a complex of crystalline and metamorphic rocks which are covered unconformably by unmetamorphosed sediments. Often the two terms are synonymous within a particular area of study, whereas in other circumstances the seismic horizon mapped and termed basement may not in fact be the base of the sedimentary succession; it may, for example, be the top of a series of lavas or sills intruded into the sediments, the top of which is the lowest horizon which can be effectively mapped by the seismic method. In some circumstances, this may, in practical terms, form the economic basement as far as exploration for hydrocarbon reservoirs is concerned. On the other hand, it is not unusual for igneous rocks such as volcanic sills, lavas and tuffs to occur as important markers within a stratigraphic sequence in circumstances where lower horizons and structures are easily identified: in this case the top of the igneous rock layer would not be termed the basement horizon. Care must therefore be taken in interpreting the use of the term basement both in well logs and on seismic maps.

Thick accumulations of sedimentary rocks occur in large scale geological structures termed sedimentary basins (see Section 7.2) and the material which forms the major component of infill of such sedimentary basins is a group of rocks known as the clastic rocks. These rocks are formed as a result of weathering of pre-existing rocks into fragments of varying size and composition followed by transportation from the area of origin to the area of deposition. Clastic rocks are thus built up of fragments and are classified according to the size and mineralogy of these grains. A commonly used particle size classification is the Wentworth Classification (see Table 7/1). Most sedimentary rocks are a mixture of grains of various sizes. The bulk of the rock may, for example, be made up of sand grains but infilling spaces between these grains may be a matrix of much finer rock fragments, and finally the rock may have been cemented by a material, usually of chemical or biochemical nature, which has impregnated the rock and bonded the grains together during the process of diagenesis. In figure 7/1 we see a typical notation used in preparation of composite logs to indicate lithological variation within a borehole section;

this diagram shows only a selection of the many rock types which may occur. The uppermost six rock types are varieties of clastic rock, the coarsest being the pebbly sandstones, grits and conglomerates (breccias are similar to conglomerates but with angular as opposed to rounded rock fragments). Sandstones are common reservoir rocks, especially if relatively uncemented and clean, that is, they do not contain a high proportion of fine-grained matrix minerals. Sandstones are of the arenaceous group of rocks in which grain sizes range from 1/16mm to 2mm, and can be graded from very fine to very coarse. Sandstones have grains predominantly of quartz. Other arenaceous rocks can be important, such as arkose rocks which contain notable quantities of feldspar grains in addition to quartz, also greywackes which are poorly sorted rocks made up of a wide range of grain material. Fine-grained sedimentary rock types are usually termed argillaceous; such rocks include clay, shales, mudstones, siltstones and marls. They are usually porous but because of their fine grain they are in most cases poorly permeable. They often provide cap

Table 7/1 Particle size classification (Wentworth)

Classification	Grade limits (diameters in mm)	Microns	Retained on mesh
Boulder	Above 256		
Large cobble	256–128		
Small cobble	128–64		
Pebble			
Very large pebble	64–32		
Large pebble	32–16		
Medium pebble	16–8		
Small pebble	8–4		5
Granule	4–2		6
Sand			
Very coarse sand	2–1		12
Coarse sand	1–1/2		20
Medium sand	1/2–1/4		40
Fine sand	1/4–1/8		70
Very fine sand	1/8–1/16	125–62.5	140
Silt			
Coarse silt	1/16–1/32	62.5–31.2	270
Medium silt	1/32–1/64	31.2–15.6	
Fine silt	1/64–1/128	15.6–7.8	
Very fine silt	1/128–1/256	7.8–3.9	
Clay			
Coarse clay	1/256–1/512	3.9–1.95	
Medium clay	1/512–1/1024	1.95–0.975	
Fine clay	1/1024–1/2048	0.975–0.487	
Very fine clay	1/2048–1/4096	0.487–0.243	

Notation used in presentation of
vertical sections in borehole logs

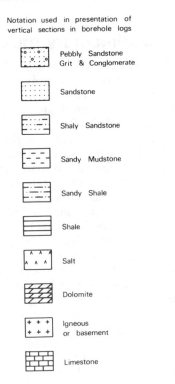

Pebbly Sandstone
Grit & Conglomerate

Sandstone

Shaly Sandstone

Sandy Mudstone

Sandy Shale

Shale

Salt

Dolomite

Igneous
or basement

Limestone

Figure 7/1 Typical lithological notation.

rock for hydrocarbon reservoirs. Shales are well laminated mudstones and siltstones are similar to mudstones with a predominance of slightly coarser grade silty material. Marls are calcareous mudstones. Such terms as shaley sandstone as shown on a borehole log usually indicate a rock with layers and partings of shaley and sandstone rock material, the bulk being a compound of both rock types.

Apart from clastic rocks, the other two main groups of sedimentary rocks to be considered are the chemical and organic deposits. Chemical deposits can be of great importance in the development of hydrocarbon-containing sedimentary basins. These deposits are formed by chemical precipitation from aqueous solutions. Organic deposits are formed by deposition of fragments of material which were originally parts of living organisms; for example, some organic reef limestones contain many shell fragments, coral remains etc. Some common rocks form through contemporaneous accumulation of organic fragments coupled with chemical deposition from solutions. Carbonates are often of this type and in petroleum geology these rocks along with sands and sandstones

form most of the world's hydrocarbon reservoirs.

The carbonate group of rocks includes limestone (main constituent calcite: $CaCO_3$) and dolomite (main constituent being the mineral dolomite: $CaMg(CO_3)_2$) as well as intermediate rocks containing both calcite and dolomite minerals such as dolomitic limestones. Other rocks in the group contain varying proportions of such impurities as sand, silt and clay and are named accordingly; for example silty limestone. Chalk is a particular variety of limestone, usually very pure, with a high percentage calcite content. As reservoir rocks, the carbonates differ from sands and sandstones in that the rocks usually have relatively low porosity and permeability. Some porosity may be original as is the case in detrital limestones but more usually in carbonate reservoirs the necessary porosity and permeability occurs as a result of post-depositional solution and fracturing.

Other chemical deposits which can form important constituents of a sedimentary basin sequence are the groups of chloride and sulphate rocks which form as a result of the evaporation of saline water. Such rocks are not important as reservoir rocks but can be important as cap-rock material by virtue of their low permeability. Rocks in this group include rock-salt (NaCl), sylvite (KCl) and carnallite ($KCl.MgCl_2.6H_2O$) and these have additional importance in that, under certain conditions, they have the properties of a plastic as opposed to an elastic solid. Deformation and flow migration of these rocks can then bring about development of structures which form traps for hydrocarbon accumulation. The sequence of rocks formed by evaporation of saline water is often termed the evaporite sequence. Typically in a sedimentary basin this might include layers of rock of the types: chemically deposited limestones, both calcitic and dolomitic, followed by gypsum ($CaSO_4.H_2O$) or anhydrite ($CaSO_4$) then, usually as a major component, rock-salt with at the uppermost end of the depositional cycle deposits of potash and magnesium salts such as sylvite and carnallite.

The last group of sedimentary rocks to be considered are the pyroclastic deposits. These are sediments which consist predominantly of volcanic ejectamenta. They can be deposited on land or in water. Lithology can range from coarse volcanic agglomerates to very fine deposits usually known as ashes or tuffs. These latter rocks are deposited from the clouds of very fine rock particles which are formed during explosive volcanic eruptions. Such clouds can distribute material over very wide areas and tuffs and ashes can act as valuable marker horizons in seismic interpretation as well as being an invaluable aid in the task of correlation between well logs in tectonically active areas. Such rocks are not generally important either as reservoir rocks or as cap material even though porous and permeable pyroclastic rocks do exist.

Most sedimentary basins are formed over very long periods of time, often with stages of both marine and non-marine deposition, and including sedimentation in both shallow water and deep water environments. These environmental changes are reflected in the type of rocks formed; cycles of sedimentation can be detected from which the history of the basin may be deduced. Many of the events which control changes in sedimentation type occur abruptly; thus changes from one rock type to another are often seen in a rock succession at distinct horizons without gradation. This does not necessarily indicate any cessation of deposition between one bed and the next. With continued development of a sedimentary basin earlier sediments become more deeply buried. Sedimentation is followed by a long period of diagenesis. Diagenetic processes are those which affect the rock during burial and bring about its compaction and cementation. These are relatively low temperature, low pressure processes (as opposed to metamorphic processes) which convert the unconsolidated sediment into a mature rock. Following diagenesis, the mature rocks will have varying physical densities and elastic moduli. In general there is a tendency for seismic velocity (and to some extent density) to increase with depth of burial. These changes are not gradual however, and as we have noted previously (p. 18) it is at the boundaries between units of rocks of significantly different velocity and density that strong seismic reflections occur. Depending on the history of the basin, a seismic reflector associated with change of lithology at a boundary between rock units can be very widespread or restricted to a limited area or areas.

7.2 Depositional features of sedimentary rocks

Only fairly large-scale features of the depositional structure of sedimentary rocks can be studied using reflection seismics. Internal structure within thin beds of rock is not generally discernible. Basins form by regional subsidence of an area and one of the results of a seismic interpretation such as described in the previous chapter is that it provides the geologist with a picture of large-scale variations in thickness of the main rock units. This gives evidence on the mechanism of subsidence (whether or not fault controlled) and where maximum subsidence has occurred, as well as detailed evidence of the present-day shape of the basin. Basins can be either symmetrical or asymmetrical and both graben and half-graben structures are common, though faulting may post-date the main period of sediment accumulation (see figure 7/2). In studying the large-scale features of a basin it is often valuable to attempt to reconstruct the shape at various stages in its history and interpreted seismic profiles across the entire area, especially if controlled by well data, give exactly the information required to undertake such a reconstruction; see, for example, figure 7/3 and Section 7.5.

Seismic sections, as well as giving data on the macrostructure of a sedimentary basin, can also give evidence on the prevailing depositional environment at various stages of its development. This evidence relates in the main to non-tectonic structural features. Seismic profiles can indicate the presence of marine transgressions and regressions; evidence of

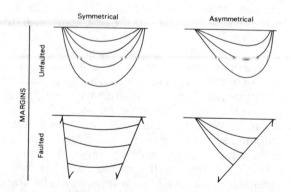

Figure 7/2 Classification of sedimentary basins (after Chapman, 1973).

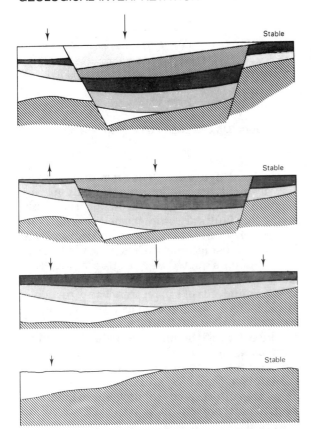

Figure 7/3 Reconstruction of the history of a sedimentary basin from seismic data. Top schematic geological section represents a seismic interpretation of present-day configuration of rock units. This structure is estimated to have developed from the earliest stage of development, the bottom section, through stages of vertical movements (indicated by arrows), sediment infill and erosion.

shallow water sedimentation is seen in such features as cyclic sedimentation and the presence of infilled river channels, deltaic deposits and coastal features such as sand bars; evidence of deep water sedimentation is seen in the presence of thick uniform accumulations of homogeneous generally fine-grained rock, in places associated with marginal accumulations of turbidites. Evidence of halokinesis (see p. 141) indicates firstly a continental depositional environment during formation of evaporites in which the basin had only restricted connection with ocean waters. Secondly, halokinesis indicates the eventual burial of the evaporites to a depth sufficient to promote mobilisation of the sodium and potassium salt formations.

Ancient desert sands can be identified through seismic structure attributable to dune bedding. Also, the presence of reefs, either of barrier type or reef-knolls, indicates a particular type of shallow marine water environment at the time of deposition. Examples of seismic sections showing a number of the sedimentary features noted above are illustrated in the case studies in Chapters 10–13; see for example the reef structure in figure 11/8 and the complex system of channels on the section in figure 12/16. The object of the above discussion is to stress the importance of co-operation between regional geologist and seismic interpreter in viewing seismic data as a means of radically improving an understanding of the stratigraphy and structure and historical development of any basin under study. This is particularly true where limited borehole data are available and any regional geological interpretation must depend on seismic evidence as an indicator of the variation in sedimentary environment. Depositional features can be recognised on seismic sections which are not evident from any study of interpreted seismic maps.

7.3 Deformation mechanism

As well as giving an indication of the depositional history of sedimentary rocks formed within basins, seismic sections also give clear evidence of any major deformation which has affected such rocks. The forces which cause deformation can be both compressional and tensional and have both vertical and horizontal components. In some circumstances gravity is an important factor in producing deformational stresses. Rocks may be deformed to produce both small- and large-scale structures; in seismic interpretation we are concerned only with the large-scale structures which can be broadly classified as being either faulting or folding. It is not uncommon however for structures to develop which combine both types; a deep seated fault for example may be draped at a higher level by a monoclinal fold. In general, rocks are more likely to fracture with the consequent development of faulting when subject to tensional or shear stresses than when subject to a primarily compressional stress though thrust faulting in particular is usually closely associated with compression and folding in stratified rocks.

7.3.1 Faults

The term faulting describes the displacement of
a body of rocks by shearing or fracturing along a
planar surface which is called the fault plane. In
many situations the fault plane is not a simple
surface but a zone of crushed rock which may
range from a few centimetres to hundreds of
metres wide depending on the magnitude and
type of fault involved. On seismic sections,
faults are identified where reflectors can be
seen to be displaced vertically, or (less usually)
by identification of zones of crushed rock. The
interpreted sections in Chapter 10 (colour
section) show numerous examples of fault
structure.

Geological interpretation of seismic data, not
only sections but also isochron and depth maps,
requires description of faulting in terms of
standard   geological   nomenclature.   This
nomenclature is now briefly summarised.
Faults are usually classified as being of the
types: normal, reverse (thrust) or strike-slip.
Figure 7/4 shows diagramatically the types of

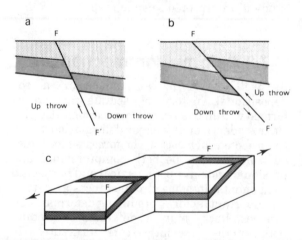

Figure 7/4 Classification of types of faults, (a) normal
fault, (b) reverse fault, (c) strike-slip fault. Arrows show
direction of movement. Γ–Γ' is the fault plane.

displacement involved. Normal faults are
generally associated with tensional stress,
reverse faults with compressional stress and
strike-slip faults with shear stress. The
directional trend of a fault is termed its strike
and the angle of inclination of the fault-plane
with respect to horizontal, its dip. Alternatively, the hade of a fault may be quoted, this
being the fault-plane angle with respect to the

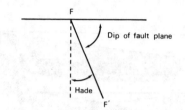

Figure 7/5 Dip and hade angles of a fault.

vertical (see figure 7/5). The amount of vertical
displacement associated with a fault at any
location is termed the throw of the fault.
Strike-slip faults are often termed alternatively
as wrench faults, tear faults or transcurrent
faults, and displacement of such a fault, if it can
be assessed, is quoted as a horizontal displacement which in some cases may amount to tens
or even hundreds of kilometres. The direction of
displacement is designated as either dextral or
sinistral, a dextral fault being one in which
displacement of the far block is to the right as
seen by an observer facing the fault and a
sinistral fault one in which displacement is to
the left (see figure 7/6). In Chapter 10, seismic
data are interpreted to provide evidence of
dextral fault movement along a major transcurrent fault.

During the process of development of a
sedimentary basin, periods of faulting are more
likely to be episodic than continuous. Nevertheless, the fact that vertical displacements of
major rock units are likely to occur contemporaneously with sedimentation is of fundamental importance to the history of any basin
development. Fault blocks which are structurally elevated, and such structures can be

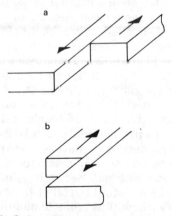

Figure 7/6 Strike-slip faults. (a) sinistral, (b) dextral.

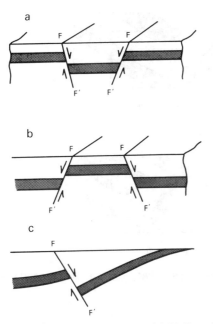

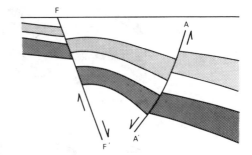

Figure 7/8 Normal growth fault F–F' with antithetic fault A–A'.

Figure 7/7 (a) Graben, (b) Horst, (c) Half-graben.

7.3.2 Folds

identified on seismic sections, are termed horst blocks whereas faulted depressions are termed grabens. An asymmetrically developed graben is often termed a half-graben (see figure 7/7). When development of any fault system is contemporaneous with sedimentation within a basin, the faulting is associated with growth structure and such structure has particular relevance to hydrocarbon accumulation (see Chapman, 1973). For example, growth faults are often associated with antithetic faulting (see figure 7/8 and the seismic section in figure 8/26). The combination of normal faulting, antithetic faulting, growth structure and 'roll-over' (that is, the development of an anticlinal trapping mechanism in association with a growth fault) can provide all the principal requirements for development of a hydrocarbon pool. Also, growth structure over horst blocks and in association with normal faulting can provide the mechanism for both hydrocarbon fluid migration and formation of traps. In oil exploration, the geological interpretation of seismic data is therefore not just a process of defining potential structural traps but also that of interpreting the structural development of a basin or province so that its history may be related to the many factors which contribute to its hydrocarbon prospectivity.

Whereas the interpretation of fault patterns, as seen on seismic sections, and the synthesis of interpreted sections into seismic horizon maps can often pose many problems, particularly those of correlation across fault lines, the identification and mapping of folds, dome structures, anticlines, synclines and monoclines is usually a relatively straightforward interpretational task. Problems can arise, however, at the stage of compiling isochron maps, particularly where structure is complex and both folding and faulting have affected the rock strata. As described in Chapter 6, it is often useful, before commencing a detailed interpretation of a set of seismic data, to prepare a map of principal structural features and trends (see figure 6/2) and such a map can be particularly helpful in establishing, as early as possible, the trends of fold structures.

Any flexure of rock strata might be termed a fold. Only large-scale folds can be detected and mapped by the seismic method. Various types of fold are shown in figure 7/9. It should be noted that the simple classification of folds into the three groups, anticlinal, synclinal and monoclinal, would need to be expanded if more deformed structures were being described than can normally be seen on a seismic section. Many folds are asymmetric; thus when horizon maps of different levels are being compared, the axes of anticlines and synclines will be seen to have different positions. When the axis of a fold is not horizontal, it is said to plunge in a particular direction, and the angle between horizontal and this axis is termed the angle of plunge. This angle will vary along the axis of a fold and may also vary between horizon levels in the fold due to variations in bed thickness along the fold

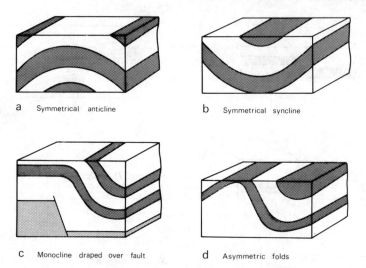

a   Symmetrical   anticline

b   Symmetrical   syncline

c   Monocline   draped   over   fault

d   Asymmetric   folds

Figure 7/9 Types of folds. (a) Symmetrical anticline, (b) Symmetrical syncline, (c) Monocline draped over fault, (d) Asymmetric folds.

axis. Doming of strata, as is often caused by salt movement through formation of diapirs, is a type of anticlinal development where there is no well-defined fold axis.

As with faulting, deformation by folding can occur at various times during the history of deposition of a sedimentary basin, in which case growth structures are likely to be apparent with thinning of rock units over anticlinal axes and thickening within synclinal axes. Recognition of such growth structure helps in dating the periods of deformation. In hydrocarbon exploration such growth structures are important in that they are likely to correlate with lateral variations in rock type and with variations in rock properties such as porosity and permeability.

7.3.3 Unconformities

When deformation occurs within a sedimentary basin contemporaneously with the long term process of its infilling by sediments, it is likely that in at least parts of the basin the process of deposition of sediments will be at times interrupted. Such interruptions can be recognised as periods of non-deposition or removal of previously deposited material by erosion. When such a period is followed by further deposition of sediments, the surface which marks the break in deposition is termed an unconformity. Unconformities can be caused by processes other than deformation of rocks within a basin,

such as major climatic changes, changes of sea-level, or tectonic activity outside the basin, in which case unconformities are recognised which, for example, can be seen as periods of basin-wide non-deposition.

On seismic sections, unconformities can be recognised easily only when the strata beneath the plane of unconformity lie at an angle to those above it (see figure 7/10). Note also the development of an unconformity as illustrated in figure 7/3 and the unconformity which can be recognised in the seismic section of figure 7/12.

7.3.4 Diapirism and salt tectonics

Some sedimentary materials, primarily salt and clay, have the property that, under certain conditions, bodies of such rocks will deform by plastic flow, and migration can take place both vertically and horizontally. Diapirs are said to have formed when this process leads to intrusion of the migrating plastic sediment body upwards through overlying strata to a level of equilibrium higher in the rock

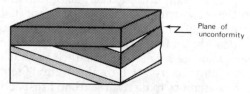

Plane of unconformity

Figure 7/10 An angular unconformity.

succession, or in some cases to extrusion at the earth's surface. Rock flow structures are most easily recognised on seismic sections where horizons can be mapped at levels below the rock units which have been deformed by migration. The structure of underlying formations will be seen to be relatively undisturbed and largely independent of the structure in rock units overlying the deformed clay or salt. In the case of salt tectonics, strata overlying the salt may be deformed into anticlines, synclines, domes, faults and overthrusts, whereas beneath the salt layer the beds can be undisturbed and flat-lying. A typical salt diapir structure is illustrated in figure 9/5. Clay diapirs are less common, and clay migration usually takes the form of flow into anticlines as a result of folding associated with compressive tectonic forces.

Compressive tectonic forces are not an essential requirement to the mechanism of salt migration. According to Trusheim (1960) many salt structures in northern Germany can be attributed to the autonomous movement of salt under the influence of gravity. These structures develop as a result of the comparatively low density of salt and its ability to flow as a result of overburden pressure alone (see Section 7.5 for an example of such flow on a seismic section). Such structures are termed halo-kinetic structures, whereas those which arise as a result of compressive tectonic forces are termed halo-tectonic.

Halo-kinetic structure can only develop where salt layers have been buried to such a depth that loading causes the salt to deform as a plastic (as opposed to an elastic) solid, and experience in northern Germany indicates that an over-burden of approximately 1000m thickness is necessary. Figure 7/11 shows the typical development of a salt stock. According to this reconstruction, salt movement has not only caused deformation of the overlying strata but has also strongly influenced the deposition of sediments during periods of structural growth. Both faulting and folding are associated with the deformation.

Salt structures occur in a wide variety of forms: stocks, domes, plugs, pillows and walls. Each structure may require detailed seismic mapping before its shape can be properly defined and in some circumstances specialised survey techniques are required to acquire good reflection data from below salt structures. Salt structures are important in oil exploration in

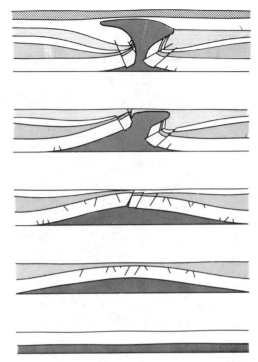

Figure 7/11 Diagrammatic development of an assymmetric Zechstein salt stock (after Trusheim, 1960).

that a wide variety of trap structures can develop in a region as a result of deformation by salt tectonics (see Chapter 9).

7.4 Outcrop geology ties

In Chapter 4 we discussed the importance of borehole data in establishing the geological significance of a seismic interpretation and how ties are made between seismic sections and well logs. Well data afford the most reliable means of providing geological control to a seismic interpretation. In situations where well data are not available or inadequate, it is necessary to attempt a geological interpretation of the seismic sections, including chrono-stratigraphic identification of principal reflectors, using such other sources of geological data as are available. One method is that of comparing the seismic sections to be interpreted with sections from other adjacent areas where good stratigraphic ties have been made with well data. This technique involves correlation of major unconformities, making

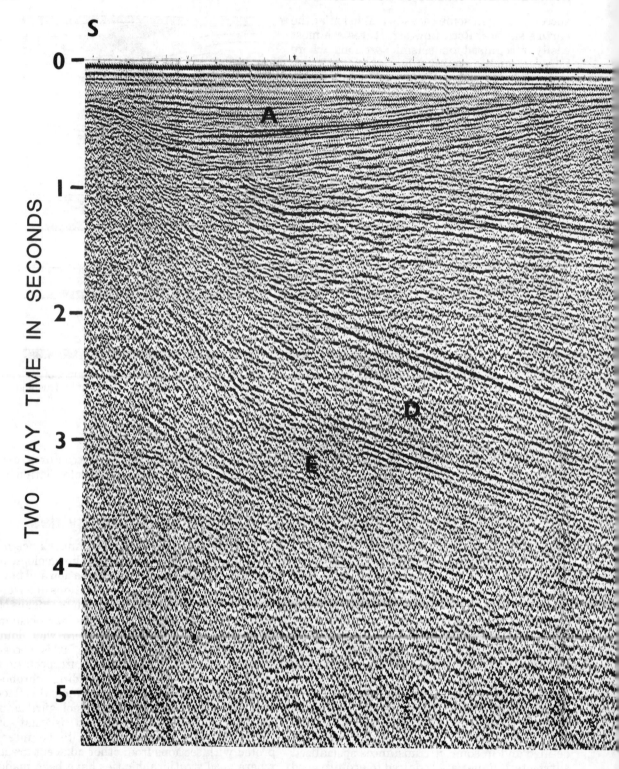

Figure 7/12 N–S seismic section across part of the St George's Channel basin (Courtesy: Phillips Petroleum; after Dimitropoulos and Donato, 1983).

N

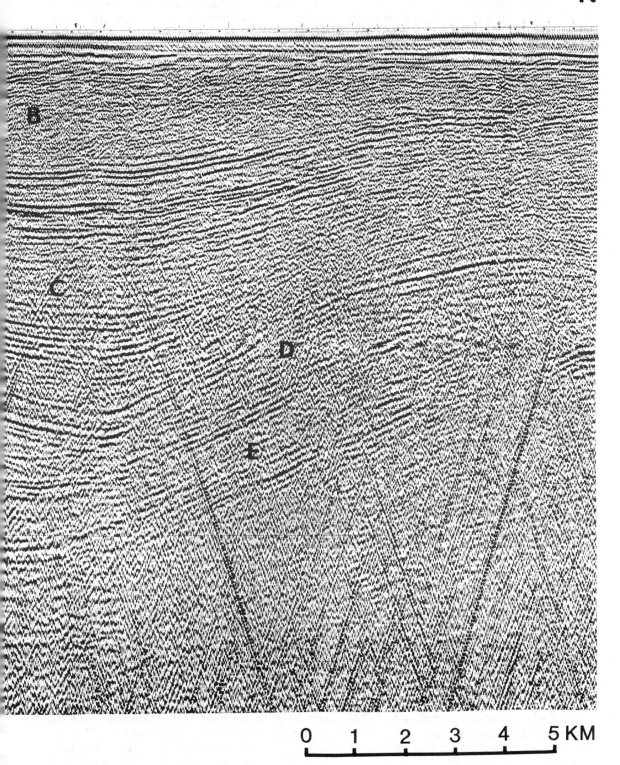

comparisons of stratigraphy as indicated by seismic character, and identifying and correlating such features as marine regressions and transgressions. To be reliable, this demands a good knowledge of the regional geology around the prospect, and in particular of the stratigraphy of the area.

Another means of obtaining geological control is through tying the seismic interpretation to the outcrop geology of an area. The efficacy of such a tie will depend on how well the outcrop geology is known and whether structure at outcrop level can be reliably extrapolated downwards to allow prediction of structure at depth. There may well be difficulties associated with poor seismic data.

On land, outcrop geology is likely to be well established and the costs of revision mapping and other additional shallow drilling are relatively low, so that outcrop geological mapping, or a study of existing maps and sections is generally accepted as an essential part of any exploration programme. At sea, the outcrop geology is usually less well known, and the costs of additional sampling and shallow drilling are high when compared with landward investigations. Exploration programmes at sea therefore do not often include any extensive survey of outcrop geology. Nevertheless, in both situations, it may be possible to add significantly to the geological validity of a seismic interpretation by careful study of, and tying to, known outcrop geology. An example of the value of establishing an interpretational tie to outcrop geology is described in Chapter 10 where, as a consequence of the study, it was furthermore found necessary to update and modify the existing geological map.

One notable difficulty in tying reflection seismic data to outcrop geology is that in the very zone where correlation is being attempted, immediately adjacent to land surface or seabed, seismic data give poor structural resolution. In the majority of seismic surveys, both acquisition and processing parameters are generally defined so as to investigate structure down to the greatest possible depths, or at least to the level of economic basement which may be at a few thousand metres below datum. Penetration to such deep horizons is often achieved at the expense of high resolution, particularly in the 0–500ms two-way reflection time zone. However, in some situations good ties to outcrop geology can be of equal benefit and less costly to

obtain than ties which could be alternatively established by drilling shallow stratigraphic test wells. Seismic surveys can in such circumstances be designed to give optimum data quality in the 0–1s two-way reflection time interval. Design of high resolution surveys is described in Chapter 3 and use of such data in Chapter 10. At sea, the results of single-channel seismic profiling can provide a valuable aid to the establishment of outcrop ties (see Chapter 8 and, in particular, figures 8/2 and 8/3).

7.5 Basin history analysis: an example

To prepare a thorough analysis of the history of development of a sedimentary basin can involve very detailed study of both seismic and well data as well as a study of regional tectonics and palaeogeography. The object here is to compare an actual seismic section with the very simple reconstruction shown in figure 7/3. The example is a N–S line across part of the St George's Channel basin in the southern Irish Sea shown in figure 7/12. The geology of the area has been described by Dobson *et al.* (1973), Barr *et al.* (1981) and Whittington *et al.* (1981). The section is reprinted from Dimitropoulos and Donato (1983) who interpret the gravity anomaly in this area in terms of a basin in which salt migration has occurred.

At seabed, Tertiary sediments occur along the entire section. To the north these are thin, and base Tertiary can be mapped as the southerly dipping reflector with maximum depth at about 1s TWT towards the south of the profile (Unit A). A major fault exists just beyond the southern end of the profile from which diffracted events can be seen on the section. The Tertiary basin is seen to be mainly controlled by hinging on this fault, the depo-centre being farther south than that of the underlying Mesozoic basin. The base Tertiary is strongly unconformable and well data from this area show the absence of Cretaceous strata. The sediments immediately underlying the Tertiary to the north and in the central part of the basin are Upper Jurassic (Unit B). These are seen to truncate to the south against the base Tertiary unconformity. A strong reflector which is at deepest just below 3s TWT is identified by Barr *et al.* (1981) as intra-Permo-Triassic. Development of the Jurassic basin is seen as largely fault controlled to the south but to the north beds thin onto a basement arch (the St Tudwall's Arch). Unit C shows no marked unconformities, but pinchout to the south against a boundary fault and possible pinchout as well as condensation of the sequence across the northerly arch. Below the intra-Permo-Triassic event a unit can be mapped (Unit D) which is seen to be thicker on the flanks of the basin than in the depocentre. This is probably the uppermost part of a salt bearing sequence, which may also include Unit E, the base of which is less well defined except in the synclinal axis at about 4s TWT. The deepest part of the basin is probably infilled with Carboniferous rocks which have been drilled on its flanks (Barr et al., 1981) but are not well defined on the seismic section.

From a study of this seismic section, it is thus possible to reconstruct in very broad terms the Mesozoic and Tertiary subsidence history of the basin. A fairly uniform sequence of Permo-Triassic beds were deposited with possibly some thinning across the northerly arch. From the evidence of this profile, the southerly fault does not appear to have exerted a major control on sedimentation. In the Jurassic an asymmetric basin formed with active fault control to the south and drape across the northerly arch structure. At some time following deposition of the Jurassic, uplift occurred to the south with erosion of Jurassic sediments, probably associ-ated with some movement on the southerly fault. Finally, a new phase of subsidence occurred with formation of the Tertiary half-graben. During the post-Triassic period of development, it is further suggested (by Dimitropoulos and Donato, 1983) that salt migrated into the flanks of the basin from the central zone of deposition.

References

K. W. Barr, V. S. Colter and R. Young, The geology of the Cardigan Bay-St George's Channel Basin, in: L. V. Illing (ed). *Petroleum Geology of the Continental Shelf of North-West Europe*. Heyden, London (1981), pp. 432–443.

R. E. Chapman, *Petroleum Geology. A concise study*. Elsevier, Amsterdam (1973).

K. Dimitropoulos and J. Donato, The gravity anomaly of the St George's Channel Basin, southern Irish Sea – a possible explanation in terms of salt migration. *J. geol. Soc. London*, **140** (1983), pp. 239–244.

M. R. Dobson, W. E. Evans and R. Whittington, The geology of the South Irish Sea. *Rep. Inst. geol. Sci. London* 73/11 (1973).

F. Trusheim, Mechanism of salt migration in northern Germany. *Bull. AAPG*, **44** (1960), pp. 1519–1540.

R. J. Whittington, P. F. Croker and M. R. Dobson, Aspects of the geology of the South Irish Sea. *Geol. J.* **16** (1981), pp. 85–88.

Chapter 8

Other Geophysical Methods

8.1 Introduction

As is stated in the introduction to Chapter 6, the interpretation of conventional reflection seismic records, has, as its first object, the provision of a series of structural contour maps which describe the subterranean topography of selected reflecting horizons within the geological strata of the area under investigation. If, for this same area, the interpreter has access to data from other types of geophysical surveys, he can use these both as a direct aid to the task of interpreting his seismic sections, and as a means of improving, or extending, his geological interpretation of the conventional seismic results.

Although, in principle, all types of geophysical data might have such an application, here we shall consider only those commonly used: shallow penetration high resolution reflection seismic profiling; seismic refraction surveys; magnetic and gravity surveys. These same methods are often used in reconnaissance of regions previously unexplored by reflection seismics as a means of, at relatively low cost, making a first evaluation of regional geological structure and economic prospectivity. Results can then be used to guide the planning of subsequent conventional seismic reflection exploration. Also, in recent years, it has become fairly common in offshore exploration for seismic survey vessels to be fitted with magnetometers and gravity meters. If gravity

and magnetic data are acquired concurrently with seismic reflection data, this can be achieved at relatively little extra cost additional to that of the seismic survey. Over both land and sea, regional magnetic surveys are usually made from aircraft, and profiles along seismic lines are best interpreted in conjunction with the results of such regional aeromagnetic surveys. Similarly, single gravity profiles along seismic sections are of limited value unless interpreted against the background of a high quality regional survey based on either a regular grid of lines (at sea) or a regular grid of stations (on land). Gravity and magnetic surveys should cover not just the immediate prospect area; local anomalies should always be interpreted against a knowledge of broad scale regional variations.

8.2 Shallow seismic profiling

High resolution, shallow penetration, reflection seismic surveys are of particular value as an aid to interpretation in offshore areas which are not blanketed by thick young superficial sediments and which have been surveyed by coring and shallow drilling to provide data on the stratigraphy of rocks cropping out at the seabed or sub-cropping close to it. Techniques have been developed which use relatively low cost equipment and which do not require computer processing of digital records; these techniques nevertheless produce well resolved seismic sections in the uppermost few hundred metres of strata beneath seabed. For a fuller description of this method see McQuillin and Ardus (1977, Chapter 4). Comparing the method with conventional seismic reflection acquisition systems, the main differences are as follows:

(i) The seismic source is of relatively low power but is designed to have a broad spectrum with the main energy in a frequency band ranging from a few hundred hertz to a few kilohertz; such devices as the sparker, boomer, pinger, airgun and water gun sources are commonly used. These devices are operated at firing rates of usually between one half second and two seconds depending on source energy and penetration required. With a ship travelling at four knots this corresponds to a pop interval (and thus sub-surface coverage) of between approximately 1.5 and 6m.

(ii) Single channel streamers are commonly used, either containing a single hydrophone (for highest resolution) or a short linear group of hydrophones.

The main components of a typical system are shown in figure 8/1. A seismic section is produced in real-time aboard ship on a graphic recorder with analogue tape-recording as a back-up and to allow subsequent replay using altered signal processing parameters.

Seismic profiles so produced offer very

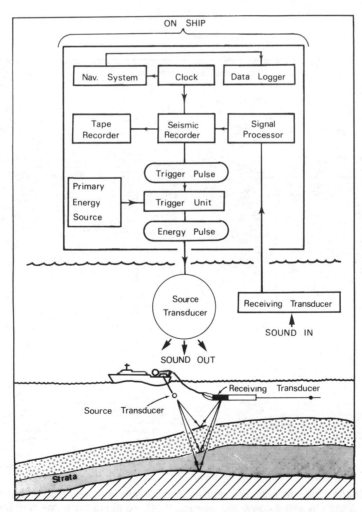

Figure 8/1 Block diagram showing the main components of a continuous sub-bottom profiling system.

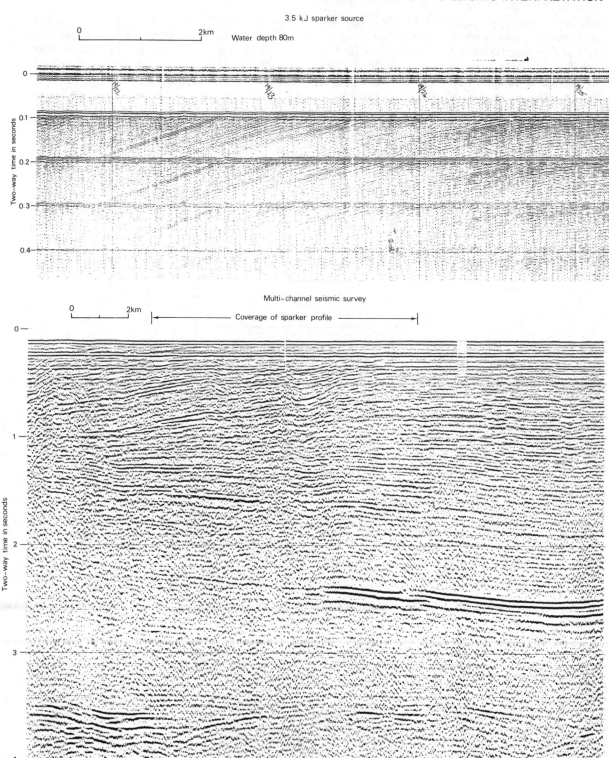

Figure 8/2 English Channel seismic reflection profiles; comparison of results obtained using a 3.5kJ sparker single-channel profiling system with a conventional seismic record along the same line. Multi-channel section obtained using gas-guns as source, 2400 m 48-channel cable and 24-fold processing. (Courtesy: IGS; surveys by IGS and Seiscom.)

limited penetration (usually only a few hundred metres) but the sections provide a very well resolved indication of structure in the zone immediately beneath seabed and it is in this zone that conventional seismic sections are usually very obscure. In Chapter 3 we discussed how to improve resolution close to seabed using a conventional digital acquisition system adjusted in terms of source frequency, streamer tow-depth and digital sampling rate. Such

specialised acquisition and processing is very costly when compared with single-channel profiling and would only be used if a combination of this with conventional acquisition proved inadequate. In figures 8/2, 8/3 and 8/4 comparisons are shown between conventional sections, high resolution multi-channel sections and sparker profiles, all from the English Channel. These comparisons illustrate how the sparker profile can be used to trace to

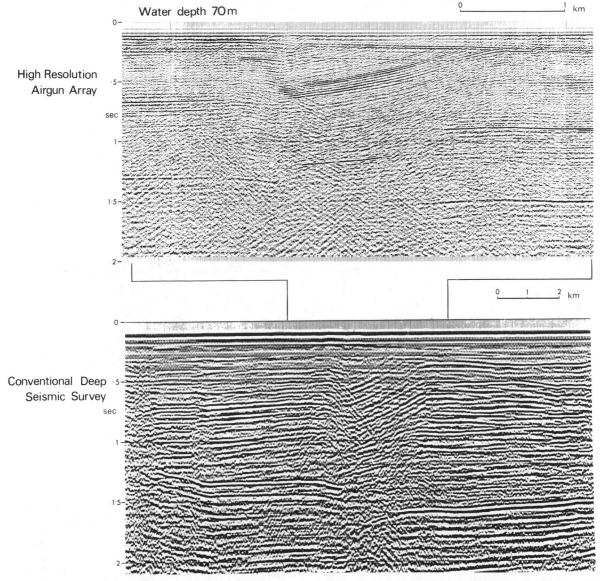

Figure 8/3 English Channel seismic reflection profiles; comparison of results of high resolution acquisition and processing with conventional seismic record. Conventional record from same survey as in figure 8/2, high resolution survey uses 600 m 24-channel cable, 2 ms sampling and processing to optimise retention of high frequencies (Courtesy: IGS; surveys by GSI and Seiscom).

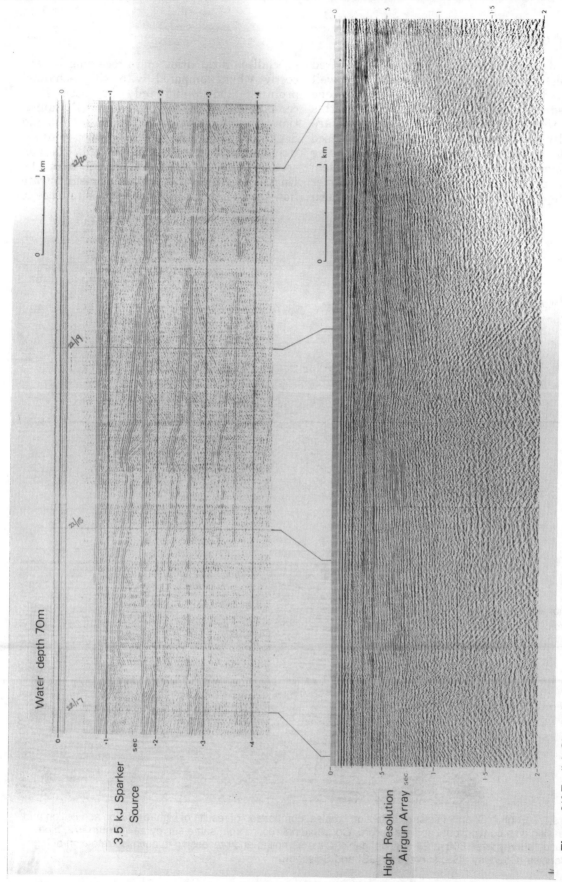

Figure 8/4 English Channel seismic reflection profiles; comparison of single channel sparker and high resolution multi-channel seismic sections (Courtesy: IGS; surveys by IGS and GSI).

exposure at seabed horizons which are well developed at depth in conventional sections but which give only an obscure indication of seabed outcrop. This link between seabed and deeper structure is also well resolved in the high resolution seismic sections in figures 8/3 and 8/4, but these sections were acquired at much greater cost than the single channel sparker sections. The drawbacks of poor penetration and the interference caused by multiples in the sparker sections are also well illustrated.

The area from which the above example is taken is one where seabed geology is fairly well established by coring and shallow drilling, results which have been interpreted in conjunction with those from a grid of sparker profiles to produce geological outcrop maps. These maps show major structures and stratigraphic boundaries. This known surface geology can then be used to provide stratigraphic identification of important reflectors seen at depth on the conventional seismic sections, a task which would have been more problematic in the absence of profiler data. Further, a study of the sparker sections gives information on the occurrence or absence of minor faulting and other small scale structures which cannot be seen on the conventional sections but which might nevertheless be of sufficient magnitude to have an important bearing on the hydrocarbon reservoir characteristics of the area.

To summarise, profiling data can be used to clarify the interpretation of conventional reflection data close to seabed, they can be used to illustrate small scale structure, too fine to be resolved by the conventional method, and they can be used to provide information on the geology of the seabed, subcrops beneath superficial sediments and the subcrop geology beneath shallow unconformities. The data, along with high resolution digital seismics, are also of value in planning well-sites and have a wide range of engineering applications offshore. It is appropriate here to discuss such engineering applications since the exploration seismologist may well on occasion become involved in the interpretation of data acquired during site investigations.

8.3 High resolution seismic applications in offshore site surveys

The purpose of a site investigation offshore in connection with the siting of a drilling or production platform is firstly to define engineeering parameters to be used in the design of structural foundations and secondly to discover any anomalous geological conditions which may be regarded as hazardous to the safety and success of the planned exercise. Seismics are used primarily in the investigations of potential hazards. What constitutes a hazard will depend on the type of structure it is planned to emplace on the seabed and is thus an engineering evaluation. Also, whether the geological conditions defined by a seismic study necessitate relocation of the planned site, or modification to structural design, or use of special drilling techniques etc. is again primarily a decision to be made by engineers. The role of the seismic interpreter is to locate and describe any anomalous geological conditions which may then require further study by shallow drilling in association with engineering tests.

The subject is too large for full treatment here, but some examples of how seismic data are used in such studies will illustrate the technique. The first example is of a 500 J sparker single-channel profile (figure 8/5) and the second (figure 8/6) is of a 15 kJ sparker 12-channel digitally processed profile displayed as both equalised and relative amplitude stacks. Both are from the North Sea.

The profile in figure 8/5 shows good resolution of data down to the first seabed multiple at 400 ms, i.e. about 200 ms TWT of section or approximately 200 m thickness of well resolved structure. The record is analogue filtered 300–1200 Hz. It should be noted that, as with all profile records of this type, there is a gross scaling distortion; a horizontal distance of 1 km is approximately equivalent to a vertical depth of 120 m of unconsolidated sediments. Thus dipping events appear very much steeper than in actuality. However, this (like a squash-plot in conventional seismics) serves to highlight structural features which might otherwise be difficult to recognise. Assuming that this record has been acquired across a site on which it is planned to establish a drilling platform, what geological features can be seen which are relevant to an engineering evaluation? The seabed can be seen to be smooth and flat but careful examination shows that there are a number of minor depressions, particularly towards the right-hand side of the record. This

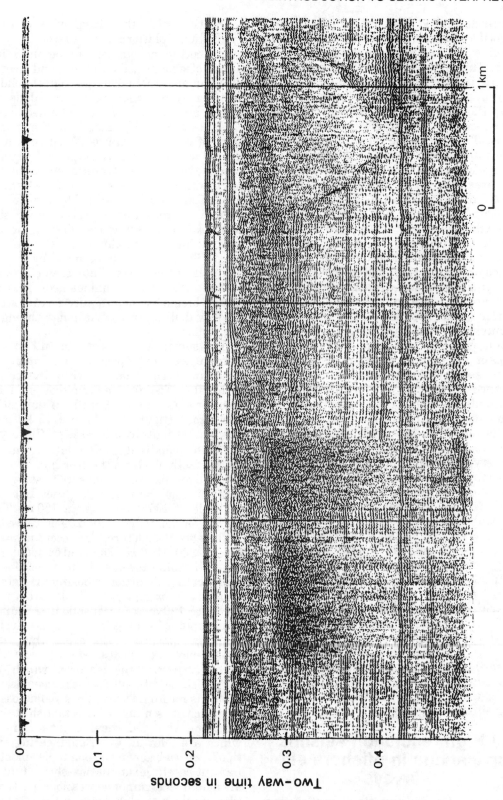

Figure 8/5 500J sparker profile from the North Sea (Courtesy: IGS).

should alert the interpreter to look at the echo-sounder records where he would see these to be of dimensions about 5–10m deep and 30–100m across. Such depressions are known as 'pockmarks' (see McQuillin and Fannin, 1979) and indicate that the area is one in which gas emission occurs at seabed, and also, that sediments at seabed are very soft and muddy. This layer of soft sediment is easily recognised by its acoustic transparency and can be quite accurately mapped as about 15m (15ms TWT) thick. The layer immediately underlying is acoustically more reflective and its character indicates a more consolidated material than that above. Two other main features are relevant to an engineering evaluation of the area. First, there is a deep channel about 2 km across and 150 m deep easily identified to the right-hand side of the record. Such a feature can prove to be problematic both in terms of foundation stability and in drilling. The second is a zone of gasified sediments towards the left of the record. The gasified zone is about 2 km across and buried at a level of about 50 m below seabed. At this depth the gasified layer is not likely to be at high pressure, but the existence of gas in the sediments will critically affect the foundation strength of this zone and should be avoided in siting any large structure. It is seen as a layer of acoustically absorbent material under which layering in the underlying sediments is masked and pull-down of reflectors can be seen due to the low velocity of the gasified zone.

In figure 8/6, a deeper and potentially more dangerous zone of gas accumulation can be identified. The main structural feature is underlain by a salt diapir. The section is of Quaternary and Tertiary sediments. At the left of the record, top Miocene occurs at about 0.92 s. The equalised stack can be interpreted to show that the salt diapirism was active through into the Quaternary, but that following uplift of the sediments, a period of collapse occurred with formation of a crestal graben. It would appear that petrogenic gas then migrated up the fault zones and has accumulated in porous zones both within the graben and more extensively in marginal layers. The gasified zones are seen as bright-spots on the relative amplitude stack. Such gas accumulations at such depths may be at high pressure and thus are likely to cause potentially hazardous drilling conditions.

8.4 Refraction seismics

In the early days of seismic exploration, refraction seismics provided an important contribution to a high percentage of hydrocarbon exploration programmes. Even as recently as the late 1950s and early 1960s the method was in vogue in many parts of the world, both on land and at sea. Improvements in the reflection technique have however demoted the refraction method to one of only occasional commercial use, though it is still widely used in studies of deep crustal structure and studies of regional tectonics, mainly as one of a range of methods employed in such studies by government or university research groups. Today, the seismic interpreter is seldom likely to be called upon to integrate the results of refraction surveys with an interpretation of reflection data. However, if refraction data are available in an area which is being subject to interpretation, these data can be a useful supplement to the results of reflection seismics. The method does also have some special applications which aim to solve certain problems less easily tackled by the reflection method; furthermore, combined refraction/reflection surveys can often be achieved at marginal extra cost if surveyed concurrently with reflection surveys, and adoption of this exploration strategy may find wider acceptance in the future with consequent development of better acquisition techniques and interpretational expertise.

Some of the purposes for which refraction surveys are conducted at present are:

1. Basement mapping for reconnaissance purposes.
2. Horizon mapping beyond the depth range of the reflection method.
3. Provision of control data for reflection interpretation to give both depth and velocity structure; such data are of particular value in areas where borehole control is sparse or absent.
4. To map horizons which underlie complex geological structures such as salt domes, shale and salt diapirs, igneous intrusions and the multiple thrust-faults of orogenic belts; often a combination of refraction survey and reflection undershooting is used to tackle such problems.
5. To determine the thickness of low velocity unconsolidated or weathered layers encoun-

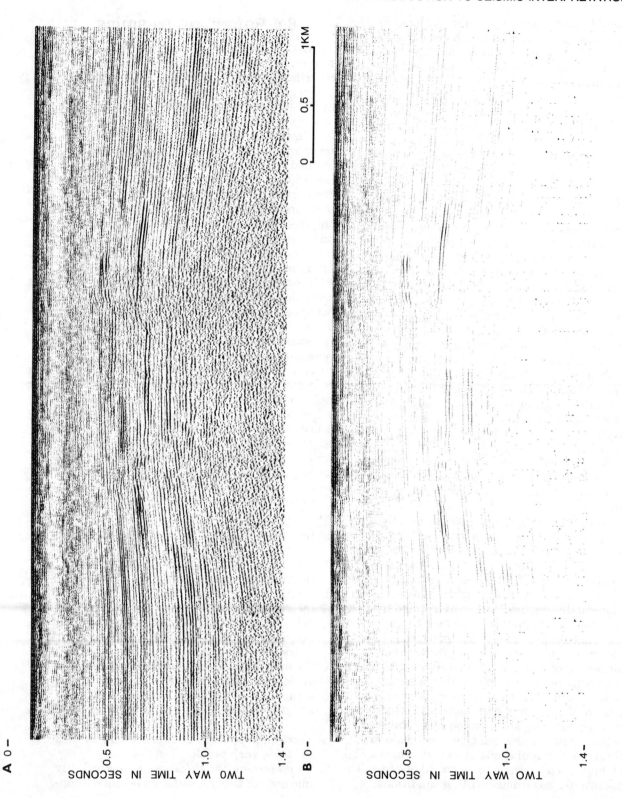

Figure 8/6 15kJ sparker seismic section; 12-fold stack. Equalised stack above, relative amplitude stack below (Courtesy: IGS).

tered in many land exploration terrains. Data so obtained are used to apply corrections during the processing of reflection seismics (see Chapter 4).

The theory of refraction seismics has been referred to earlier in connection with weathered-layer surveys undertaken during land seismic reflection work. To briefly recapitulate, refraction of seismic waves is governed by the simple law, Snell's Law, illustrated in figure 8/7 which states $V_1/V_2 = \sin\alpha/\sin\beta$ where V_1 and V_2 are the seismic velocities either side of a boundary plane. Refraction exploration depends on an effect whereby acoustic energy is propagated along the plane of the boundary, with emission of energy into the media both sides of the boundary. Seismic waves so propagated are termed head waves. The effect occurs under the condition that $\beta = 90°$, in which case α is said to be the critical angle α_c such that $\alpha_c = \sin^{-1}(V_1/V_2)$. In the simplest type of refraction shooting, reversed in-line profiling, an interpreted section can be built up which should correlate well with a seismic reflection profile, but with the advantage that refracting horizons are defined in terms of depth, not two-way reflection time. Other types of refraction shooting include fan-shooting and broadside shooting (see later).

In practice, the methods of analysing and processing refraction data can be quite complicated unless the geological conditions are fairly simple, i.e. horizontal layers of isotropic rock media and in the absence of velocity inversions.

Often sites can be selected in a prospect area, or field programmes designed, which can minimise actual deviations from these conditions, particularly in reconnaissance work or in situations where the principal aim is to gain data on velocity information and depth control to aid reflection interpretation. For the interpretation of refraction data over complex structures, computer methods have been developed, as well as graphical aids. Such methods are described in the text book, *Seismic Refraction Prospecting*, published by the Society of Exploration Geophysicists (Musgrave, 1977). Here, only a very simple case will be treated, that of a simple three layer structure as shown in figure 8/8 where $V_1>V_2>V_3$ and the uppermost two layers have a thickness d_1 and d_2. Let us assume a shot-point at position O with receivers (groups of geophones or hydrophones) spaced at intervals along the x axis. In refraction exploration, it is the time between shot instant and first arrival of acoustic energy through the rock layers which is of principal use. Receivers close to the shot-point will receive, as first arrivals, waves which have passed directly through the uppermost layer, velocity V_1, these first arrivals being followed by reflections from deeper layers. Receivers at increasing distance from the shot-point will receive, as first arrivals, seismic waves refracted as head waves of velocity V_2 or V_3, depending on distance along the x axis. At the more distant receivers, arrivals from layers 2 and 1 will follow the first arrival refracted waves from layer 3, along with reflected waves, multiples and other refractions, giving a complicated signal. However, with processing, it is usually possible to detect and display the primary signals from the upper layers, as well as the first arrival signal. A time–distance graph of first arrival times has the characteristics shown in figure 8/8. From this graph it is possible to extract data on both the seismic velocities and thickness of the rock layers. Slopes of straight line segments of the graph can be measured, these being equal to $1/V_1$, $1/V_2$, and $1/V_3$. If the intercepts of these lines through the time axis are plotted, then the first segment passes through $t=0$, the second segment through t_1 such that:

$$d_1 = \frac{V_1\, t_1}{2 \cos\alpha_1}$$

and the third segment through t_2 such that:

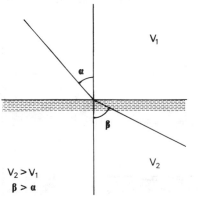

V_1

α

β

V_2

$V_2 > V_1$

$\beta > \alpha$

Figure 8/7 Refraction of seismic waves according to Snell's Law.

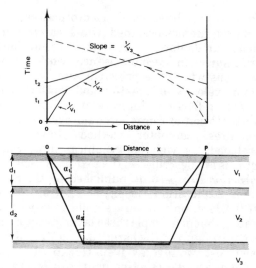

Figure 8/8 Refraction of seismic waves in a three-layer earth.

$$d_2 = \frac{V_2 \left[t_2 - t_1 \left(\dfrac{\cos \alpha_3}{\cos \alpha_1} \right) \right]}{2 \cos \alpha_2}$$

where $\alpha_1 = \sin^{-1} (V_1/V_2)$, $\alpha_2 = \sin^{-1} (V_2/V_3)$ and $\alpha_3 = \sin^{-1} (V_1/V_3)$.

These relationships can be adjusted and extended for dipping interfaces and for more than three layers, but with increasing complexity of the mathematical relationships.

A profile as described above, will not by itself detect the presence of dipping events or other deviations from the simple horizontal layer case. If, however, the profile is reversed, that is, if the shot-point is removed to a point P, and the same spread of receivers is used, then an exactly symmetrical time–distance graph should result as shown by the dashed lines in figure 8/8. Any deviation from this symmetry can be used as a measure of dip between layers (for example, see Slotnik, 1950).

In practice, surveys are usually designed so that data acquisition provides a set of overlapping reversed profiles. A typical layout for a long land profile is shown in figure 8/9. Shot-holes are positioned at SH0, SH1, SH2 etc. and geophone groups G0, G1, G2 . . . G50 etc. Depending on target depth, it may be necessary to use spreads extending to 40–50 km or more from each shot-point. Considering the shot at SH2, it can be seen that both up-dip (to the east) and down-dip data (to the west) will be observed from this shot position.

In figure 8/10 the system is shown which is commonly adopted for refraction surveys at sea. The seismic source can be either a conventional reflection seismic source, such as an airgun or gas-gun array, or, in the case of long range experiments, it is often necessary to use explosive charges at distances beyond that at which good arrivals can be detected using the relatively low energy reflection sources. In the system illustrated a sonobuoy is used as a receiver-transmitter; the hydrophone can be either suspended in the sea from a free-floating or anchored buoy, usually on an elastic suspension designed to dampen out sea wave motion, or it can be laid on the seabed attached to an anchored buoy. The buoy contains an amplifier and radio transmitter which transmits seismic signals to a shipboard receiver. Acoustic signals are displayed either on a multichannel oscillograph or on a graphic recorder as well as being recorded on magnetic tape for subsequent processing. Shot instant is

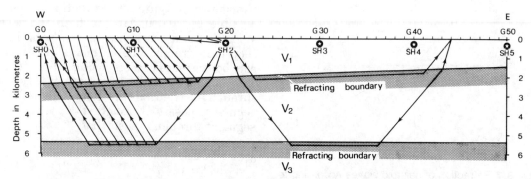

Figure 8/9 Layout of reversed in-line refraction survey on land.

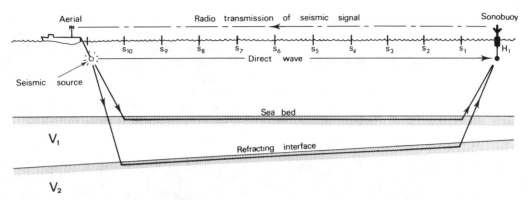

Figure 8/10 Layout of seismic refraction experiment at sea.

recorded, and from the display it is possible to detect, as well as the first arriving seismic signal, a later strong high frequency signal which is the acoustic impulse which has travelled directly through the seawater from source to hydrophone. Thus measurements can be made of interval times: shot instant to first arrival and shot instant to direct-wave arrival. This second time interval divided by the velocity of sound through seawater gives the distance between shot-point and sonobuoy. Such a measurement is particularly important if the sonobuoy is free-floating and is likely to be drifting in the sea current or by force of wind in which case its position will vary over the duration of the profile acquisition period. An alternative marine acquisition system which is much more costly in use but which can give much higher quality data utilises a two-ship system. In this case a conventional digital reflection acquisition system can be used, appropriately adjusted to accept low frequencies. The survey vessel tows its cable away from a fixed shooting location, shots being fired at intervals such as to allow stacking of data. At the end of the profile the shooting vessel occupies the end of line location and the survey vessel re-runs the line in the opposite direction so that the profile is fully reversed. In practice a number of shot locations might be used, with the recording ship sailing a profile up to, then away from, each shooting location. Thus a series of overlapping reversed profiles is built up.

In figure 8/11 an example is shown of an interpretation of the results of a marine refraction experiment carried out in the Irish Sea (Bacon and McQuillin, 1972). Results indicate a layer of approximately 1.5 km thickness of velocity 3.46 km/s overlying rocks of velocity 5.07 km/s. Geologically, these results suggest that 2 km of Permo-Triassic rock overlie a Palaeozoic basement.

In figure 8/12 a refraction seismic section is shown from a land seismic survey. On this section the high velocity basement refractor (γ)

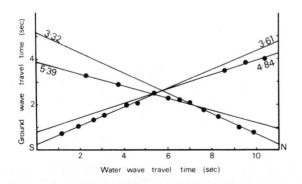

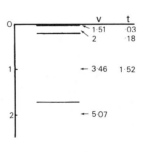

Figure 8/11 Results of a seismic refraction experiment in the Irish Sea (from Bacon and McQuillin, 1972).

crosses a fault. Also shown is the interpretation of a set of travel time curves using the wavefront method over 45 km of profile. In this particular study, velocity information for rocks overlying the refractor (γ) was based on reflection seismics, though low velocities can be detected as both first arrivals and as late refracting arrivals in the 4–7 s interval. The

application of refraction seismics to an investigation of a complex subsalt tectonic problem in northern Germany is illustrated in figure 8/13. These travel time curves and the depth presentation are from part of a 300 km refraction line, the presentation giving a good example of the use of Thornburgh's wavefront method (Thornburgh, 1930). Other refraction

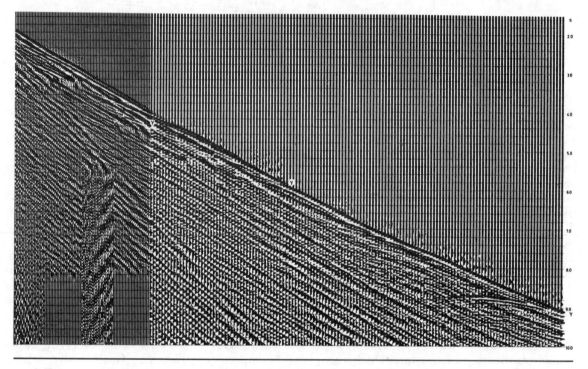

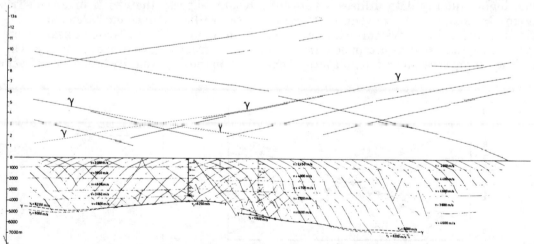

Figure 8/12 Refraction seismic section (above) with fault indication in the basement refractor (γ). Travel time curves (below) of a 45 km long refraction line with depth interpretation of the basement refractor (γ) using the wavefront method (Courtesy: Prakla-Seismos).

seismic techniques include fan-shooting, broadside (arc) shooting, radial shooting and combined reflection and refraction surveying. Fan-shooting is a well established technique which was devised principally to detect and map salt domes within a relatively low velocity section. Firstly, a calibration profile is established in the area of study where the seismic section is known to be normal, i.e. no salt domes are present. Then a number of fan-arrays are shot (see figure 8/14) and early arrivals (time-leads) are plotted which indicate where ray paths have passed through the high velocity salt domes.

Broadside (arc) shooting is similar to fan-shooting but has more general application. A central shooting location is used and detector stations are located on circle segments. Different radii may be used, values being gauged so that the various refractors being investigated will provide first-arrival signals in the sections produced. An example of broadside shooting is shown in figure 8/15 where the object of study was fault detection at depths where reflection data might provide inconclusive evidence. Radial shooting is similar to arc shooting but in this case a geophone is lowered into a well to the level of an identified refractor, then a series of shots are fired in an arc to investigate structure associated with the refractor.

An interesting application of combined reflection/refraction surveying has been described by Prakla-Seismos (Prakla-Seismos Report no. 1/76). The method depends on utilising geophone spreads already in place during reflection profiling and by additional shooting obtaining broadside sections. An

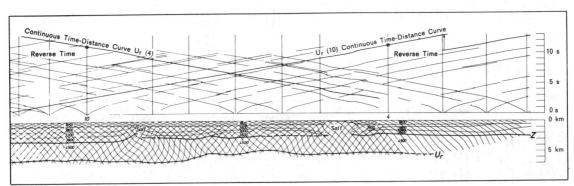

Traveltime curves and depth presentation for a part of a 300 km refraction line. The pre-Permian horizon U_r can be traced over long distances (Presentation using Thornburgh's wavefront method).

Z = Base of Zechstein

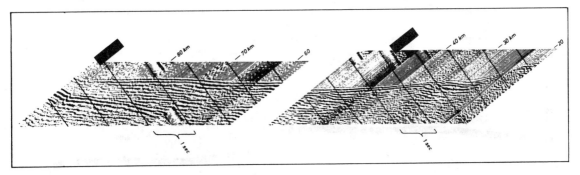

Corresponding seismogram sections; shot-receiver distances up to 50 km, in some cases up to 90 km have been used.

Figure 8/13 Application of refraction seismics to a subsalt tectonic problem in a deep salt dome basin in northern Germany (Courtesy: Prakla-Seismos).

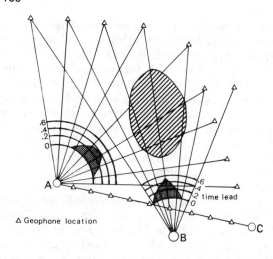

△ Geophone location

Figure 8/14 Layout of refraction fan-shooting survey. Profile A–C is shot as a reversed refraction profile to establish a calibration of the normal time-distance curve for the area. Shaded area shows approximate outline of salt dome responsible at depth for the plotted time leads (After Nettleton, 1940).

example of the method from a survey carried out in the South German Molasse Basin is shown in figure 8/16. The method can only be applied if a good refractor is present. In the example shown, the location of a fault is traced from its position on the left-side refraction offset line, through its position on the reflection profile, to its position as indicated on the right

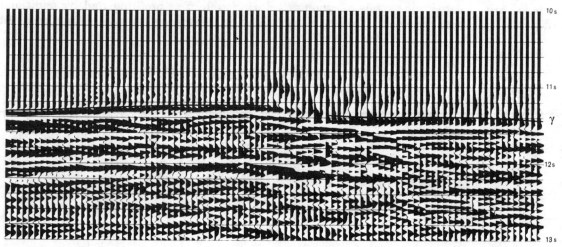

Seismogram section

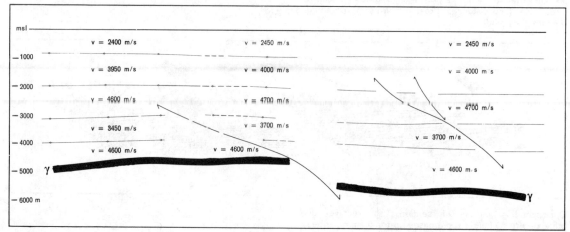

Depth presentation

Figure 8/15 Results of a broadside-refraction survey. Offset between shots and receiver = 5 km (Courtesy: Prakla-Seismos).

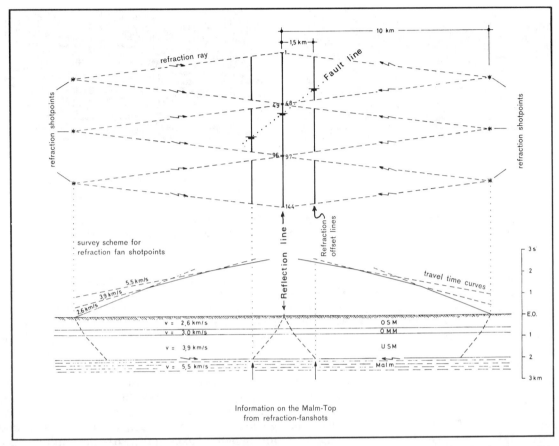

Figure 8/16 Combined reflection-refraction survey. Any fault which displaces the Malm refracting horizon will be detected in both the reflection section and in the refraction offset lines (Courtesy: Prakla-Seismos).

side refraction offset line. The method can under some circumstances give a more positive indication of faulting than is seen on the reflection record as well as giving a determination of fault strike. An example of a section showing fault detection by this method is shown in figure 8/17.

8.5 Magnetic method

An aeromagnetic survey is an economical method of undertaking a geophysical reconnaissance of any relatively unexplored region by providing data on broad scale structural trends, the positions of faults, the distribution of shallow and deep crystalline basement, as well as on the occurrence and distribution of volcanic rocks within sedimentary basins. Such information can be invaluable in planning more

costly seismic reflection exploration work which will aim to detect defined economic targets. Magnetic survey results are furthermore useful to the seismic interpreter even at the stage of detailed interpretation of a close grid of seismic lines. He may have access to both the results of an aeromagnetic reconnaissance survey as well as magnetic profiles surveyed along actual seismic lines, whether on land or at sea. We are mainly concerned here with how such data can be used by the interpreter. Later, in Chapter 10, the Moray Firth Case History illustrates how the results of an aeromagnetic survey can be used as an aid to seismic interpretation, particularly in the early stages of exploration of an area when seismic coverage is open and few or no drilling results are available.

The basic principle underlying magnetic

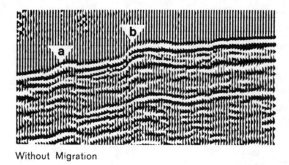

Without Migration

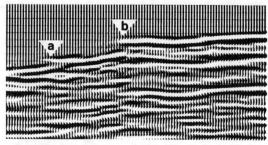

With Migration

Two faults (a, b) detected by broadside shooting. The refraction events were corrected to standard distances of
30 km.

Figure 8/17 Detection of faults by broadside-refraction shooting. This example shows how migration processing can be applied to refraction data to improve definition of the fault structures (Courtesy: Prakla-Seismos).

surveying is that some rock types, notably though not exclusively of igneous origin, are naturally much more highly magnetised than others. These bodies of magnetised rock produce a magnetic field which is apparent as a local perturbation of the earth's magnetic field.

Magnetic surveys therefore aim to detect anomalous magnetic field variations associated with geological structure; magnetic anomalies are determined by comparing measured magnetic field values with either a local or whole-earth reference field. Most modern survey work uses equipment which measures the total field; the earth's main field varies from approximately 0.4 oersted at the Equator to 0.6 oersted at the poles. The unit of measurement in most common use in magnetic exploration work is termed the gamma and 1 gamma is equal to 10^{-5} oersted. Most present-day surveys are referenced against the International Geomagnetic Reference Field (IGRF), which is defined in terms both of the main field and of secular variations. Thus, for any locality on the earth's surface, and for a particular epoch, a reference field value can be calculated. Secular variations are sufficiently small for it to be acceptable to assume no significant secular change over the period of a normal survey programme (say up to 6 months). A map showing contoured values of the differences between magnetic field measurements and the reference field values is called a magnetic anomaly map. Magnetic profiles are sometimes shown on top of seismic sections and anomaly values for these are usually similarly derived. Figure 10/4 shows a magnetic anomaly map of the Moray Firth area. This map is derived from the results of an aeromagnetic survey flown at 1000 ft above sea or ground level with a line grid spacing of 2 km × 10 km. Instrument accuracy is approximately 1 gamma and the results are usually contoured at a 10 gamma interval.

Geological interpretation of magnetic data usually aims as a first objective to indicate the depth of burial of magnetised rocks. Major geological features such as faults, anticlinal and synclinal axes may also be determined. As a further refinement, model structures are derived with computed magnetic fields which give a good fit to the observed magnetic anomaly fields. These geological models can then be compared with results of other geophysical surveys and assessed in terms of geological probability.

Interpretation maps are often produced by aeromagnetic survey companies which are based on automated or semi-automated procedures for handling the large quantity of data involved. Fourier analysis techniques are widely used; short wavelength anomalies are produced by magnetic rock bodies occurring near ground level whereas long wavelength anomalies originate at deeply buried levels. In figure 8/18 part of such a map is shown. It can be useful in seismic interpretation to compare such a map with interpreted isochron or isopach maps for the same area. If the seismic data allow mapping of depth to economic basement, it would be interesting to see if this correlates well with depth to magnetic basement. The magnetic map will also indicate if volcanic or

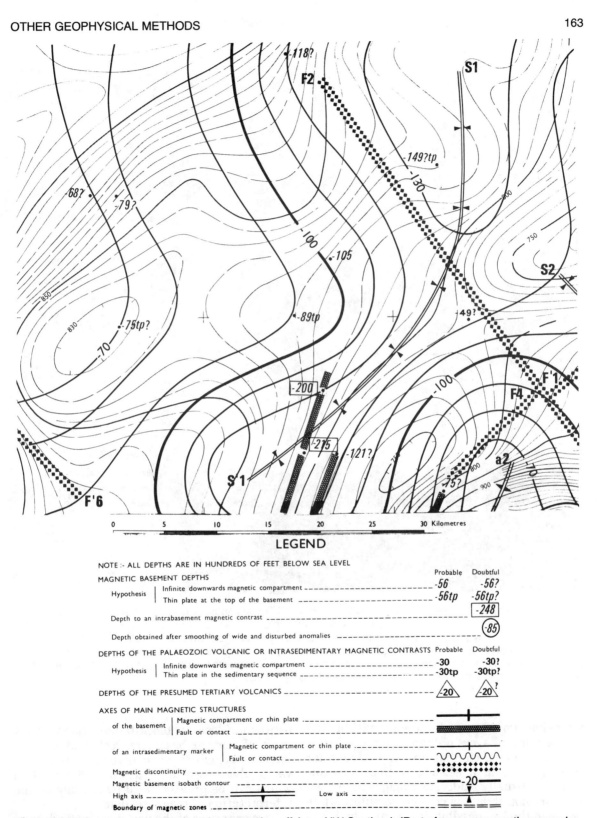

Figure 8/18 Magnetic anomalies and interpretation offshore NW Scotland. (Part of an aeromagnetic survey by Hunting Geology and Geophysics, UK.)

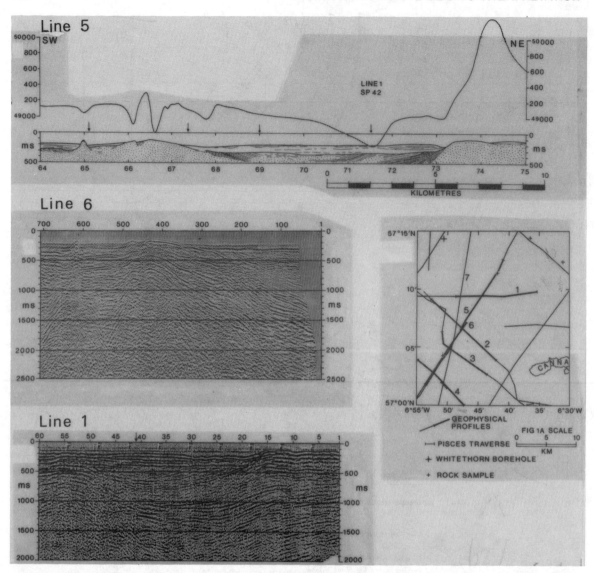

Figure 8/19 Geophysical study of a small basin of presumed Tertiary sediments in the Sea of the Hebrides. Map shows positions of geophysical profiles, a Pisces submersible traverse, shallow borehole locations and rock sample sites. Line 5 was surveyed using magnetometer and sparker; the sparker interpretation is shown beneath the magnetic profile. Line 6, a 12-fold stacked seismic section, overlaps the south-western part of line 5 as shown. Line 1, a 24-fold stacked section (shot by Forest Petroleum UK, intersects the basin, and an interpretation of this section is shown in figure 8/20 (After Smythe and Kenolty, 1975).

intrusive igneous rocks occur within the mapped area. Layers of high velocity volcanic rocks can in some circumstances appear similar to evaporites in a seismic section. Volcanic rocks are usually moderately magnetic, evaporites are not, thus it is possible to differentiate between the two possibilities. Similarly, it can be important to differentiate on seismic sections between ridges or block-faulted horst structures of normal basement rock and igneous intrusive structures such as volcanic necks and vents, dyke swarms, etc.; a differentiation which can be greatly aided by a study of and interpretation of magnetic data.

Two examples of the use of magnetic data as an aid to the interpretation of seismic results are discussed now and in both cases a qualitative interpretation of the magnetics is sufficient. The first example is derived from the Sea of the Hebrides, NW British Continental Shelf (Smythe and Kenolty, 1975). In this area Mesozoic sediments are in places overlain by high velocity (presumed Eocene) lavas. Because of generally poor reflector quality and the presence of numerous igneous intrusives, seismic data quality is poor. Economic basement in the area comprises Pre-Cambrian and Lower Palaeozoic rocks. Prior to the survey, Tertiary sediments of post-Eocene age were unknown in this area; however, seismic sections indicate in some localities up to in excess of 1 s two-way time of low velocity sediments with a character unlike that of the Mesozoic section. A post-lava age was confirmed by showing that in at least one locality, thick sediments of this type overlay a thickness of Eocene lavas. In figure 8/19 two seismic sections are shown over one of which a marine magnetometer profile was obtained. Also shown is a magnetometer profile across the axis of the basin, along a line which has also been interpreted using shallow profiling (sparker) records. Figure 8/20 shows eventual interpretation of Line 1. In this case it was the interpretation of a combination of conventional seismic data, seismic profiling data, magnetic data and seabed sampling which led to a plausible and scientifically significant geological interpretation. The particular contribution of the magnetics was to indicate and allow mapping of lavas at seabed, such areas being characterised by short wavelength, high amplitude anomalies. Also, the anomaly patterns of Lines 2 and 5 are more consistent with the interpretation of a basin underlain with

lavas than with a possible alternative interpretation of the seismic data, that the area with lavas absent at seabed represented an area where these had been removed by subsequent erosion in the Quaternary and deposition of superficial sediments on top of the Mesozoic.

As a second example of integrated interpretation of seismic and magnetic data, a structure is illustrated which was discovered during an early (1968) seismic survey of the southern Irish Sea (Bullerwell and McQuillin, 1969). The seismic section (figure 8/21) indicated a strongly positive feature which was difficult to interpret as it was only crossed by one seismic line in a very open survey grid. No other seismic or drilling data were available in the area and a number of interpretations were possible: that the structure was associated with an igneous intrusion; that it was associated with a basement ridge; or, that it was a diapiric salt structure. The fact that the structure had no associated magnetic anomaly strongly suggested that the latter interpretation was correct.

The mathematical basis of quantitative interpretation and model fitting techniques are beyond the scope of this book. However, as a very approximate guide to estimating depth of burial of magnetic rock bodies, the interpreter can abstract profiles normal to the trend of mapped magnetic anomalies, then measure the half-amplitude width of prominent deflections. This horizontal distance equates to a maximum depth value to the source of the magnetic anomaly. To illustrate the effect, computed magnetic anomalies over two simple structures are shown in figure 8/22 with varying depths of burial; the size of the magnetic body and its magnetisation are kept constant. For the

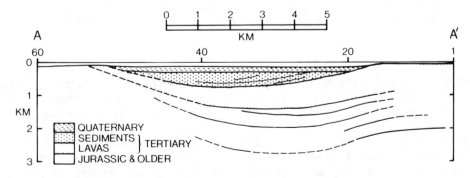

Figure 8/20 Interpretation of seismic line 1 of figure 8/19.

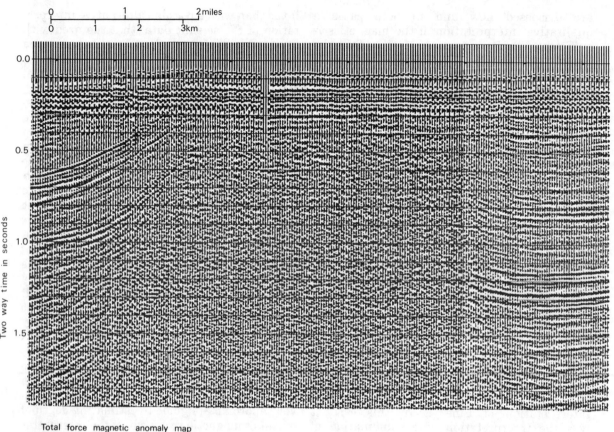

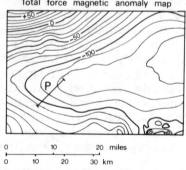

Figure 8/21 Seismic section from the Cardigan Bay area of the southern Irish Sea. Magnetic map (inset) shows absence of a magnetic anomaly associated with the high structural feature, marked: P. Magnetic contours are at 10 gamma interval (Courtesy: IGS).

vertical cylinder example, the half-amplitude width varies from 3 km at 2 km depth burial to 7.5 km for 5 km depth burial; in all cases an overestimate.

Application of this method to an actual interpretation is illustrated in figures 8/23 and 8/24. Figure 8/23 shows an aeromagnetic anomaly map of the North Minch area of NW Scotland. This map shows an area of low magnetic gradients intersected by a strong linear magnetic feature. Geological interpretations suggest that a Mesozoic basin lies between the landward basement area of the Outer Hebrides and the mainland of Scotland. The linear magnetic feature is assumed to be associated with an igneous dyke of presumed

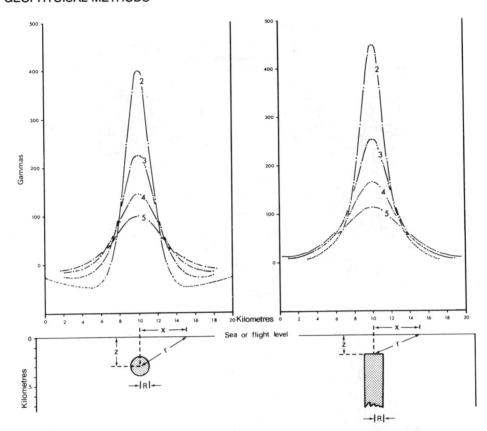

Figure 8/22 Computed total field magnetic anomalies over simple structures with vertical magnetisation. On the left, over a horizontal cylinder, on the right, over a vertical cylinder. Curves 2–5 represent different depths of burial (Z) from 2–5 km.

Tertiary age, one of a large suite of dykes which were intruded in the early Tertiary during a period of intense igneous activity in NW Scotland. Figure 8/24 shows two marine magnetic profiles across the feature compared with seismic reflection profiles. Study of the magnetic profiles indicates a depth to top of the dyke of 1600 m in profile A and 2700 m in profile B. These figures compared with depth values of 1750 m and 3150 m obtained from the seismic sections using a velocity of 2000 m/s for the sediment cover under which the dyke is buried.

8.6 Gravity methods

Gravity surveying can in many respects play a similar role to that of magnetic surveying within any exploration programme based principally on the interpretation of the reflection seismic data. Gravity surveys are often made as a means of obtaining, at relatively low cost, a geophysical reconnaissance of any relatively unexplored region. On land, the method is fairly straightforward and operations utilise compact easily transported instruments which are simple to use. At sea, the situation is more complex, and the whole operation of gravity surveying demands very much higher technical resources, both in terms of manpower and instrumental hardware. Attempts have also been made to survey gravity from aircraft (particularly helicopters), but it is only very recently that systems have been developed which are able to monitor craft position, course, velocity and vertical acceleration to an accuracy which allows gravity computation at a level of accuracy which is acceptable for reconnaissance survey work. The results of regional gravity surveys, both on land and offshore, allow delineation of boundaries of sedimentary basins as well as gross estimates of

Figure 8/23 Total force aeromagnetic map of the North Minch area off NW Scotland (between the Scottish mainland and the Outer Hebrides). High frequency large amplitude anomalies over land areas are associated with Pre-Cambrian basement. The sea area is characterised by generally low amplitude anomalies and gentle magnetic gradients due to the offshore development of Mesozoic basins underlain by Torridonian sediments. The prominent near-linear anomaly which intersects the basin is associated with an igneous dyke, probably of Tertiary age. (Adapted from IGS aeromagnetic map.)

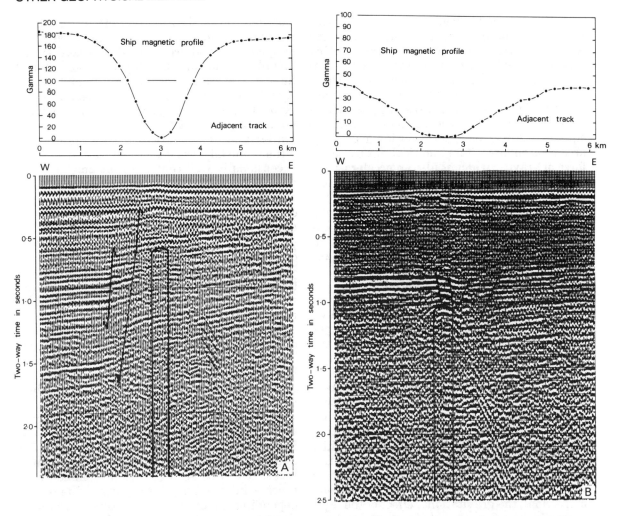

Figure 8/24 Comparison of magnetic with seismic profiles from the North Minch. Note single-point type diffraction patterns due to reflections from top and margins of dyke intrusion, also the minor faulting associated with dyke emplacement. (Seismic profile A, courtesy IGS; seismic profile B, courtesy BP and Gas Council; marine magnetic survey by IGS.)

total thickness of infill within such basins. Combined interpretation of gravity and magnetic data can give improved reliability in such estimates. Thus areas can be selected on the basis of regional gravity and magnetic surveys for further investigations by the reflection seismic method. However, our main concern here is to consider how the interpreter of seismic data can use gravity results at that stage of his interpretation where he is using a grid of seismic lines in a, by then, defined prospect.

The laws of gravitational attraction are similar to those of magnetic attraction; in both cases we are concerned with potential field theory. In gravity studies it is Newton's Law of Gravitation which describes the nature of attraction between bodies of matter. This force of attraction is expressed in the equation:

$$F = \frac{G \, m_1 m_2}{d^2}$$

where F is the gravitational force, d is the distance between the point masses, m_1 and m_2, and G is the gravitational constant which has been experimentally determined as 6.670×10^{-8} cm^3/g s^2.

As with magnetic fields, we are dealing with an inverse square law of attraction. If the earth were a perfectly homogeneous non-rotating sphere in empty space, then the gravitational pull would be uniform over its entire surface. This is not the case; the earth is not a sphere but an ellipsoid of revolution; a site at either pole is closer to the earth's centre than one at the equator. Also the earth's revolution leads to an acceleration force of an identical nature to the force of gravity, which is also an accelerational force. The sum of these main effects on variation of gravity over the earth's surface, as measured at sea-level, is that gravity at the earth's poles is approximately 0.5% higher than at the equator. A reference field can be defined for the earth which takes account of these effects and that used in modern gravity survey work is the 1967 International Gravity Formula.

From the results of gravity surveys, anomaly values are computed (as with magnetic surveys) which are in this case the differences between calculated reference gravity field values and observed gravity values corrected for the effects of topographic variation. The anomalies of interest to us have wavelengths less than about 100 km and are due to variations in density of the rocks from place to place within the crust. Anomaly maps and profiles can be used to study geological structure involving rock units which exhibit contrasting rock density. In computing gravity anomaly values, allowance is made for variation in height of observation (the free-air correction), the gravitational effect of the layer of rocks between the observation site and sea-level (Bouguer correction) as well as for the gravitational effects of surrounding hills and valleys (the terrain correction). After all such corrections are applied, anomaly values are contoured and/or profiles prepared. Such maps or profiles (in terms of Bouguer anomaly values) are in exploration work usually expressed in milligals. One milligal (mGal) is 1/1000Gal (from Galileo) which is an acceleration of 1cm/s². The earth's main (reference) field varies from approximately 978Gal at the equator to 983.2Gal at the poles. Land surveys are made to an accuracy of better than 0.1mGal and results are contoured at one milligal intervals. Measurements at sea (except where seabed gravity meters are used, and this is a special application of marine gravity work in which data are treated in a way almost identical to land data) are made using instruments mounted on stabilised platforms linked to computer control systems which aim to compensate and correct the accelerations brought about by the ship's motion. For a further discussion see McQuillin and Ardus (1977). Accuracies obtainable are less than for land surveys. A good marine survey may have an average cross-tie error of approximately 1mGal and results are usually contoured at either 5mGal, or if the survey has been exceptionally well controlled, at 2mGal intervals. Surveys made from aircraft or helicopters are likely to suffer even greater inaccuracies especially in application of corrections, and an anomaly accuracy of 2–4 milligal is probably the best achievable with present-day technology.

Before any attempt can be made to interpret the patterns of gravity anomalies associated with a particular area, some estimate must be made of local contrasts in rock density between different structural units. Using such estimates of density contrast, model geological structures can be tested to find the most plausible disposition of rock units which will produce, by computation, the same anomalous gravity field pattern as has been observed in the actual gravity survey. This process of interpretation is well illustrated in the Moray Firth Case History in Chapter 10 of this book. In Table 8/1 a list is shown of the densities of some common rocks. It can be seen that some common rock types exhibit a fairly large range of densities, depending mainly on porosity variation. Furthermore, one of the more difficult aspects of

Table 8/1 Common densities

	g/cc
Oil	0.90
Fresh water	1.00
Sea water	1.03
Unconsolidated sand	1.95–2.05
Boulder clay	1.90–2.10
Porous sandstone	2.00–2.60
Rock salt	2.10–2.40
Granite	2.55–2.65
Quartzitic sandstone	2.60–2.70
Compact limestone	2.60–2.70
Gneisses	2.70–3.00
Basalt	2.70–3.10
Basic intrusive rocks	2.80–3.20
Ultrabasic intrusive rocks	2.80–3.30

gravity interpretation is that even low-porosity crystallised basement rocks exhibit a similar wide density range, due in this case to differences in mineralogical composition. Average density of the earth's upper crust is approximately 2.7g/cc. Thus it can be seen that large granite intrusions (density 2.55–2.65g/cc) will cause reduced gravitational attraction, that is, a negative gravity anomaly or gravity low. Similarly, a massive basic intrusion or a thick layer of basalt lavas will cause a gravity high.

As an aid to seismic interpretation, gravity data are mainly used in direct comparison with seismic sections to test the plausibility of a particular seismic interpretation wherever this is problematical. Fault structures can be modelled to test seismic correlation across major discontinuities. Estimates of the density of seismic basement can sometimes be used to interpret its nature; for example, a seismic basement horst might consist of Devonian sandstones of relatively low density (a potential oil or gas reservoir) or a non-porous basement of higher density (a structure with no reservoir potential). Gravity modelling can be used to test such alternative hypotheses. Rock salt has a relatively low density, and large structures involving thick salt intervals will have associated gravity anomalies. Salt is also a high seismic velocity material and in areas of complex salt movements, it is often the case that reflection data show only poor penetration beneath salt layers. Again, gravity modelling can be used to test a range of possible seismic interpretations where these are based on unreliable seismic evidence. Another use of gravity data is in the study of deep structure within sedimentary basins to depths beyond that penetrated by available seismic data. In such situations it is possible to evaluate the 3-dimensional gravitational effect of the entire sedimentary infill as mapped from a seismic interpretation. This gravitational effect can then be compared with the observed gravity field to obtain an indication of the main elements of underlying structure. In this way, structural features can be detected which merit further study by the seismic method now optimised for deep penetration in these areas of maximum interest.

To illustrate the analytical method of gravity interpretation we shall use here a fairly simple example. The formula from which a gravity

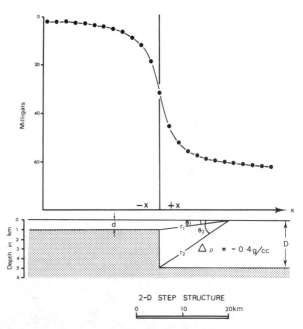

Figure 8/25 Calculated gravity profile across a model step structure.

profile across a fault or step structure can be calculated is given by

$$\Delta g = 2G\Delta\rho\, x \ln\left[(r_1/r_2) + D(\pi - \theta_2) - d(\pi - \theta_1)\right] \text{mGal},$$

where $\Delta\rho$ is the density contrast and d and D are the depths to top and bottom of the step, $(D - d)$ being the throw of the fault if it is a simple fault structure; see figure 8/25 for definition of r_1, r_2, θ_1, θ_2, and x. It can be seen from the above that the total gravity change across the structure is given by

$$\Delta g = 2\pi G\, \Delta\rho(D - d) \text{mGal}.$$

That is, for the structure illustrated in figure 8/25, a gravity change of 67mGal. In figure 8/26 the direct comparison is shown between a gravity profile along a major fault in the Moray Firth and a seismic section along the profile, and in figure 8/27 a gravity profile is fitted to a model fault structure. The relationship between seismic interpretation and gravity interpretation is clearly demonstrated.

As was stated early in this discussion, it is important to have knowledge of density contrasts associated with geological structures being interpreted. One of the most common

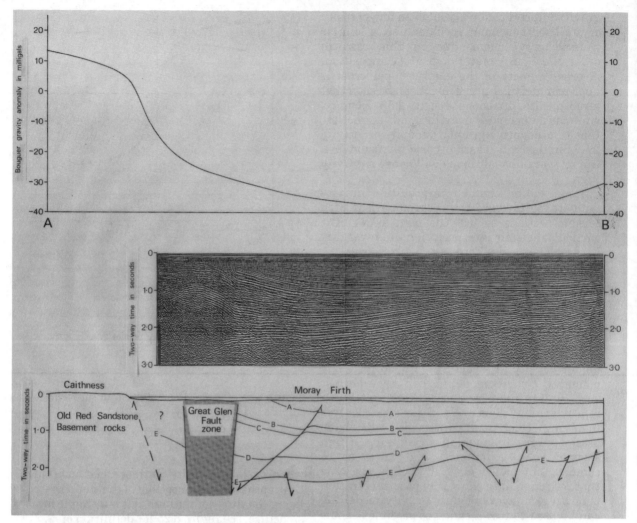

Figure 8/26 Comparison of gravity profile and seismic section across a major fault (the Great Glen Fault) in the Moray Firth, off NE Scotland. For detail of seismic section see figure 10/8(g).

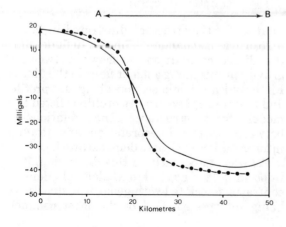

methods of determining rock density values is by laboratory measurement of the water-saturated density of rock samples, either collected from land or seabed exposures, or from cored boreholes. Alternatively, a very useful estimate of rock density can be obtained from geophysical well logs. The gamma–gamma log, or, as it is often referred to, the compensated density log, is a calibrated profile of bulk

Figure 8/27 Computed (dots) and measured gravity profiles across margin of the Moray Firth sedimentary basin. Model is a 4 km step identical to that illustrated in figure 8/23.

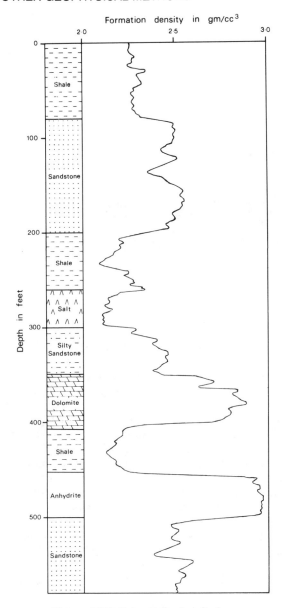

Figure 8/28 Schematic density log.

density. A schematic density log is shown in figure 8/28. The principle of the method, its operation and interpretation of results are described in Chapter 5.

If borehole data are sparse or not available, then seismic velocities derived from velocity analyses can be used as a means of determining density information. Such determinations are less precise than sample or borehole logging tests but do have the advantage that lateral variations in density within a large sedimentary basin can be accounted for. The basis of determining density values from seismic velocities is empirical and depends on experimental data relating P-wave velocity measurements in sedimentary rocks to laboratory measurements of saturated density. Figure 2/7 shows the relationship as determined by Gardner, Gardner and Gregory (1974).

It can be seen then that there is an interplay between gravity interpretation and seismic interpretation which can allow the seismic interpreter to utilise the results of seismic surveys to define the interpretation of gravity data, and by so doing, test the results of the original seismic interpretation. An example of the value of comparing seismic refraction and gravity survey data is described by Bacon and McQuillin (1972).

References

M. Bacon and R. McQuillin, Refraction seismic surveys in the North Irish Sea. *Jl. geol. Soc. Lond.*, **128** (1972), pp. 613–621.

W. Bullerwell and R. McQuillin, Preliminary report on a seismic reflection survey in the southern Irish Sea. *Rep. Inst. geol. Sci. London* (1969), no. 69/2.

G. H. F. Gardner, L. W. Gardner and A. R. Gregory, Formation velocity and density – a diagnostic basis for stratigraphic traps. *Geophysics*, **39** (1974), pp. 770–80.

R. McQuillin and D. A. Ardus, *Exploring the Geology of Shelf Seas*. Graham and Trotman Ltd., London (1977).

R. McQuillin and N. G. T. Fannin, Exploring the North Sea's lunar floor. *New Scientist*, **83** (1979, July 12), no. 1163.

A. W. Musgrave, *Seismic refraction prospecting*. Society of Exploration Geophysicists, Tulsa, Oklahoma (1967).

L. L. Nettleton, *Geophysical prospecting for oil*. McGraw-Hill, New York (1940).

M. M. Slotnik, A graphical method for the interpretation of refraction profile data. *Geophysics*, **15** (1950), pp. 163–180.

D. K. Smythe and N. Kenolty, Tertiary sediments in the Sea of the Hebrides. *Jl. geol. Soc. Lond.*, **131** (1975), pp. 227–233.

H. R. Thornburgh, Wavefront diagrams in seismic interpretation. *Bull. AAPG*, **14** (1930) pp. 185–200.

Chapter 9

Seismic Stratigraphy and Hydrocarbon Detection

A seismic trace is regarded as the superposition of reflections from the many places where acoustic impedance (the product of velocity and density) changes. This concept pictures a seismic wavelet being reflected back at each acoustic-impedance change, the reflection sign and amplitude being proportional to the sign and fractional magnitude of the change. The equivalent mathematical operation is to convolve the seismic wavelet shape with the earth's reflectivity, which is simply a 'log' of the acoustic impedance plotted as a function of two-way travel time; this concept constitutes the 'convolution model' of a seismic trace (figure 9/1).

In conventional seismic interpretation, we assume that acoustic impedance changes are parallel to the bedding so that mapping the arrival time of reflections gives a picture of structural relief, and that interruptions in the reflection continuity indicate faulting or other structural features. For many years reflection interpretation involved mainly determining arrival times with little regard to amplitude or waveshape variations.

Most seismologists realise that reflection signals often are small compared with noise and that changes in the reflection waveshape may merely indicate changes in the superimposed noise. However, where noise can be attenuated sufficiently, waveshape changes indicate variations in the earth's reflectivity, that is, stratigraphic changes. Distinguishing between waveshape changes which are caused by noise and those indicating stratigraphic changes is not obvious. Consequently stratigraphic interpretation of seismic data requires good data quality and involves some artistry. The effects of structural complications may override effects of stratigraphic variations even in good data areas. However, in many areas, seismic stratigraphy can add important geologic understanding and enhance the likelihood of discovering hydrocarbon accumulations.

Stratigraphic interpretation of seismic data is done on different scales: (1) on a regional basis to distinguish between different depositional systems and determine the environments of deposition, aspects of which are called 'seismic-sequence analysis' or 'seismic-facies analysis': and (2) on a local basis to examine individual reflections or groups of reflections to determine where and what stratigraphic changes are; this aspect is called 'reflection character analysis'. Hydrocarbon accumulations sometimes produce evidence of their presence in seismic data and its study is also part of reflection-character analysis.

9.1 The nature of seismic reflections

Interfaces where the acoustic impedance changes are generally much closer together

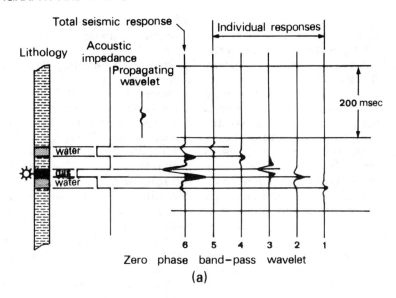

(a)

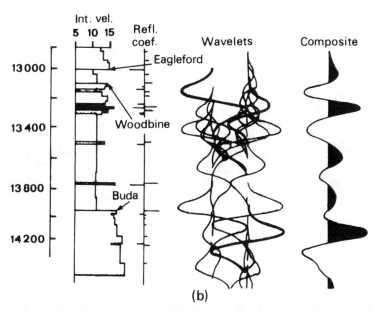

(b)

Figure 9/1 Convolutional model of seismic trace. (a) Each acoustic impedance change produces its own reflection and the result is their sum. (After Schramm *et al.*, 1977.) (b) More realistic example with the component reflections overlapping so that there is no one-to-one correspondence between events and single interfaces. (After Vail *et al.*, 1977.) Note that the waveshapes differ in these two examples.

than the seismic wavelength so that the reflections from nearby interfaces interfere. Usually interfaces are nearly parallel and contrasts remain nearly the same between seismic data sample points: consequently the interface composite remains nearly the same and shows the same attitude as the interfaces. However, there may be no simple correspon-

dence between reflection events and individual interfaces, and a change in a reflection waveshape can result from a change in any of the component reflections. Seismic reflections generally indicate time lines in the geologic sections. 'Time lines' are relict surfaces of the solid earth (time 'surfaces' would be more accurate). Time lines agree with 'stratal surfaces': laminae seen in cores and fine-structure correlations between nearby wells, because strata were deposited on what was the surface of the solid earth at some past moment in time. The lithology may change as one follows along a time line, just as one may encounter different lithology at various points on the Earth's surface today. Most stratal surfaces are too closely spaced to give rise to distinctive individual reflections. Successive stratal surfaces generally parallel each other because the surface relief does not change much between their times of deposition. Hence the interference pattern resulting from component reflections from nearby stratal surfaces parallels the individual stratal surfaces. The result is that reflections show the same attitude as stratal surfaces (or time lines) even though they may not be identified with any specific stratal surface.

Diagrams of depositional patterns (such as figure 9/2(a)) often show facies lines which indicate changes in lithology (here indicated by the 'shale to basal sand' line). Seismologists used to argue that reflections should follow such facies lines because these are where changes in acoustic impedance should be expected, rather than time lines where one does not necessarily expect a change in lithology. However, observations overwhelmingly indicate that reflections parallel time lines rather than facies lines as drawn in figure 9/2(a). The seeming contradiction results because facies lines are usually based on inadequate sampling (the only information often being from wells which are an appreciable distance apart). With more complete information, the picture usually shows facies lines paralleling time lines over most of their length, although crossing over the lines (as shown in figure 9/2(b)). Looked at in detail, since facies boundaries mainly parallel time lines, the contradiction between the attitudes of reflections and facies lines disappears.

Unconformities in a sense are also time lines, in that they were (at least locally) the surface of the solid earth at some past moment of time. Sometimes they represent many superimposed time lines, appreciable time gaps during which conditions likely changed. Hence it is probable (though not certain) that there is an acoustic impedance contrast across the unconformity and an associated seismic reflection. In fact, unconformities are often among the best seismic reflectors although the nature (amplitude and sometimes polarity) of an unconformity reflection may change as different lithologies become juxtaposed along the unconformity (see for example the seismic section in figure 9/17(c)). In addition there is often an angular relationship between reflections on opposite sides of an unconformity and this tends to make unconformities stand out more clearly on a seismic section. A consequence is that reflections from unconformities are often among the easiest and most obvious reflections to pick and identify. This is important for seismic sequence analysis because the picking of unconformities is the key to isolating seismic

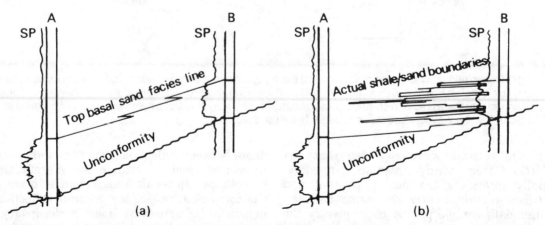

Figure 9/2 The nature of facies lines. (From Sherriff and Geldart, 1983; after Vail et al., 1977.) (a) Facies line as it might be drawn based on data from two wells about 17 km apart; (b) redrawn based on added information from many intervening wells.

sequences. It is also important for stratigraphic trap detection because most are associated with unconformities in some way.

9.2 The formation of hydrocarbon reservoirs

Usually the objective of seismic work is the discovery of hydrocarbon accumulations. To see how seismic data contribute to such discovery, it is necessary to understand the formation of hydrocarbons and the factors involved in their accumulation.

Hydrocarbon reservoirs are found within sedimentary basins in which rock sequences have accumulated to a sufficient thickness to allow diagenetic maturation of those rocks which are the original source of oil and gas, the source rocks. Maturation of the organic material in the source rocks demands a high enough temperature (possibly between 100–200°C) for certain chemical changes to produce petroleum fluids. In addition, a high pressure is needed to expel these fluids from the source rocks; this pressure results from burial under other rocks (the overburden). The fluids may then eventually become trapped in oil and/or gas reservoirs.

It is important to establish in the first place whether conditions within a sedimentary basin were likely to have been capable, at some time during its history, of producing sufficient organic source material as well as the correct environment for primary production and migration of hydrocarbon fluids. Secondly, it is necessary to establish whether suitable source and reservoir rocks actually occur, or, in the absence of definitive borehole data, to assess the likelihood of such occurrences from an evaluation of palaeoenvironmental considerations. Thirdly, the structural history of the basin is studied to assess whether or not any excessive tectonic activity might have resulted in the loss of previously formed hydrocarbons. Finally, the relationship is assessed between present structural features and those associated with earlier phases of basin development, hydrocarbon generation, tectonic activity and hydrocarbon migration. The principal elements of the hydrocarbon-forming process can be summarised as follows:

1. Development of a sedimentary basin.
2. Deposition of source rock material; normally organic clays deposited in an anaerobic environment, or in some cases, peaty formations which will form residual coal deposits as well as providing source material for natural gas and/or petroleum.
3. Deposition of porous and permeable reservoir rocks, usually either sandstones or carbonates. Some carbonate reservoirs owe their porosity and permeability to diagenetic and tectonic modification of original sediments. It is important that reservoir rocks are interbedded with impermeable potential cap rocks.
4. Burial and diagenesis of source and reservoir rocks leading to formation of hydrocarbons and primary migration of fluids into porous rocks.
5. Formation of traps by differential subsidence or relative sea-level changes causing pinch-outs of reservoir rocks interbedded with impermeable rocks, or by subsequent tectonic basin modification.
6. Migration of mature hydrocarbons as a result of basin downwarping or tectonic activity and increases in pressure due to increased depth of burial. Petroleum fluids have a lower specific gravity than water and migrate upwards through the influence of gravity; however, hydrodynamic and structural effects may induce considerable lateral migration, and in some cases even downward migration.
7. Capture of migrating fluids in traps of porous and permeable reservoir rock capped by impermeable strata which seal the reservoir and inhibit further migration, thus leading to oil/gas pools in the reservoir.

It should be noted that the above summary is very simplified; processes associated with the origin and migration of petroleum fluids are not fully understood and are still the subject of some controversy. The history of a real sedimentary basin is usually complex including numerous phases of marine and non-marine deposition associated with cycles of marine transgression and regression; pulses of tectonic activity may occur at different times during its development leading to major unconformities as well as dislocation and folding of the strata; other structures such as major growth faults which often control sedimentation across horst blocks and at basin margins, may have developed more continuously.

9.3 Hydrocarbon traps

Traps can be broadly classified as of three types; structural traps, stratigraphic traps and combination traps (partly structural and partly stratigraphic). Not every hydrocarbon-bearing structure fits neatly into this classification but the exceptional types are of relatively minor economic importance. In every hydrocarbon trap the hydrocarbons are contained within a body of reservoir rock which is porous and permeable and escape is prevented, both upwards and laterally, by enclosure of the reservoir in impermeable strata, sometimes called the cap rock. If both oil and gas are present, gas will occupy pore space in the highest part of the trap. Below this will be a layer of oil and below the oil layer pore space will be water-filled. The contact between gas and oil is termed the gas/oil contact or GOC, that between oil and water the oil/water contact or OWC. If only gas occurs, the only contact will be between gas and water, a gas/water contact or GWC. These contacts are usually horizontal or nearly horizontal. In some circumstances water flow in the reservoir, or the effects of varying capillarity, can render this contact substantially non-horizontal, but such circum-

stances are relatively uncommon.

Structural traps are formed by post-depositional deformation of rock strata with development of folds, domes, faults and unconformities. Stratigraphic traps are formed as a result of lithological variations in strata which are related to environmental conditions at the time of deposition. Lenticular sand bodies and reef structures are among the more important stratigraphic traps. Within an oil province, exploration for such traps is often of increasing importance as time passes and all the obvious structural traps have been drilled. Combination traps include elements of both stratigraphic and structural control; within this group, traps associated with unconformities and with salt movements are among the more important.

Potential structural traps can be located on seismic horizon or isochron maps. They are characterised by closed structures as mapped on, or near, the top of a reservoir rock formation. For the trap to contain oil or gas it is necessary for the closed structure to be sealed by a suitable cap rock. Figure 9/3 shows typical structural traps as mapped on two-way time or depth-converted horizon maps and as seen on geological sections through each structure.

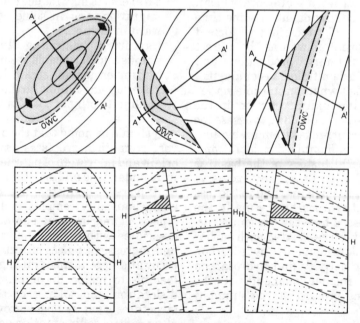

Figure 9/3 Schematic structure traps showing (above) contours on a horizon H with geological sections AA' (below). On the left, a structure over a closed dome; in the centre, a faulted anticline with closure; on the right, a fault-controlled closure in beds having homoclinal dip.

Seismic sections along A–A' might appear similar to the schematic geological sections but modified in various ways (for example, by diffractions, energy coming from the side etc. and usually without any indication of the oil/water contact, although such contacts sometimes show, as in figure 9/16). An important characteristic of structural traps is that they are often capable of providing oil and/or gas production from a number of horizons. If the rock succession contains several potential reservoir intervals, more than one of these might be sealed and thus a number of pools might be stacked, greatly enhancing the value of the field. Such a characteristic is particularly true of closed domes, but can also be the case with other types of structural traps.

To locate stratigraphic traps from seismic data, the quality and resolution required is usually greater than that required for detection of structural traps. A further problem is that of identifying the reservoir interval. A borehole sited on a structural high will usually penetrate a thick succession of beds, all of which have closure. With stratigraphic traps, such as pinch-outs for example, different prospective intervals will lap against older formations to

give potential traps at locations well-separated laterally. With such reservoirs as sand lenses, channel sands, etc. there may be very little seismic evidence at all for the trap. Reef carbonates are exceptional in that they are subject to direct location by the seismic method, as subsequently described. Figure 9/4 shows schematically three important types of stratigraphical traps. The method of location of the sand lens would be to identify the fact that the reflector H was not continuous throughout the prospect and that it had an areal form akin to that of a known type of sand body such as a sand bar or buried channel. A map of the occurrence of H combined with an isochron map of the top of the body, or of a strong reflector immediately above H could then be used to locate an exploration well.

In figure 9/5 are shown a number of combination traps. The first is a typical unconformity trap in which a productive sandstone reservoir is overlain by impermeable clays above the unconformity. Closure depends on the one hand on a plunging unconformity and on the other on a fault which seals off the highest part of the closure. This type of trap can be directly detected by seismic mapping. The

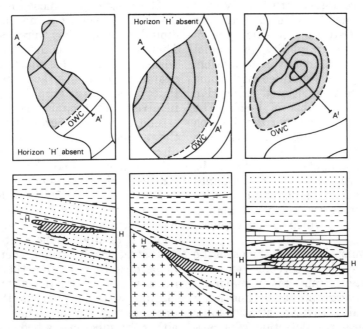

Figure 9/4 Schematic stratigraphic traps showing (above) contours on horizon H with geological sections AA' (below). On the left, a lenticular body of sand; in the centre, a pinch-out closing against a basement high; on the right, a reef.

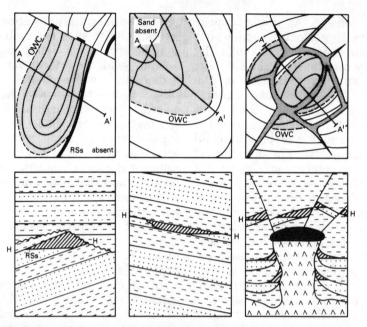

Figure 9/5 Schematic combination traps showing (above) contours on horizon H with geological sections AA' (below). On the left, an unconformity trap; in the centre, a plunging anticline sealed up-dip by a change in lithology; on the right, a complex salt-dome structure.

second example shows a permeable reservoir sealed up-dip by change of lithology to impermeable rock, combined structurally with a plunging anticline to give an arcuate oil/water contact. Detection of this type of reservoir depends on an ability to predict lithological change in the reservoir formation. The final example is a simplified salt dome structure with which is associated productive reservoir sands at a number of levels. The contour map shows the form of horizon H which is subject to complex doming and subsidence with attendant faulting leading to the development of a number of disconnected pools. Mapping complex structures around salt domes is a complicated task requiring high quality seismic data; nevertheless, seismic mapping can identify closures on which to site exploration wells.

Many explorationists believe that most of the still-to-be-discovered hydrocarbons will be found in stratigraphic traps. Many major oil and gas fields are stratigraphic traps, and many structural traps also involve stratigraphic variation. In general the detection of stratigraphic traps is more difficult and challenging than the detection of structural traps; it seems reasonable to conclude that a larger fraction of

them have been missed to date and therefore remain to be found.

The discovery of most known stratigraphic traps has involved some element of 'serendipity', implying they were discovered by 'luck'. This point of view is discouraging to one who wishes to explore for them. A more complete examination of the facts, however, indicates that most stratigraphic discoveries were involved with recognition of anomalies, and the serendipity aspect merely involved the fact that the anomalies were misinterpreted so the discoveries involved erroneous premises. With this thought in mind, the deliberate search for stratigraphic traps is not so discouraging.

Rittenhouse (1972) divided stratigraphic traps on the basis of whether they were or were not associated with unconformities and Halbouty (1972) divided them into three classes: (1) 'true stratigraphic traps' where lateral changes in permeability provide the trapping mechanism without any association with an unconformity surface; (2) traps above or below an unconformity surface, where the trap involves either the subcrop of a bed at the unconformity or the onlap of a bed onto the unconformity; and (3) palaeogeographic traps associated with palaeorelief, the palaeorelief usually being at

COLOUR SECTION

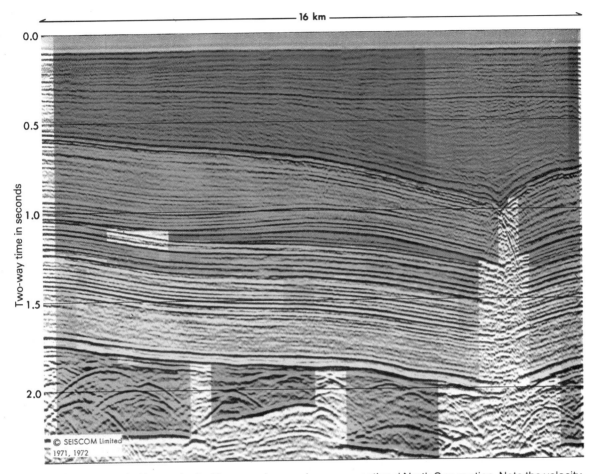

Figure 4/36 Colour coded interval velocities superimposed on a conventional North Sea section. Note the velocity gradient reversal at approximately 1.15 s in the centre of the section. Colour code as follows: computer interval velocity in ft/s: pale blue—5000; dark blue—5000–8500; dark green—8500–10500; pale green—10500–12000; yellow—12000–13000; orange—13000–15000. (Courtesy: Seiscom).

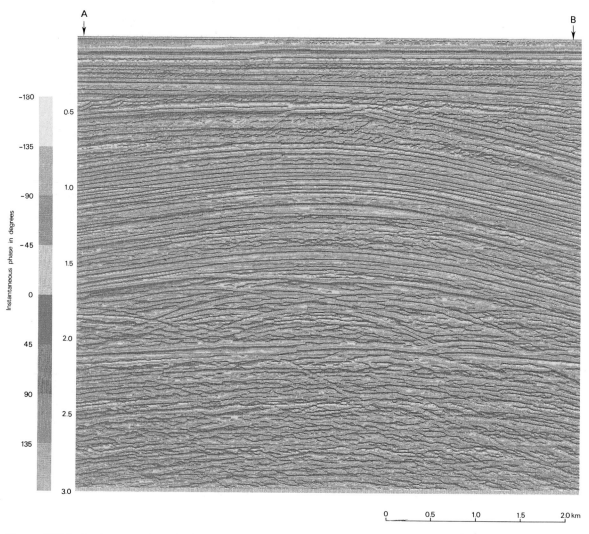

Figure 4/41 Instantaneous phase in degrees: same section as in figure 4/40.

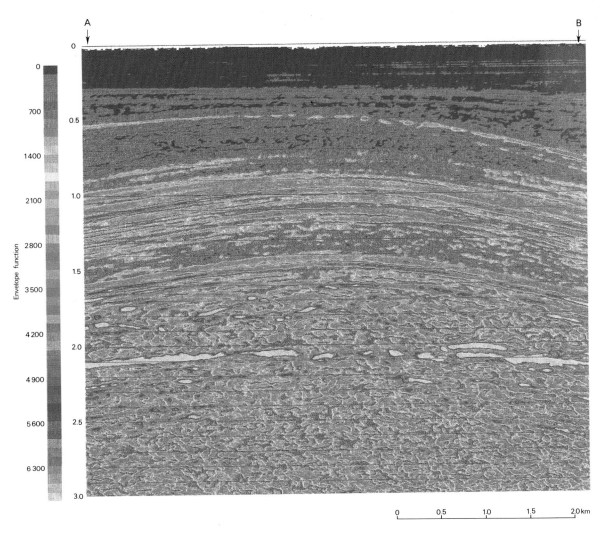

Figure 4/42 Envelope function: same section as in figure 4/40.

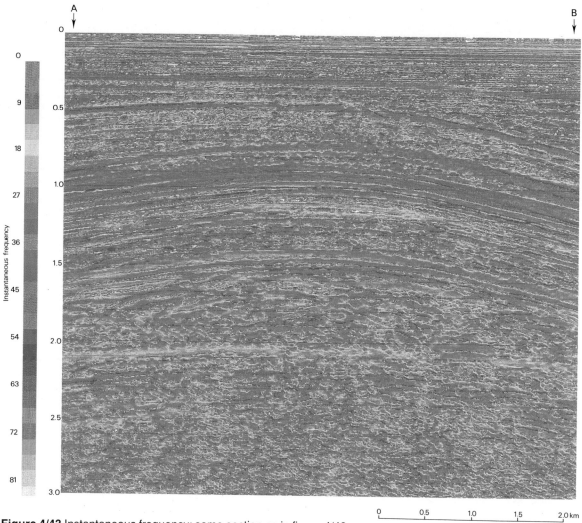

Figure 4/43 Instantaneous frequency: same section as in figure 4/40.

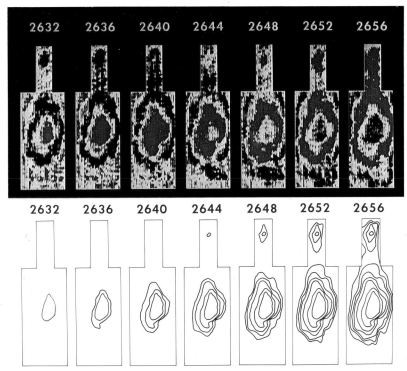

Figure 4/44 Seiscrop sections, 4 ms apart, from Peru. The raw contour map (lower right) is constructed by successively circumscribing the red event on each section (after Brown, 1983).

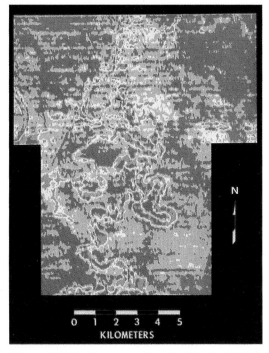

Figure 4/45 Seiscrop section at 196 ms over whole of prospect area showing meandering river channel. (After Brown *et al.*, 1982.)

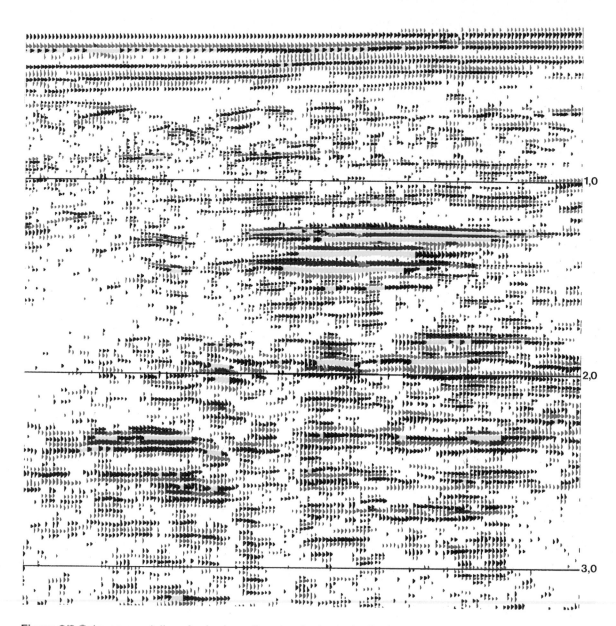

Figure 9/9 Colour presentation of seismic section showing both amplitude and polarity of reflections. Amplitudes below 2 are supressed; positive amplitudes from 2 to 6 are blue, negative black; positive amplitudes above 6 are red, negative yellow. (Courtesy: Prakla-Seismos.)

COLOUR SECTION

ORIGINAL FIELD

140   90   40 μ SEC./FT.

1/4 MILE

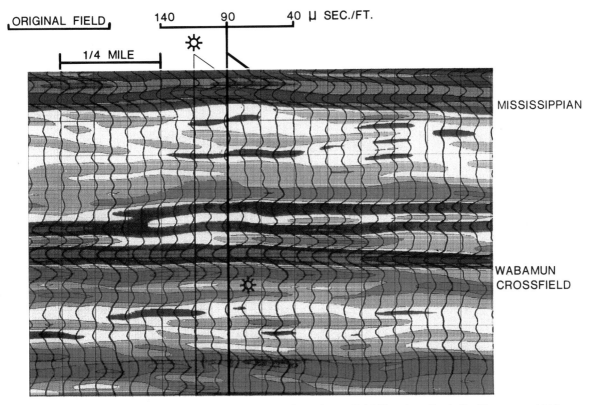

MISSISSIPPIAN

WABAMUN
CROSSFIELD

Figure 9/15 Seismic log colour presentation of section through gas field (Courtesy Technika Resources Ltd.).

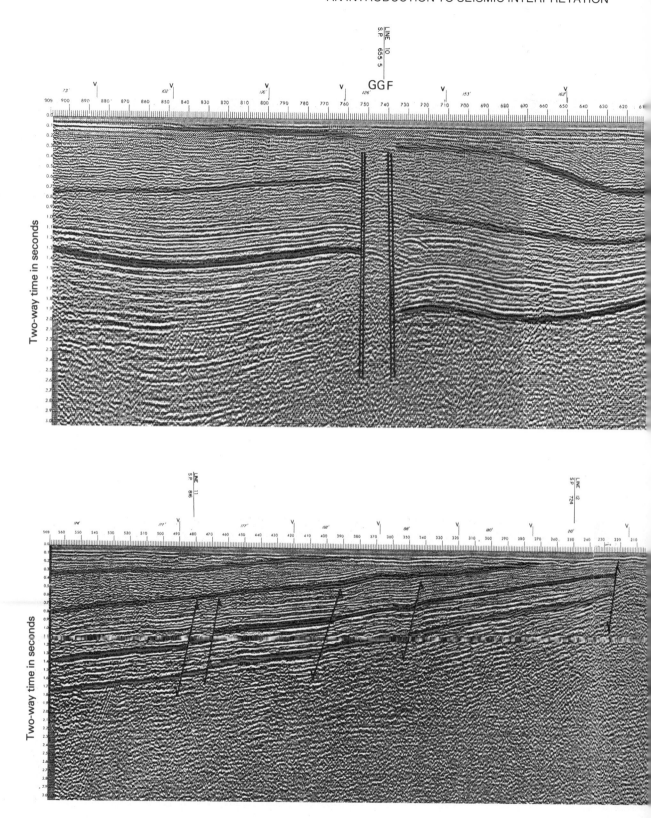

Figure 10/8 Selected sections from Moray Firth seismic survey, letter V on top of record signifies location of a velocity analysis. Vertical scale is 0–3 s two-way reflection time. Reflections are colour coded: A — upper red; B — upper green; C — blue; E — lower red; F — lower green. Arrows on faults indicate direction of throw. On the northwestern parts of lines 4 and 5 the dashed coloured lines indicate horizons as interpreted prior to the survey of line 14. Otherwise dashed lines indicate an uncertain pick. GGF marks the location of the Great Glen Fault zone (courtesy: IGS records; Seiscom survey).

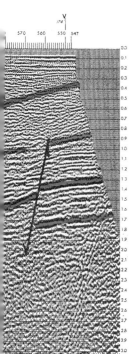

Figure 10/8(a) Northwestern part of line 3.

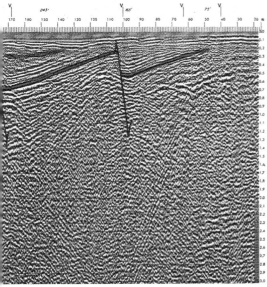

Figure 10/8(b) Southeastern part of line 3.

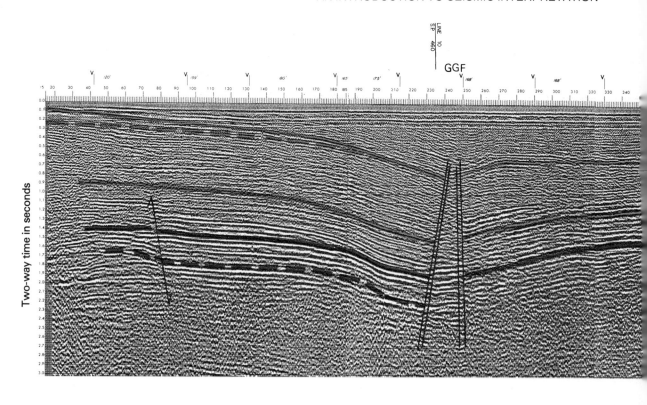

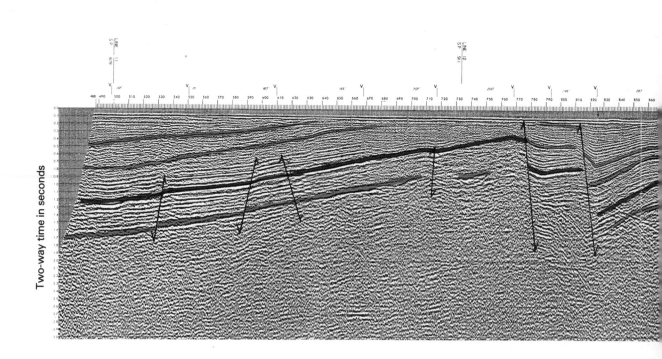

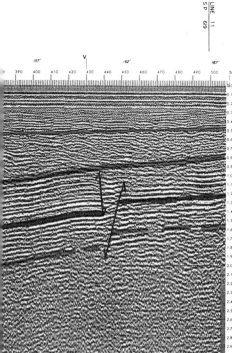

Figure 10/8(c) Northwestern part of line 4.

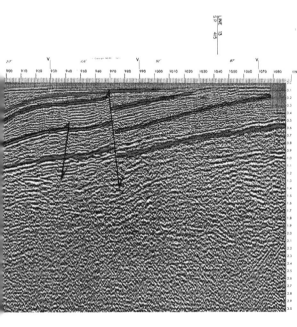

Figure 10/8(d) Southeastern part of line 4.

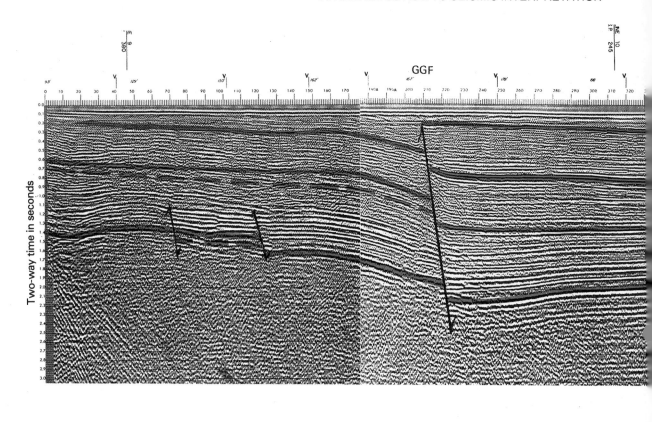

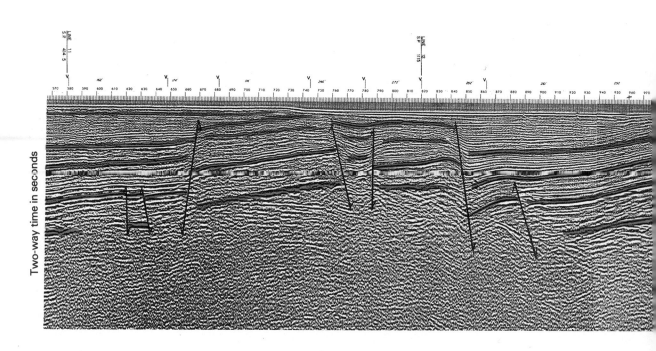

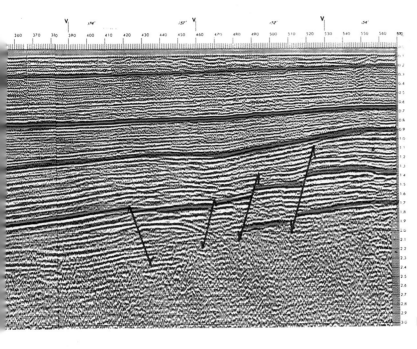

Figure 10/8(e) Northwestern part of line 5.

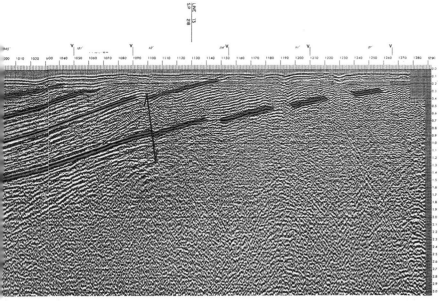

Figure 10/8(f) Southeastern part of line 5.

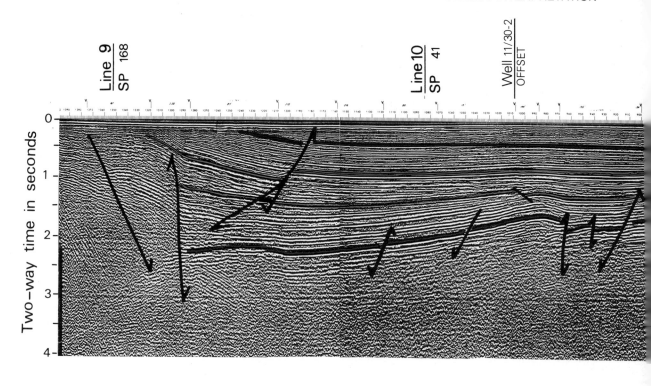

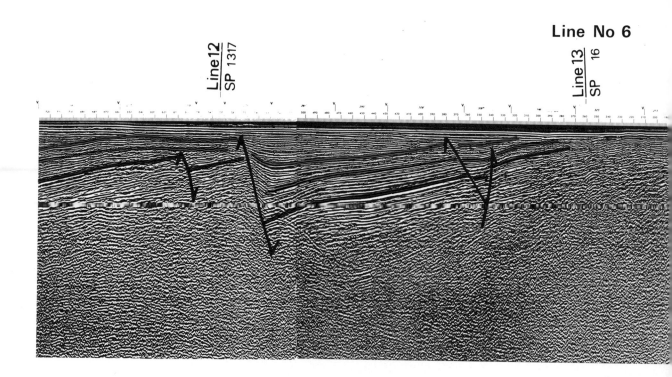

Line No 6

Figure 10/8(g) Northwestern part of line 6.

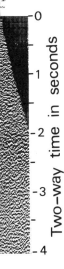

Figure 10/8(h) Southeastern part of line 6.

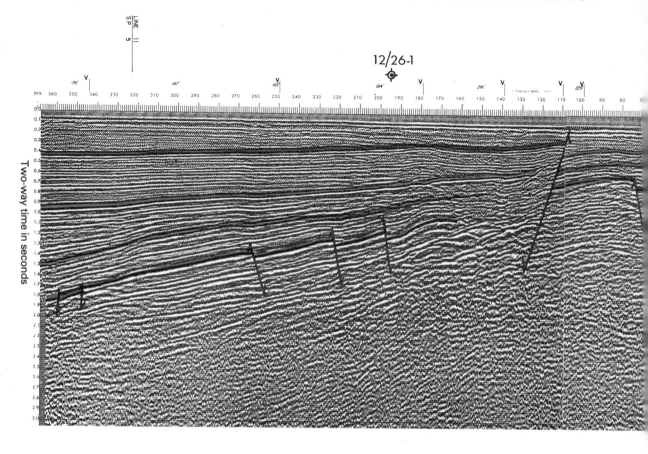

Line No 10

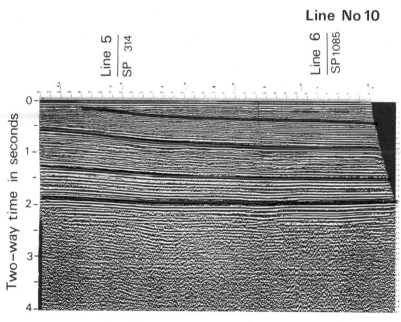

Figure 10/8(j)
Northeastern part of line 10.

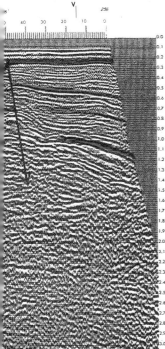

Figure 10/8(i) Line 7 showing location of well 12/26-1.

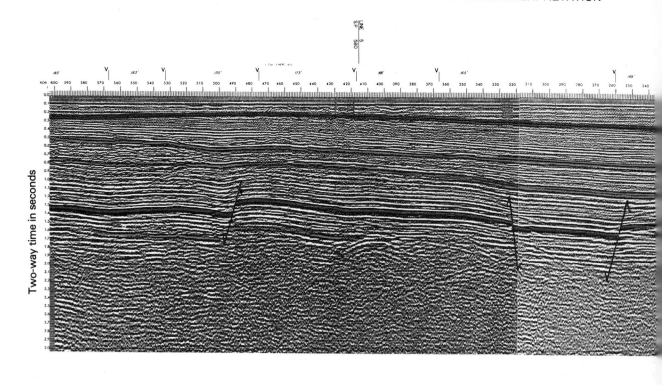

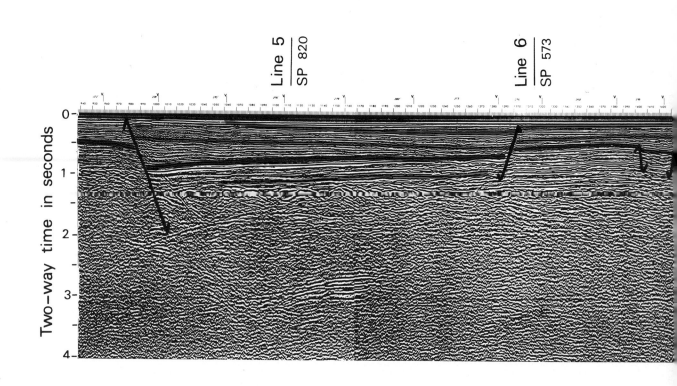

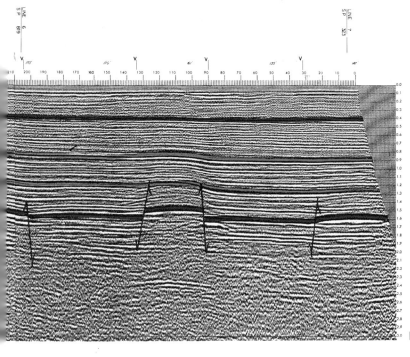

Figure 10/8(k) Northeastern part of line 11.

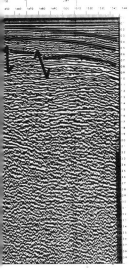

Figure 10/8(l) Northeastern part of line 12.

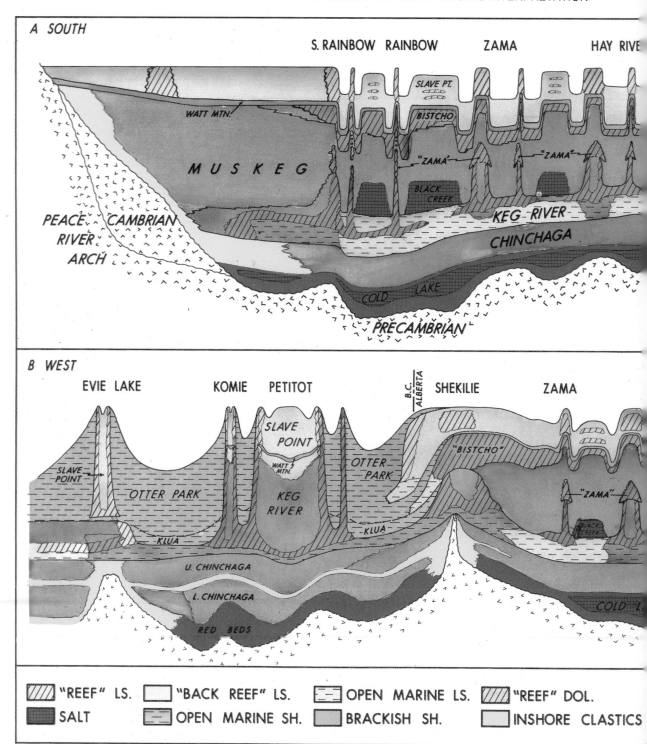

Figure 11/4 Regional stratigraphic/structural sections. Datum is top Middle Devonian adjusted to late-Upper Devonian to reflect shale compaction and solutioning of the Black Creek Salt. Lines of section are shown in figure 11/2 (after McCamis and Griffith, 1967).

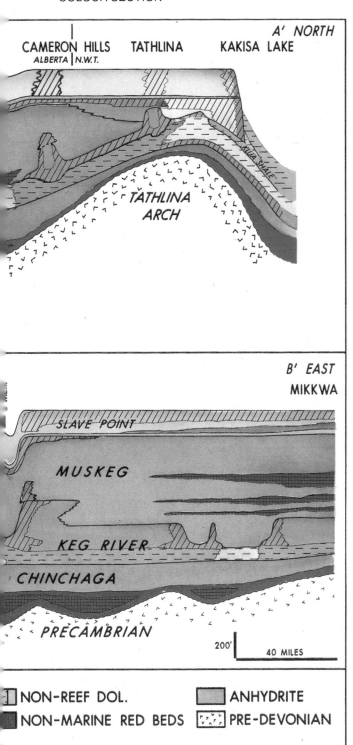

A' NORTH

CAMERON HILLS TATHLINA KAKISA LAKE
ALBERTA | N.W.T.

KLUA SHALE

TATHLINA ARCH

B' EAST
MIKKWA

SLAVE POINT

MUSKEG

KEG RIVER

CHINCHAGA

PRECAMBRIAN

200' 40 MILES

NON-REEF DOL.   ANHYDRITE
NON-MARINE RED BEDS PRE-DEVONIAN

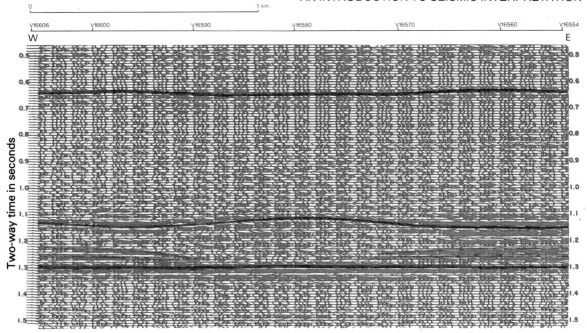

Figure 11/11 Tehze seismic line T108-51. Reflectors coded as follows: top red — Wabamun; middle red — Slave Point; green (peak) — Top Black Creek Salt; blue (trough) — Bottom Black Creek Salt; bottom red — Base Cold Lake Salt. (Courtesy: Aquitaine.)

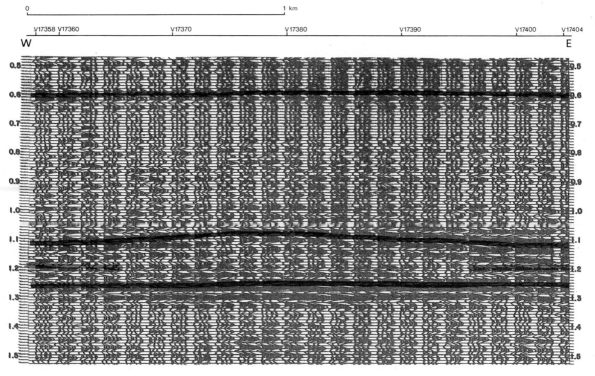

Figure 11/12 Tehze seismic line T108-44 (Courtesy: Aquitaine).

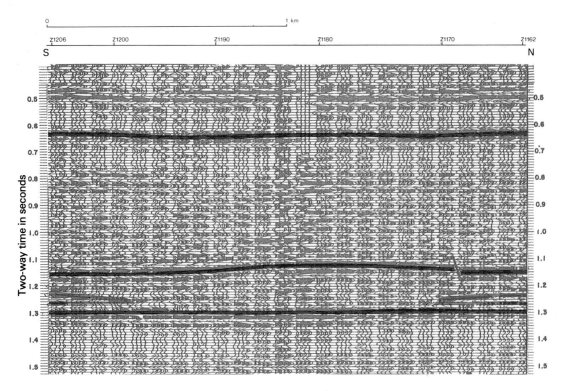

Figure 11/13 Tehze seismic line R9-32 (Courtesy: Aquitaine).

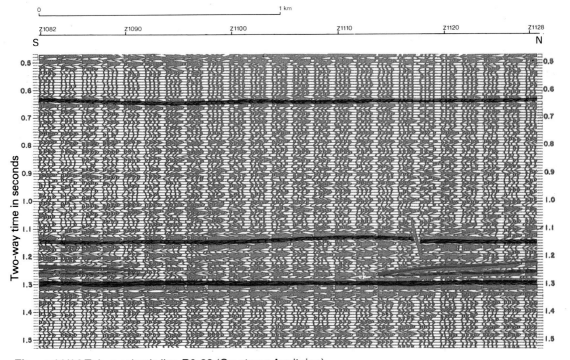

Figure 11/14 Tehze seismic line R9-33 (Courtesy: Aquitaine).

2-27-108-9 W6M

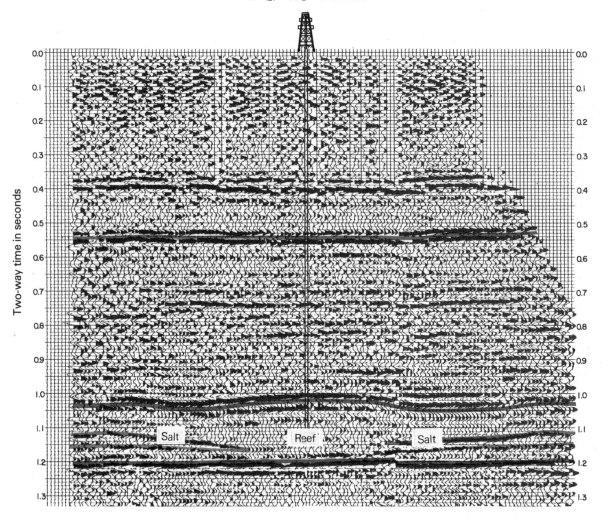

Figure 11/17 Tehze reprocessed (1976) seismic section of line T108-51.

Facing page
Figure 11/18(a) and (b) Rainbow 'AA' Pool interpretation. Seismic lines are flattened at 1.2 s resulting in pull down or up in places where the basement is high or low;
(a) E–W line T 110-33 with sonic log and synthetic seismogram aligned with the Slave Point reflector. A velocity/ check shot survey, if run, would have provided control on the low velocity shale section between the Wabamun and Slave Point carbonates, resulting in a more precise NMO curve and elimination of the mis-ties (Sonigram: Courtesy Digitech).
(b) N–S line R6-22.

COLOUR SECTION

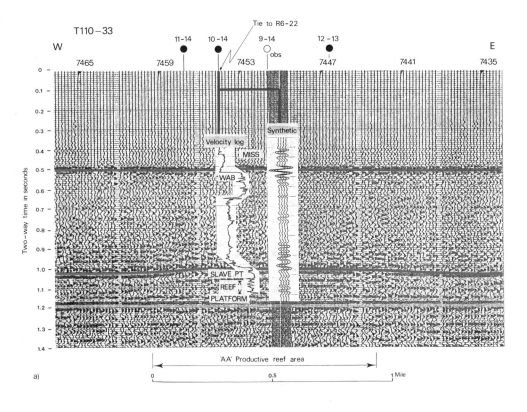

a)

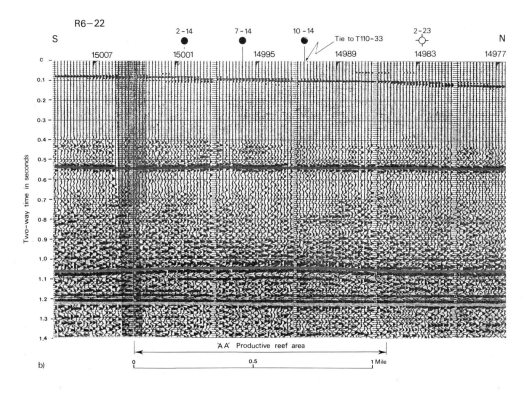

b)

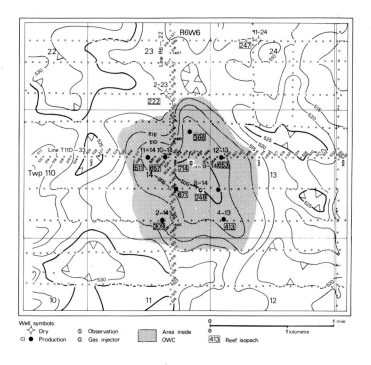

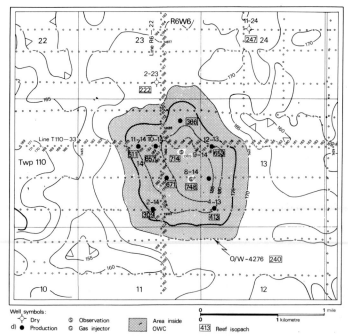

Figure 11/18 (c) and (d) Maps of two way time interval between the three main reflectors shown in figures (a) and (b) (Mapped from lines T110-33 and R6-22; remaining values supplied by Mobil Canada, but not posted). (c) Wabamun–Slave Point: probably more accurately reflects the NW–SE lineated reef underlying this isochron. (d) Slave Point-Base Cold Lake Salt; this isochron contains the reef and also suggests a NW–SE trend. Both maps suggest a bifurcation at the NW end. Reef isophachs and estimates (above the platform) are also shown.

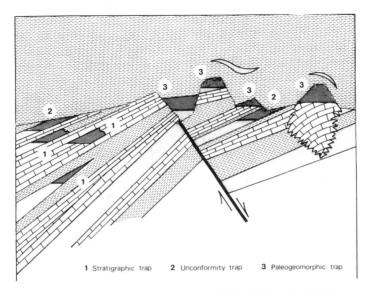

Figure 9/6 Types of subtle traps. (After Halbouty, 1972.)

an unconformity. These are illustrated in figure 9/6. The frequent association of stratigraphic traps with unconformities (which usually show clearly in seismic data) is encouraging with regard to the deliberate search for stratigraphic traps.

9.4 Evidence of pinch-outs, reefs and channels

Pinch-outs such as illustrated in figure 9/4 are often detectable on seismic records. Figure 9/7 shows a seismic section from the southern Irish

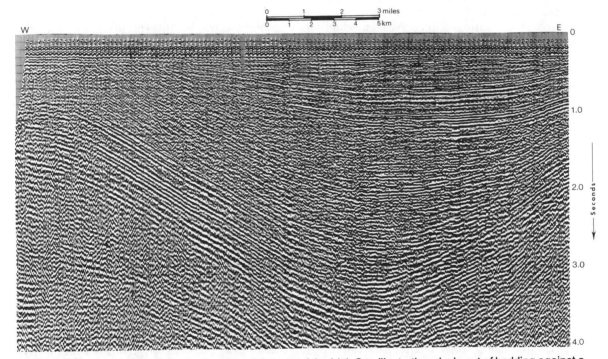

Figure 9/7 Seismic section from the Cardigan Bay area of the Irish Sea illustrating pinch-out of bedding against a structural high (Courtesy: IGS).

Sea and a number of intervals can be seen to thin out against a structural high to the east. The main problem is one of resolution and actual definition of the feather edge. A further complication can be that lateral sealing of such a reservoir can be due to lithologic change, not closure.

Reefs are one type of palaeogeomorphic trap which are often fairly evident in seismic data. Reefs often grow on topographic highs or at a shelf edge and these underlying features may be visible in the seismic data. A reflection (and diffractions) is often obtained from the surface of a reef where there is appreciable lithologic contrast. The interior of a reef usually is a non-reflection region. Reefal material often has a different velocity from the surrounding sediments resulting in velocity anomalies underneath them. A reef usually is less compactible than surrounding sediments so there may be a differential compaction anomaly above the reef. Reefs often separate different depositional environments and reflection patterns are apt to differ on opposite sides of the reefs. This abrupt change of pattern can aid in finding reefs and also can make it clear which side is fore-reef and which back-reef.

The schematic reef structure shown in figure 9/4 is a simplified version of the type of structure described in Chapter 11. A good example of a seismic section across a productive reef is shown in figure 11/8 (p. 218).

Channels are a type of palaeogeomorphic anomaly often evident in seismic data, especially when they involve appreciable relief. The material infilling a channel often differs from that of the material into which the channel was cut. Sometimes channel-fill material constitutes the reservoir (channel sands, for example) and sometimes channel-fill clay provides a trapping barrier. Figures 12/5 and 12/16 (pp. 234 and 245) show seismic sections across a well developed Miocene channel system. In this case the oilfield under discussion occurs at a deeper level but the ability to map channels using seismics is nevertheless well illustrated. Channels may also produce velocity anomalies in underlying reflections (see the case history in Chapter 12). The effects on underlying reflections may be rather complicated and structurally misleading because the nature of the channel fill may change laterally.

Focht and Baker (SEG 49th Annual Meeting) described a very successful seismically control-led 86 well exploration and development programme in channel sands of the Lower Cretaceous of east central Alberta, Canada, which resulted in a 78% commercial natural gas success ratio. They also addressed the pitfalls of misinterpretation in changing stratigraphic and lithologic environments involving coals, shale-filled channels, etc. Figure 9/8 is a seismic section through two gas-topped sand-filled channels.

9.5 Processing and display techniques

Stratigraphic interpretation is based on more subtle aspects of the seismic data than structural interpretation and hence is often possible only in good record areas which do not involve major structural deformation. In general, processing requirements are not especially different from those for structural interpretation. Processing for both types of interpretation objectives requires that variations due to recording or processing be kept to a minimum, that noise be attenuated as much as possible, that the effective wavelet length be as short as possible and that data be migrated so that reflection events indicate the correct locations of reflectors.

For stratigraphic interpretation one wants to retain relative amplitude information (which is sometimes destroyed in order to emphasise weak reflections). To rephrase, the major requirements for recording and processing are: (1) noise should be minimised and (2) conditions be kept as nearly constant as possible, in order to maximise confidence that observed waveshape variations signify changes in the deep earth rather than noise, or processing or recording variations.

Display requirements for structural and stratigraphic objectives often differ significantly. Stratigraphic interpretation requires being able to see waveshape detail which requires that the vertical scale of sections be expanded. At the same time geographical perspective is required in order to sense the 'normal' waveshape so that one can distinguish between consistent and erratic waveshape changes; this requires that the horizontal scale be compressed. The result is considerable vertical exaggeration, often of the order of five times and sometimes much larger. Vertical

COLONY CHANNEL

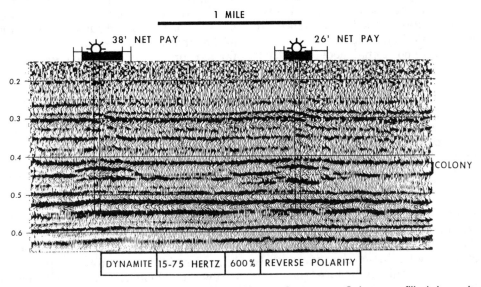

Figure 9/8 Two of many successful by seismically located Lower Cretaceous Colony gas-filled channel sands. The amplitude of the gas anomaly peaks can be related to the gas pay thickness with an average 80% accuracy. Note also the drape due to differential compaction above the resistant sand bodies. (After Focht and Baker, *SEG 49th Annual Meeting*, figure 16.)

exaggeration, however, tends to make structural interpretation difficult.

Displays of seismic attributes (measurements of aspects of the seismic data) are useful in making it easier to see where waveshape changes occur. Amplitude of the envelope of the waveforms, instantaneous frequency, apparent polarity, interval velocity, and other quantities are sometimes displayed. Complex trace analysis (Taner *et al.*, 1979), and other techniques are sometimes used to calculate attributes for display. Colour encoding of attribute measurements superimposed on the seismic section is an especially effective display method (see, for example, figures 4/41, 4/42 and 4/43 and 9/9, colour section). Probably the most common special display is to decrease the display amplitude (relative amplitude information having been preserved as nearly as possible) so that only the strongest events stand out (see figure 4/39). Another effective display is to simply reverse the polarity of a variable-area display. Most special displays do not fundamentally increase the information content and one can argue that they are not necessary because what they show can already be seen on other displays. However, one tends to

see different aspects of the data with different displays and they are valuable even if they merely make it less likely that significant features are missed.

9.6 Seismic facies analysis

Any sort of interpretation must be based on conceptual models relating causes and their effects. One interprets observations according to the model whose effects they appear to match. Seismic stratigraphic interpretation employs models of depositional systems and deduces how such systems would show up in seismic data. Such an interpretation process lacks uniqueness, that is, it is always possible that models other than those considered might provide equally good explanations of the observations.

Among the reflection characteristics which are possible stratigraphic indicators are those features which geophysicists call 'character', features such as relative amplitude, frequency content, wavelet length, and waveshape detail. To these we add reflection abundance, reflection continuity and reflection pattern. Lateral changes, that is, changes as one follows the same band of reflections from trace to trace across an area, are of especial importance. If the lateral sequence of patterns fits our model of a likely lateral sequence of depositional environments, we have added confidence in stratigraphic predictions based on it.

As examples of depositional models and their reflection characteristics consider the following:

(a) Marine depositional conditions are uniform over a larger area than those on land, so continuity of reflections and constancy of reflection character indicate marine deposition.

(b) Sediments formed in the deep sea by the raining of debris should have roughly the same thickness and character regardless of pre-existing topography. Thus units which are laterally uniform and drape over topography with little change indicate fine-grained pelagic or hemipelagic sediments.

(c) Once a turbidity current slows it loses its carrying ability and drops its load rather quickly. Turbidites are also apt to be less compactible than adjacent sediments. Consequently turbidites should be characterised by being relatively reflection-free or having a chaotic reflection pattern, but the overall pattern should be a fairly local mound with a mounded top. The courses leading one turbidite flow to an area will likely lead others to the same area, so turbidite mounds are apt to be stacked.

(d) Tilting contemporaneous with deposition should cause intervals to thin and some to become too thin to produce reflections. Thus reflection events should die out on a more-or-less random basis, as opposed to dying out systematically as at an unconformity. This pattern is called 'divergence'.

(e) Meandering stream deposits vary horizontally, often abruptly, so reflections should be discontinuous. Organic matter may accumulate in stagnant oxbow lakes and other places to subsequently become peat and coal beds, usually with appreciably smaller acoustic impedance (lower velocity and density) than nearby deposits. Thus this depositional environment should be characterised by discontinuous patterns including occasional strong reflections.

The concept that sea level variations control depositional systems is applied in seisismic stratigraphic interpretation. If relative sea level ('relative' allows for both change in absolute sea level and subsidence or emergence of the land) were to rise, deposition would occur farther up on the land masses and show in seismic data as a pattern of onlapping reflections called 'coastal onlap', as diagrammed in figure 9/10(a). If relative sea level remains constant, deposition will prograde seaward and produce a 'toplap' pattern, as diagrammed in figure 9/10(b). If relative sea level should fall abruptly the location of deposition will shift seaward and show as a seaward shift in the onlap pattern (figure 9/10(c)). These concepts form the heart of the interpretation procedures enunciated by Vail *et al.* (1977). They recognise that the amount of material available for deposition is also important in determining the pattern. Others (Brown and Fisher, 1980) emphasise material availability rather than sea level variation as the primary control aspect, and the changed emphasis produces some changes in interpretation.

Vail *et al.* postulate that during the relatively long time of a gradual fall of sea level much section would be exposed to erosion (figure 9/10(d)) and hence would be removed above a major unconformity. They interpret seismic patterns as showing the features of figures 9/10(a), (b) and (c) but do not observe the features of figure 9/10(d). Consequently they deduce that the rise and fall of sea level is asymmetric, long periods of gradual sea level rise being interrupted by occasional stillstands and short periods of sudden sea level fall. Furthermore, they profess to see the same patterns recurring at the same time worldwide. Their conclusions are summarised in the eustatic level chart shown in figure 9/11. It relates sudden sea level falls to an absolute time scale. If the pattern of sea level rises and falls in a given area based on study of seismic sections can be correlated with the eustatic level chart, then absolute age dates can be assigned to the unconformities which bound seismic sequences.

The hypothesis of worldwide sea level rises and falls implies mechanisms which can remove large amounts of water from the oceans rather quickly. Continental glaciation during the Pleistocene may be one such mechanism. However, the hypothesis poses many problems and many geologists and geophysicists do not

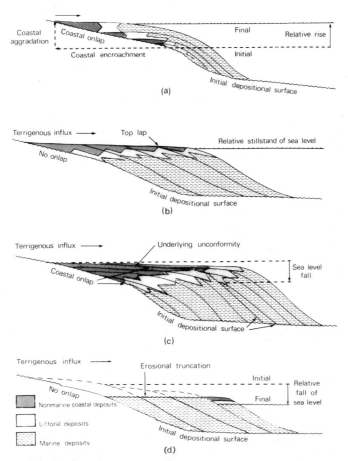

Figure 9/10 Patterns resulting from sea-level changes. (a) Gradual rise in sea-level; (b) stillstand of sea-level; (c) two periods of gradual rise of sea-level interrupted by sudden sea-level fall; (d) gradual fall of sea-level. (After Vail *et al*., 1977).

accept it. Nonetheless, correlation with the eustatic level chart has permitted age dating of seismic events in virgin basins in several instances and subsequent drilling has supported the age dates. The eustatic level chart is used as a working hypothesis even by those with misgivings about its significance.

9.7 Conclusions from seismic velocity

One might expect seismic velocity to be an important tool in stratigraphic identification and sometimes it is. The problem with velocity as a diagnostic is that the velocities associated with various lithologies (see figure 2/6) encompass wide ranges which overlap each other considerably. The reason for the broad

ranges is that other factors are involved, especially porosity and the maximum depth to which a rock has been buried (which often determines the porosity). Relatively low velocities, in the 1500–3000 m/s range, generally indicate shales or sandstones and high ones, in the 4500–6000 m/s range, carbonates or evaporites, but the intermediate range of roughly 3000–4500 m/s could indicate either. However, in specific situations where one knows the age, depth or other circumstances about the rock, the ranges for different lithologies can be narrowed considerably. Sandstones usually have slightly higher velocity than shales; from velocity data one can sometimes predict sand probability within a fairly broad interval on a statistical basis, whereas predicting for a specific sample is not much better than random. Coal and peat often have low velocity. Salt can

Global Cycles of Sea Level Changes

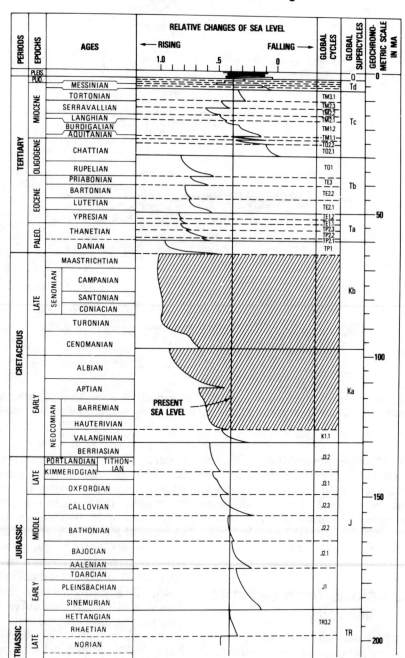

Figure 9/11 Eustatic level chart relating worldwide sea-level rises and falls to the absolute time scale. (From Vail *et al*., 1977).

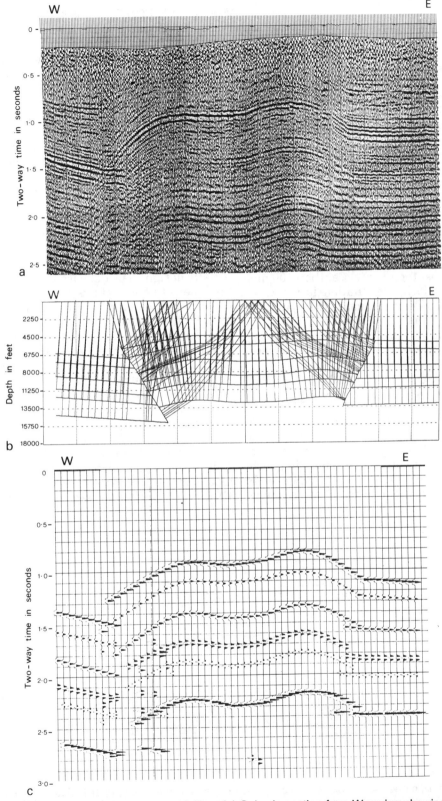

Figure 9/12 Raypath seismic computer modelling. (a) Seismic section from Wyoming showing faulting and diffractions. (b) Geological model with computed raypaths, including diffraction paths. (c) Synthetic seismic section which gives good fit to original section. The good fit between original and synthetic tends to confirm the validity of the geological model of a reverse-thrusted anticlinal horst. (Courtesy: GeoQuest International.)

sometimes be identified by its velocity but often it is embedded with other materials which result in altering the velocity measurement. Cementation and diagenesis usually result in increasing the velocity. Younger and unconsolidated rocks generally have lower velocities than older and consolidated rocks. However, even where velocity magnitude is not diagnostic of lithology, lateral changes in velocity may be used to indicate changes.

Velocity values are usually determined for the interval between two horizontal parallel reflectors by using Dix's equation (given in Chapter 4, Section 4.6). Since the calculation is based on the difference between measurements which involve uncertainties, if the two reflections are too close together (closer than 100 ms), the interval velocity calculation is apt to be unreliable. The Dix equation is not applicable to velocity layering which is not horizontal and parallel. Velocity is also determined by inversion to seismic logs, as discussed below.

9.8 Reflection character analysis and hydrocarbon indicators

Reflection character analysis concerns trace-to-trace changes in the waveform of a single reflection event or groups of events. Its implementation involves two modelling techniques, one direct and the other inverse. In direct modelling the effects of an assumed

model are calculated and in inverse modelling one attempts to determine the model from observation of effects. The first is the technique of synthetic seismogram manufacture and the latter of seismic log manufacture.

Synthetic seismograms have already been discussed and several examples shown (figures 4/34, 5/10, 5/13, 11/22 and 13/18). Synthetics are used to understand and demonstrate seismic wave principles and the effects of recording and processing methods. They are used in various models to model structural and velocity complications (see figure 9/12), stratigraphic changes (figure 9/13) and the effects of hydrocarbon accumulations.

In their stratigraphic applications, one usually designs a model based on sonic log data (and density log data where available) and convolves the reflectivity calculated from the model with a wavelet which often is extracted from the actual data. If the synthetic seismogram matches actual seismic data closely enough, one then varies the model in ways one suspects the stratigraphy may vary. The waveshape changes which such variations produce are then used as a guide to interpreting observed waveshape changes. Clement (1977) tells of using synthetic seismograms to define the reflection characteristics of channel sands. This led to the discovery of channel-sand reservoirs; it also led to a failure occasioned by a stratigraphic change which had not been

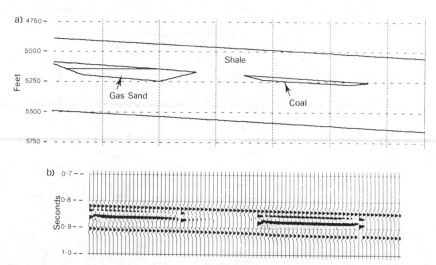

Figure 9/13 Seismic modelling. (a) Model of a gas sand (left) and a coal seam (right) enclosed in a shale unit. (b) Synthetic section showing the similarity between the effects of the two geological models. (Courtesy GeoQuest International.)

anticipated but which produced the same characteristics as the channel.

One conclusion from synthetic seismogram studies is that wavelength is not a very sensitive indicator of bed thickness for beds thinner than about one quarter wavelength: this is evident in figure 13/18. Many hydrocarbon reservoirs are within this thickness limitation. Amplitude changes can sometimes be used to determine thicknesses in this thin-bed region (Neidell and Poggiagliolmi, 1977).

An interesting model relating to the ampli-

tude mapping of hydrocarbon accumulations is given in figure 9/14 showing the effect of gas stored in an originally water-filled storage reservoir in West Germany. The gas produces a bright spot and an underlying dim spot. The amplitude ratio of the bright-spot and dim-spot reflectors is mapped at times with different amounts of gas fill. The amount of fill can be determined from a plot of amplitude ratio against thickness of gas-saturated sand.

Seismic logs or synthetic sonic logs were introduced in Section 4.10.2. The production of seismic logs implies that amplitude variations

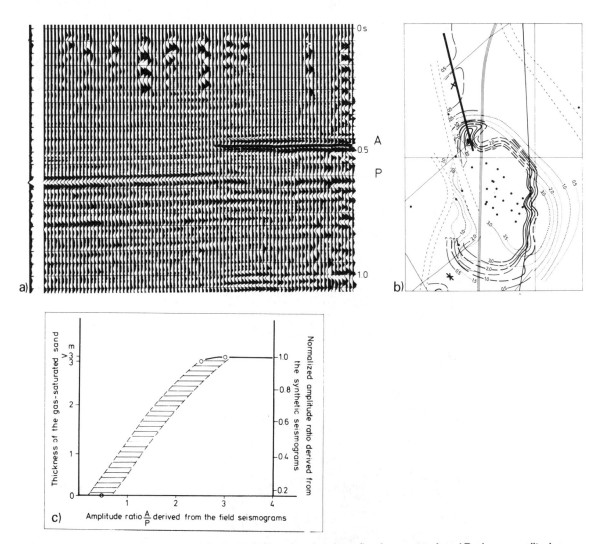

Figure 9/14 Amplitude mapping. (a) Seismic section showing the reflection events A and P whose amplitudes were measured. (b) Amplitude ratios plotted at different dates with contour values converted to gas-saturation thickness from the data in (c). (c) Graph of normalised amplitude/gas-saturation thickness versus thickness of saturated gas sand. (Courtesy: Prakla-Seismos).

have a one-to-one correspondence relating them to variations in the earth's reflectivity, so the data input into the inversion process must have preserved amplitudes faithfully and must also be free of noise. The reliability of a seismic log generally depends on how closely these conditions are met. Inversion also requires knowledge of quantities which cannot be determined from the seismic data alone in order to give magnitudes to the resulting velocity values. Where seismic data are good and sonic logs are available in nearby wells to provide calibration, seismic logs may match sonic logs which have been filtered to the same bandpass (figure 4/38(b)).

The colour section in figure 9/15 (colour Section) shows how seismic logs were used to extend the boundary of a known Upper Devonian carbonate gas field in western Canada. On this section the synthetic sonic traces are velocity contoured and the high velocities associated with carbonates are coloured blue, the darkest

blue indicating the highest velocity. The Crossfield interval is one such high-velocity interval. A known gas field in a porous zone of the Crossfield existed to the left of the section but its limit based on a dryhole had been placed at the fifth trace from the left. This study indicated an extension of the porous productive zone (the lower velocity due to the gas porosity being indicated by the light blue colour) towards the right which was subsequently proved by drilling.

The presence of gas within the pore space of a rock lowers the velocity compared with water in the pore spaces. In poorly consolidated rocks this lowering of velocity can be appreciable. Oil in the pore spaces also lowers the velocity but usually by only a small amount. Since gas or oil usually is present only in the limited area of a trap, the lateral change in velocity produces effects which sometimes can be used to indicate the trap. Gas and oil also have lower density than water and hence their presence lowers the

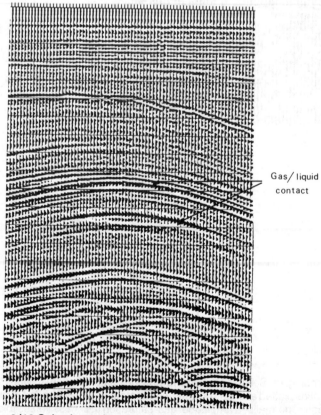

Figure 9/16 Seismic section showing gas/liquid contact in reservoirs.

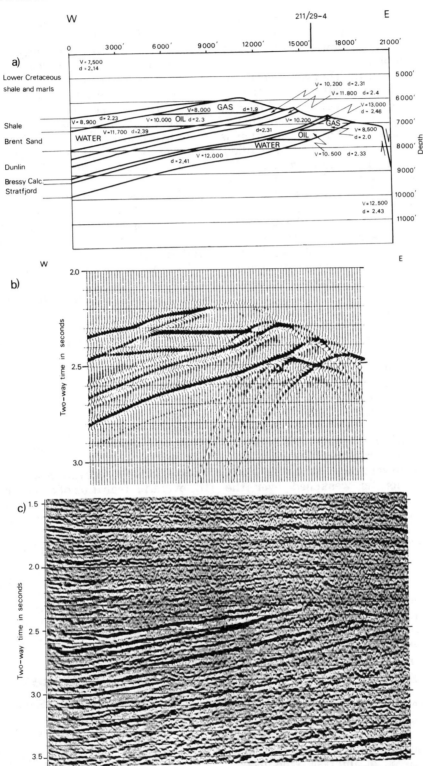

Figure 9/17 Model section across Brent Field, North Sea. (a) Model with velocities and densities. (b) Synthetic seismic section; note (i) dim-spot where Brent sand is truncated by unconformity, (ii) bright-spot and flat-spot for oil/water contact in Brent sand, (iii) bright-spot for oil/water contact in Brent sand, (iv) polarity reversal from shale/Brent-sand gas to shale/Brent-sand oil, and (v) concealed Stratfjord fluid interfaces due to diffractions at the various subcrops at the unconformity. (Courtesy: GeoQuest International.) (c) Migrated zero-phase Brent field section. (Courtesy: Shell Exploration Co. UK.)

bulk density of the rock, but the changes in density are usually small. However, the effect of density changes can substantially alter acoustic impedance values.

The major effect of the changed velocity on seismic data is a changed reflection coefficient. In Tertiary clastics a water-filled sand usually has a velocity not very different from that of encompassing shales so that its reflection is usually weak. Where the sand is gas-filled, however, the contrast is increased and so the reflection becomes stronger. The local increase in amplitude is called a 'bright spot'. Gas-induced bright spots usually have negative polarity for the reflection from the top of the reservoir. Both the top and base of the reservoir sand are generally reflectors and the reflection composite may not have clear polarity, depending on the thickness. Bright spots can be seen in figures 9/9, 9/16 and 9/18.

Where reservoirs are thick and include a fairly sharp gas-liquid contact a reflection may be obtained from this horizontal reflector to show as a 'flat-spot' (see figures 9/16 and 13/17). Sometimes the contrast between a reservoir and the encompassing rock is positive when water fills the pore space but negative when gas fills them, in which case a reversal of polarity may be associated with the reservoir. In another situation, the contrast with water in the pore spaces is large and filling the pore spaces with gas lowers the contrast, in which case the reflection is weakened over the reservoir to form a 'dim spot'. Thus hydrocarbon indicators differ depending on the local circumstances. One possibility, of course, is that the changes are too small to be obvious and hydrocarbon indicators are not seen everywhere.

Synthetic modelling showing dim-spots, bright-spots, flat-spots and polarity reversal is shown in figure 9/17, based upon the Brent Field in the North Sea. A comparison with field results is also shown. Although data quality is good, the features predicted by the model are not all clearly identifiable on the seismic section.

The presence of a very small amount of gas affects the velocity as much as complete gas saturation so hydrocarbon indicators are sometimes seen where there are not enough hydrocarbons to be of interest. One use of seismic data is to provide advance knowledge of gas pockets which may constitute a hazard in drilling even though they may not contain large volumes of gas. Of course, other changes to a rock can change the reflectivity and give effects which look like hydrocarbon accumulations. Local buildups of carbonate (old oyster reefs) sometimes create such anomalies in the US Gulf Coast (Taner *et al.*, 1979).

Other types of anomalies may be associated with hydrocarbon accumulations. There is often a lowering of frequency content immediately under an accumulation (see figure 9/9). Events below an accumulation are sometimes weakened. Small amounts of gas leaking from an accumulation may cause a degradation of reflection quality. Sometimes accumulations are thick enough to produce velocity anomalies underneath them.

The use of shear waves, a technique still largely experimental, as a means of hydrocarbon detection was mentioned in an earlier chapter.

Hydrocarbon indicators were applied in the case history of Chapter 13 (Section 13.3). The interpretation of hydrocarbon indicators is much like that of other features. The case is strengthened where evidences coincide. Some schemes attempt to quantify various types of evidence and combine them to provide a better indicator than any one by itself, as in the Hi-Scan display shown in figure 9/18. However, ambiguities and alternative explanations are always possible. Interpretations should rely on the overall integration of evidence and the best picture is that which explains the data in the most consistent manner. Thus the conjunction of evidence of a structural trap with hydrocarbon indicators builds a much stronger case than either by itself.

An example of how both lithology and pore fill may be predicted is shown in figure 9/19. Where good data are available, comparison shows a good match between predicted results and actual results from drilling.

References

L. F. Brown and W. L. Fisher, *Seismic stratigraphic interpretation and petroleum exploration.* AAPG Continuing Education Course Note Series, No. 16 (1980).

W. A. Clement, A case history of geoscience modelling of basal Morrow-Springer sandstone. Watonga-Chickasha trend, Oklahoma. *AAPG Memoir*, **26**, pp. 451–76.

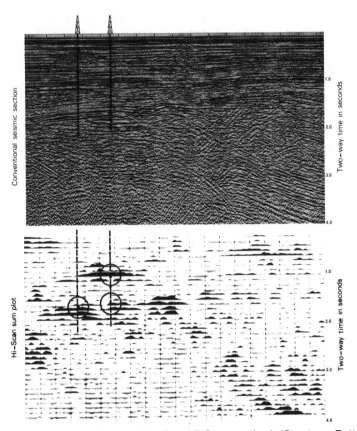

Figure 9/18 Automated hydrocarbon indicator detection, Hi-Scan method. (Courtesy: Petty-Ray Geophysical.)

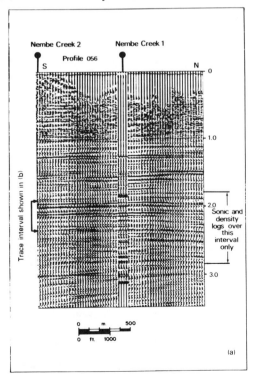

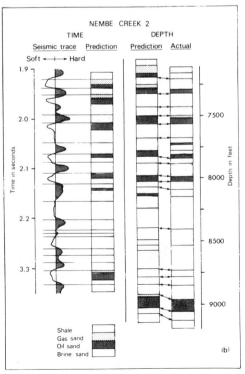

Figure 9/19 (a) Synthetic trace from Well No. 1 (repeated four times for clarity) set into a north-to-south impedance profile. The match is good over the interval with both sonic and density logs. (b) The lithology and pore fill prediction compared with results obtained in Well No. 2 (after Nelson, 1980).

M. T. Halbouty, Rationale for deliberate pursuit of stratigraphic unconformity and palaeogeomorphic traps. *AAPG Memoir*, **16** (1972), pp. 3–7.

N. S. Neidell and E. Poggiagliolmi, Stratigraphic modelling and interpretation – geophysical principles and techniques. *AAPG Memoir* **26** (1977), pp. 389–416.

P. H. H. Nelson, Role of reflection seismic in development of Nembe Creek Field, Nigeria. *AAPG Memoir* **30** (1980), pp. 565–576.

G. Rittenhouse, Stratigraphic trap classification. *AAPG Memoir* **16** (1972), pp. 14–28.

M. W. Schramm, E. V. Dedman and J. P. Lindsey, Practical stratigraphic modelling and interpretation. *AAPG Memoir* **26** (1977), pp. 477–502.

R. E. Sheriff and L. P. Geldart, *Exploration seismology*, vol. 2, Cambridge University Press (1983).

M. T. Taner, F. Koehler and R. E. Sheriff, Complex trace analysis. *Geophysics*, **44** (1979), pp. 1041–63.

P. R. Vail, R. M. Mitchum, R. G. Todd, J. M. Widmier, S. Thompson, J. B. Sangree, J. N. Bubb and W. G. Hatlelid, Seismic stratigraphy and global changes in sea level. *AAPG Memoir* **26** (1977), pp. 49–212.

Section 2
Case Studies

Chapter 10

Moray Firth Case Study

In this chapter we will review the exploration history of the Inner Moray Firth sedimentary basin with particular reference to a regional seismic survey conducted in this area by the Institute of Geological Sciences in 1972. Both borehole information and surface geological maps provide geological control. Valuable information is also obtained from gravity and aeromagnetic maps. Some high resolution seismic data are also described which were of particular value in developing a model invoking crustal extension and transcurrent fault movements as the mechanism of basin formation.

10.1 Basin exploration history

The possibility that the Moray Firth might be underlain by a large sedimentary basin of predominantly Mesozoic rocks has been evident for many years. The Firth is bounded by mainly Old Red Sandstone rocks with an intermittent fringe of Permian to Jurassic rocks, a knowledge of which led Arkell (1933) to postulate the basin's existence. However, when Collette (1958) published results of a gravity survey showing a large negative anomaly in the Firth, he favoured the idea that this was due to the existence of a large body of buried granite. Donovan (1963) argued afresh the case for the Mesozoic sedimentary basin as the most likely

cause of the gravity low; yet, even in the mid-1960s, following oil and gas discoveries in other parts of the North Sea, the area was not being investigated as of having hydrocarbon potential. In the late 1960s and early 1970s the true nature of the basin became apparent. Seabed sampling and some shallow boreholes drilled by IGS (Chesher et al., 1972) confirmed the presence of Permo-Triassic to Upper Cretaceous strata, and, also, regional geophysical surveys included a much more detailed gravity survey interpreted by Sunderland (1972, see below) as indicating 4 km of Mesozoic rocks in the deepest part of the basin. The IGS seismic survey conducted in October 1972 confirmed Sunderland's gravity interpretation (McQuillin and Bacon, 1974; Chesher and Bacon, 1975). In parallel with IGS studies, commercial interest developed at a considerable pace, although it was not until September 1977 that a commercial oil field was eventually discovered.

The history of commercial exploration can be traced through the stages of seismic exploration, licensing and drilling (see figure 10/2 for location of licence blocks and wells drilled). The first seismic surveys were conducted in the outer part of the basin in 1965, and in the inner part by IGS in 1972. By the late 1970s a dense grid of data existed. The history of licensing (see figure 10/2) has been as follows:

Round 1 (up to 1965): Nil.

Round 2 (1965):   Blocks 12/21, 22, 23, 24, 26 and 29. 18/1, 2, 3 and 4.

Round 3 (1970):   Blocks 12/25 and 8/5.

Round 4 (1971/72):   Block 11/30. Surrendered blocks 12/21, 22, 23 (part), 26 and 29 (part).

Round 5 (1967/77):   Nil. All remaining blocks surrendered except 11/30 and parts of 12/30 and 18/5.

Round 6 (1978/79):   Nil. Part of 11/30 surrendered.

Round 7 (1980/81):   Blocks 11/25, 12/21, 22, 23, 24, 27 and 29.

Round 8 (1982/83):   Nil.

Drilling began in 1967. Hamilton investigated a structure on the Inner Moray Firth central

ridge (see below) with well 12/26–1, through which IGS line 7 (figure 10/8, colour section) passes, which was dry. At about the same time well 12/23–1 was drilled to be followed by 12/21–1 and 12/21–2; all were dry in terms of commercial quantities of hydrocarbons. Other wells drilled in the area are shown in figure 10/2. The only proven oilfield (at time of writing) is the Beatrice Field which was discovered in September 1976. The group of wells in Block 11/30 were drilled to investigate this field (intersected by IGS line 6) which has estimated recoverable reserves of about 160 million barrels (see Linsley *et al.*, 1980). The gross-pay column of over 800 ft thickness extends from Lower to Upper Jurassic, see figure 1/3 (p. 5). Development plans were approved in 1978 and involve transportation of the oil via a pipeline to Nigg Bay (see figure 10/2). In late 1982 Burmah Oil Company hit gas in block 12/27 and at time of writing appraisal drilling is being planned.

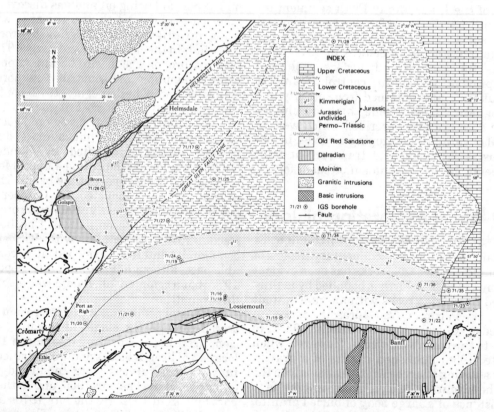

Figure 10/1 Solid geology of the Moray Firth as mapped before the IGS (1972) seismic survey was completed (after Chesher *et al.*, 1972).

10.2 Surface geology

As noted above, on land, the Moray Firth is mostly bounded by thick sequences of Old Red Sandstone and older rocks, but outcrops of Mesozoic rocks occur in a narrow coastal strip; Permian to Jurassic rocks are found in the Lossiemouth area and Triassic to Jurassic rocks crop out between Golspie and Helmsdale and at Ethie and Port an Righ (figure 10/1). At the time the seismic survey to be described here was commissioned by the Institute of Geological Sciences, the state of knowledge of the surface geology was as shown in figure 10/1.

We shall see later that normal faulting controls the structure of this part of the Moray Firth. Since it is not easy to identify faults on shallow sparker sections, and since most of the faulting, as we shall see, does not generally affect the shallowest horizons, little was known of the fault patterns in the basin prior to the deep seismic survey. On land, the major fault of the area is the Great Glen Fault, a wide fault zone along which transcurrent movement is believed to have occurred in Middle Old Red times, though the amount and direction of shift are controversial. Later phases of movement have also been postulated and these will be discussed in Section 10.7. This major fault zone is thought to continue in a north-north-easterly direction across the Moray Firth to link up with known faults on the Shetland Isles. The only other major fault known on land was the Helmsdale Fault running along the north-

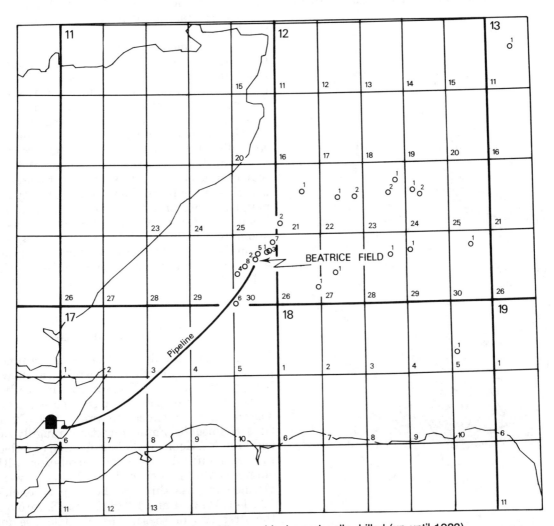

Figure 10/2 Location of licence blocks and wells drilled (up until 1983).

eastern coast of the Moray Firth and down-faulting Mesozoic strata against Old Red Sandstone.

10.3 Gravity and aeromagnetic maps

When the Moray Firth gravity low was first discovered (with a very wide spacing between observations, resulting in a generalised version of figure 10/3), it was attributed to an extensive granitic body. Doubt was cast on this interpretation by the aeromagnetic survey (figure 10/4), which showed that over the gravity low the magnetic anomalies were of long wavelength and low amplitude, consistent with the presence of a thick sedimentary cover over magnetic basement. The mapping of the surface

geology described above confirmed the sedimentary basin interpretation. An approximate depth of the basin can be obtained by modelling a profile across it. Figure 10/5 shows such a model (Sunderland, 1972). First, a linear regional field was removed, based on the observed Bouguer anomalies at either end of the profile. A density contrast of 0.4g/cc was assumed between the Mesozoic rocks of the basin and the underlying Old Red Sandstone and the Dalradian, since published determinations of density from land outcrops in the area gave a mean value of 2.3g/cc for the Permian and Mesozoic rocks and 2.7g/cc for the Old Red Sandstone and Dalradian rocks. Exceptionally, the density contrast near A (figure 10/5) was taken to be only 0.3g/cc; in this region the Mesozoic sequence is not present and the density contrast is found within the Old Red Sandstone.

The two-dimensional model calculated to fit the observed anomalies then shows that along AB the depth of the basin does not exceed 4 km, though owing to the uncertainties in regional field and density contrast this value is probably accurate only to about 0.5 km. The basin is clearly fault-controlled, as indicated by the steep dip (50°) deduced for the southern margin of this model. Further steep gradients indicate major faults on the northern and north-western margins of the basin. More recent gravity and magnetic interpretations (Dimitropoulos and Donato, 1981) show that a residual gravity anomaly which correlates with a magnetic anomaly over the central ridge is strong evidence that a granitic body underlies this ridge.

10.4 Seismic sections

To investigate the deep structure of the inner part of the Moray Firth, the seismic grid shown in figure 10/6 was shot. The survey is basically a 10 km × 13 km grid, with the more closely spaced lines intended to cross the Great Glen-Helmsdale Fault direction roughly at right angles. Geological identification of reflectors was initially based on correlation with outcrop data. However, lines 7 and 8 were planned to pass through the deep boreholes 12/21–1, 12/22–1 and 12/26–1. Line 8, through the first two of these, was of little use for identification because the Great Glen Fault

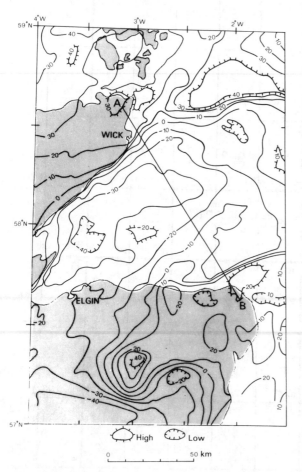

Figure 10/3 Bouguer anomaly map (courtesy IGS).

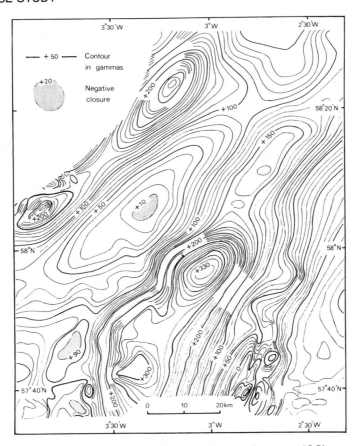

Figure 10/4 Aeromagnetic anomaly map (courtesy IGS).

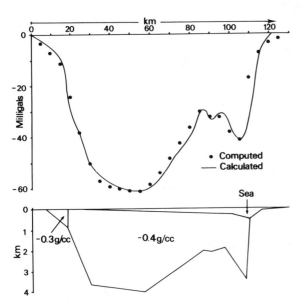

Figure 10/5 Interpretation of gravity profile (after Sunderland, 1972). For location see figure 10/3.

passes too close to the tie to the rest of the survey. When information on borehole 12/26–1 was released, however, it showed that the identifications from outcrop geology were fairly accurate (figure 10/7). Line 14 was shot after the interpretation of the rest of the data, when it became apparent that correlation across the Great Glen Fault (which runs roughly along line 10) was difficult, so that the northern extremities of lines 1 to 6 were not clearly tied together; this line was positioned to run north of and roughly parallel to the fault in order to solve this problem.

Selected seismic sections are shown in figure 10/8, together with their interpretation. The events selected for mapping were as follows:

Horizon A: Intra Lower Cretaceous. This is a continous shallow event in the eastern part of the area and is clearly defined except where it approaches seabed outcrop.

Horizon B: Base Lower Cretaceous. This is a very strong and continuous event especially in the eastern part of the area.

Horizon C: Originally described as mid-Jurassic, now known to be within Upper Jurassic. The event is not as strong as B or E but continuity is fairly good. The interval B–C is mainly lacking in continuous reflection events.

Horizon E: Base Jurassic. (A horizon D lying just above this event was considered for mapping but rejected because it showed very similar structural trends to E.) This is the strongest deep reflector, with an easily recognised, well-defined character; it can be followed reliably over most of the area, although correlation across the Great Glen Fault is poor. There are many smaller faults which are very clearly defined at this level.

Horizon F: Intra Permo-Triassic. This deep event can be mapped only in the southeastern part of the area. Correlation across faults is poor and, although deep events are visible on some other lines, the uncertainty of correlation is too great to make possible any reliable extension of the mapping.

Identification of these horizons is illustrated in figures 10/8 and 10/9. To illustrate the

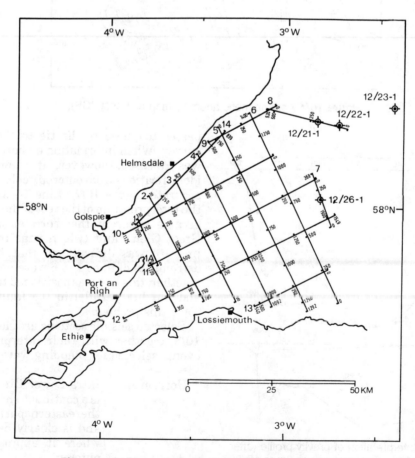

Figure 10/6 Plans of Moray Firth IGS seismic survey.

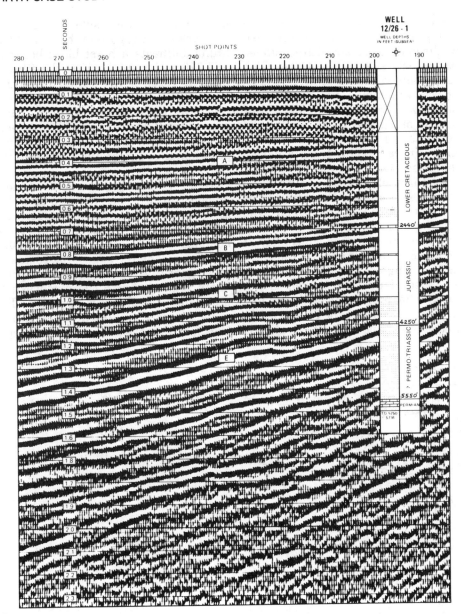

Figure 10/7 Correlation of seismic horizons and well stratigraphy (well section courtesy Hamilton, IGS seismic record, Seiscom survey).

difficulty of correlation across a major fault such as the Great Glen Fault, figure 10/8 shows, for those parts of the line north of this fault, both the original picks based on seismic character and velocity information, and the revised picks after line 14 became available, enabling us to tie together all these line segments; in some cases drastic revision was required. Figure 10/9 shows how the outcrop geology from IGS shallow boreholes fits into the structure of line 3. Once the identifications have been made, the seismic data suggest modifications to the outcrop geology map, of which the most conspicuous example is the southward extension of the Lower Cretaceous outcrop as a result of the zone of faulting with downthrow to the southeast near line 12, which is an obvious feature of lines 4 and 5.

Returning to figure 10/8, we can now comment on the details of the interpretation.

Lines 3 to 6 are a series of cross-sections of the basin as it progressively deepens to the east. Line 3 shows the main basin structure bounded on the northwesterly side by the Great Glen Fault (hereafter abbreviated GGF). The southeasterly flank is broken by a number of small faults at the horizon E level, with downthrow towards the axis of the basin. At SP100 and SP200 there are faults with downthrow to the southeast producing a shallow basin separate from the main basin. On line 4 the structure is similar, but the throw of the GGF appears to be very small; it is marked by a discontinuity at the deepest part of the basin. Again the southeasterly flank of the basin is broken by small faults giving rise to poor continuity of horizons C, E and F. At SP775 and 820 there are major faults throwing down to the southeast. Across these latter faults correlation is not absolutely certain but is probably fairly reliable for Horizons B and E. On line 5 the GGF is again seen as a major fault, and reflector quality is poor on the upthrown side. Reflection quality of horizons B to E is good in the main basin. Major faults delineate a horst block between SP660 and 770 with a sub-basin to the southeast of it. Structure seen on line 6 is very similar to that of line 5. The Great Glen Fault zone here features a well developed antithetic fault. This section crosses the Beatrice Oil Field. The approximate location (offset) of well 11/30–2 is shown. The pay-zone is in Lower Jurassic sands just above red reflector E.

Line 7 has been included mainly to demonstrate the borehole tie to well 12/26–1 and shows a fairly simple structure with all horizons rising south-easterly to a fault-bounded uplift zone. Line 11 shows the structure along the axis of the basin. There is a broad gentle arch in the southwest followed by a fairly regular dip to the northeast; however, much minor faulting is apparent. The north-eastern parts of 10 and 12 are included principally for the purpose of the interpretation

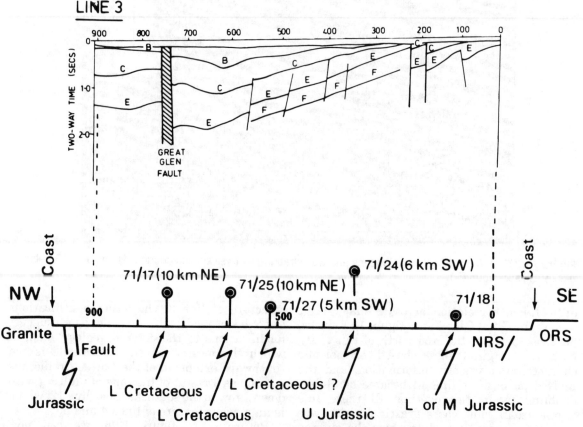

Figure 10/9 Correlation of line 3 with outcrop geology. IGS borehole identifications are shown.

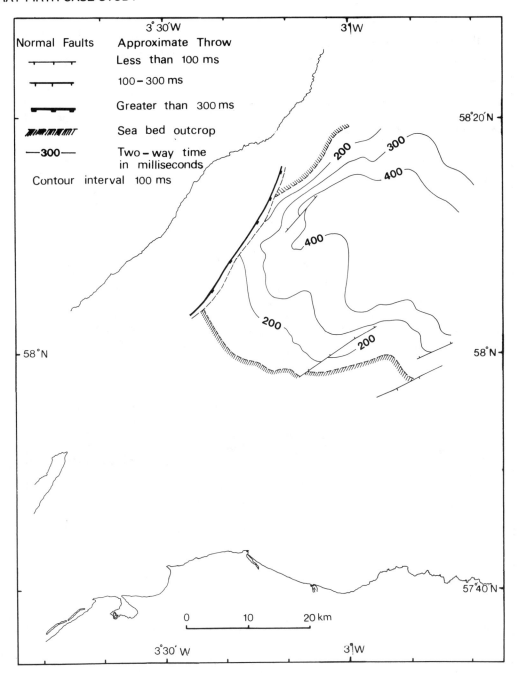

Figure 10/10 Two-way time map of horizon A, intra-Lower Cretaceous (figures 10/10–10/12 are adapted from Chesher and Bacon, 1975).

exercise described in Chapter 6.

10.5 Seismic maps

Two-way time maps for horizons A, B and E are shown in figures 10/10 to 10/12. For maps of horizons C and F see Chesher and Bacon (1975). In general the connection of major faults from line to line is straightforward but the minor faults cannot be treated so reliably and have

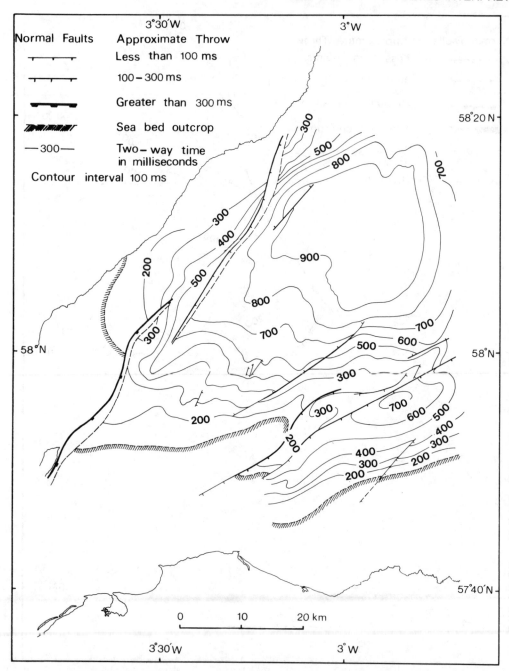

Figure 10/11 Two-way time map of horizon B, base Lower Cretaceous.

simply been assigned trends similar to those of the major features. The increase in complexity of structure with depth is readily apparent. Horizon A (figure 10/10) shows a simple basin structure, faulted out against the GGF in the northeast and against a zone of uplift to the southeast; it may, however, be present again to the southeast of this zone, beyond the uplift zone at the end of line 7. At horizon B level (figure 10/11), the main basin and southern sub-basin are already clearly defined, although only the major faults extend up to this level. A zone of uplift trending northeasterly through the main basin is also apparent. The deeper structure of horizon E (figure 10/12) confirms these major features of the basin, with the uplift

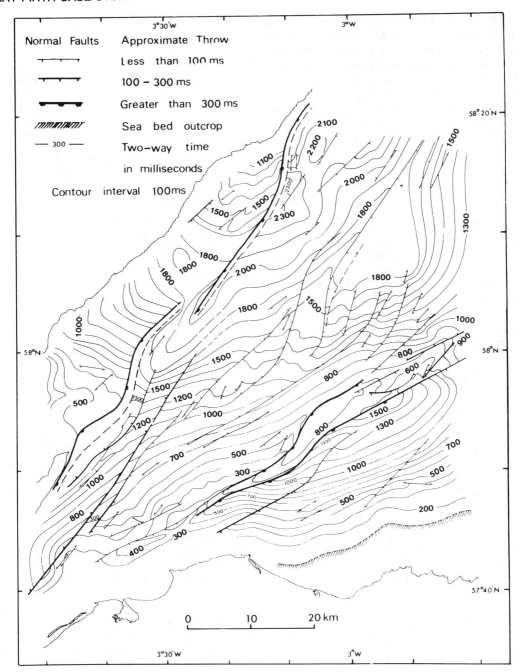

Figure 10/12 Two-way time map to horizon E, base Jurassic.

zone in the main basin appearing as a horst block.

It should be noted here that the interpretation of the seismic sections is based on a study of the IGS survey alone. If integrated with the results of more recent surveys of the area, more detail could be incorporated into the interpretation of minor faults and in places seismic picks would be modified.

10.6 Conclusions to be derived from the IGS survey

The IGS survey showed that there are hydrocarbon prospects in the area. Not only is the Jurassic sequence, productive elsewhere in

the North Sea, well developed, but there was a
reasonable hope that more detailed work would
reveal anticlinal closures over fault blocks in
the basin. Since the IGS survey was completed a
number of commercial surveys have been
undertaken in the area and a much denser
seismic coverage has become available to
offshore operators. A series of wells (11/30–1 to
4) have delineated the Beatrice Oil Field. This
feature appears as a fault block on horizon E
(figure 10/12). The IGS seismic grid spacing is,
however, too great to permit precise delineation
of the trap. Figure 10/13 shows a structure map
of the Beatrice Field.

Also, it should be noted that the IGS survey
greatly increased knowledge of the general
geology of the area. Not only could a more
precise outcrop geology map be prepared, but
much information on the detailed pattern of
faulting was obtained (see figure 10/14). In
particular, the Great Glen Fault was traced
across the area and shown to be sinuous and
discontinuous.

10.7 Development of the Inner Moray Firth Basin

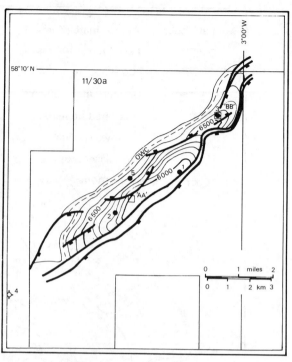

Figure 10/13 Structure map indicating depth to top of
pay, Beatrice Field (after Linsley *et al.*, 1980).

A model for the development of the basin is
described by McQuillin *et al.* (1982). This
invokes crustal extension and dextral displacement of the Great Glen Fault as the
principal mechanisms. The reader is referred to
the above paper for a full treatment. Here it is
sufficient to say that the evidence which proved
critical was that provided by a Horizon
Exploration seismic survey using high resolution digital acquisition techniques. This survey
was carried out in 1979 using a sparker source.
The high resolution and close line spacing of the
survey allowed a detailed interpretation of the
geological structure of the Great Glen Fault
zone in the uppermost 2–3 km of sediments (see
figure 10/15 for line layout). Interpretation was
mainly at the level of Horizon B, the
Jurassic-Cretaceous boundary. Figure 10/16
shows an example of the data where vertical
displacement of the GGF is relatively small and
the zone is marked by a compressional
structure, which within an extensional environment is indicative of transcurrent movements in the deeper, more rigid basement.
Further evidence comes from an analysis of the
details of the deformation pattern in Jurassic
and Cretaceous rocks along the fault zone. The
main elements are shown in figure 10/15. Of

particular relevance is the fault segment A–A'
which is seen as a 'scissor fault'. Theoretical
studies by Chinnery (1961 and 1965) predict the
development of this type of fault pattern in
association with dextral displacement. On the
basis of these observations, McQuillin *et al.*
suggest that the basin developed by crustal
extension between northerly and southerly
boundary faults; these are seen as high gravity
gradients in figure 10/3, the southerly crustal
block being driven to the southwest through
dextral movement along the GGF. The model
requires about 8 km dextral displacement. This
produces a basin (in the absence of crustal
thinning) of mean subsidence 2–3 km which
can be compared with the geological section
across the basin, interpreted along IGS line 6,
shown in figure 10/17.

References

W. J. Arkell, *The Jurassic system in Great Britain*.
Oxford University Press, Oxford (1933).

J. A. Chesher, C. E. Deegan, D. A. Ardus, P. E. Binns
and N. G. T. Fannin, IGS marine drilling with
m.v. Whitethorn in Scottish waters, 1970–71, *Inst.
geol. Sci. Rep.* no. 72/10 (1972).

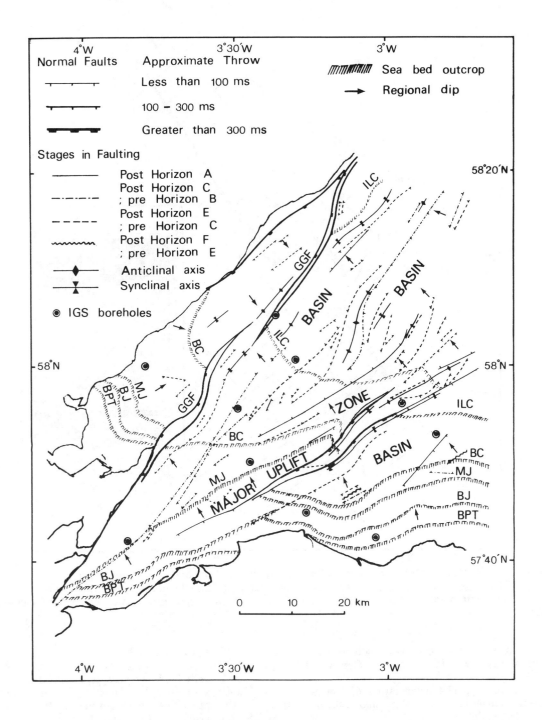

Figure 10/14 Structural summary map.

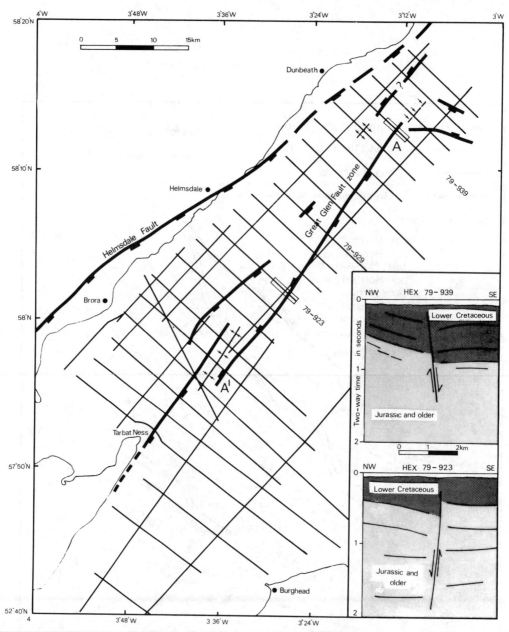

Figure 10/15 Location map showing Horizon Exploration seismic lines and interpreted positions of faults and flexures which affect the Jurassic-Cretaceous boundary along the Great Glen Fault zone. Inset shows diagrammatic interpretation of lines 79–939 and 79–923 where these cross the fault AA' (after McQuillin *et al.*, 1982).

J. A. Chesher and M. Bacon, A deep seismic survey in the Moray Firth, *Inst. geol. Sci. Rep.* no. 75/11 (1975).

M. A. Chinnery, The deformation of the ground around surface faults. *Bull. Seismol. Soc. Am.*, **51** (1961), pp. 355–372.

M. A. Chinnery, The vertical displacement associated with transcurrent faulting. *J. Geophys. Res.*,

70 (1965), pp. 4627–32.

B. J. Collette, *Gravity expeditions 1948–1958*, vol. 5, part 2. Delft University Press, Delft (1960).

K. Dimitropoulos and J. A. Donato, The Inner Moray Firth central ridge, a geophysical interpretation. *Scot. J. Geol.*, **17** (1981), pp. 28–38.

D. T. Donovan, *The geology of British Seas.* Univ. Hull Pbl. (1963).

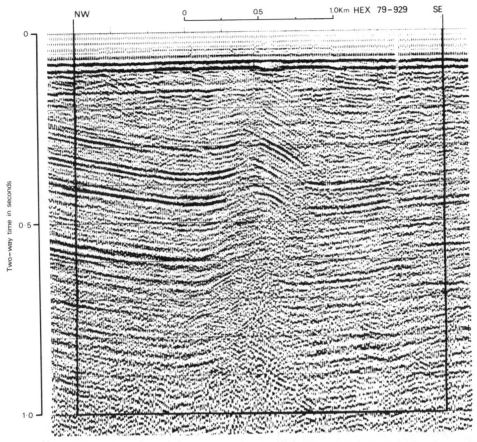

Figure 10/16 Part of seismic line HEX 79–929 where this crosses the fault AA′ shown in figure 10/14 (after McQuillin *et al*., 1982. Seismic section courtesy: Horizon Exploration).

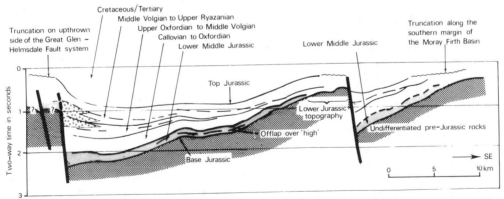

Figure 10/17 A geological section across the Moray Firth along IGS line 6 (Personal communication: Dr C. Deegan, IGS).

P. N. Linsley, H. C. Potter, G. McNab and D. Racher, The Beatrice Field, Inner Moray Firth, UK North Sea. *AAPG Mem.*, no. 30 (1980) pp. 117–129.

R. McQuillin and M. Bacon, Preliminary report on seismic reflection surveys in sea areas around Scotland, 1969–73. *Inst. geol. Sci. rep.* no. 74/12 (1974).

R. McQuillin, J. A. Donato and J. Tulstrup, Development of basins in the Inner Moray Firth and the North Sea by crustal extension and dextral displacement of the Great Glen Fault. *Earth Planet Sci. Lett.*, **60** (1982), pp. 127–139.

J. Sunderland, Deep sedimentary basin in the Moray Firth. *Nature*, **236** (1972), pp. 24–25.

Chapter 11

Rainbow Lake Case History

The purpose of this chapter is to describe how, in 1965 and subsequently, the relatively new concept of Common Depth Point reflection seismic technique was effectively used to contribute to the discovery and development of reservoirs in a series of stratigraphic traps which ultimately yielded recoverable reserves of an estimated 2.2 billion barrels of oil and 1.5 trillion cubic feet of gas, and which were collectively known as the Rainbow Lake Fields.

11.1 Location

The Rainbow Lake area (figure 11/1) is situated in the extreme northwest corner of the Province of Alberta in Canada, 25 miles east of the boundary with British Columbia (120°W longitude) and 100 miles south of the boundary with the Northwest Territories (60°N latitude). It is over 400 miles northwest of the provincial capital, Edmonton, and about 600 miles from Canada's oil capital, Calgary.

11.2 Regional setting and stratigraphy

The Rainbow area is situated towards the northern end of the Western Sedimentary Basin or Interior Plains (figure 11/2). Economic basement is composed of Pre-Cambrian metasediments and igneous rocks.

From the edge of the exposed Pre-Cambrian

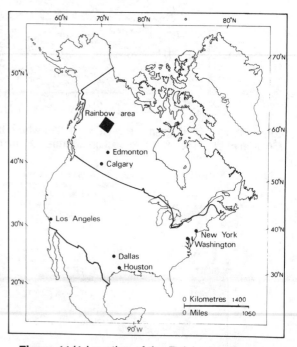

Figure 11/1 Location of the Rainbow Lake area.

shield, the basement dips very gently south-westwards for about 500 miles, then rapidly disappears deep beneath the Rocky Mountains where Palaeozoic and Pre-Cambrian rocks are thrust over the accumulated 15 000–20 000 ft of Mesozoic and Palaeozoic sediments.

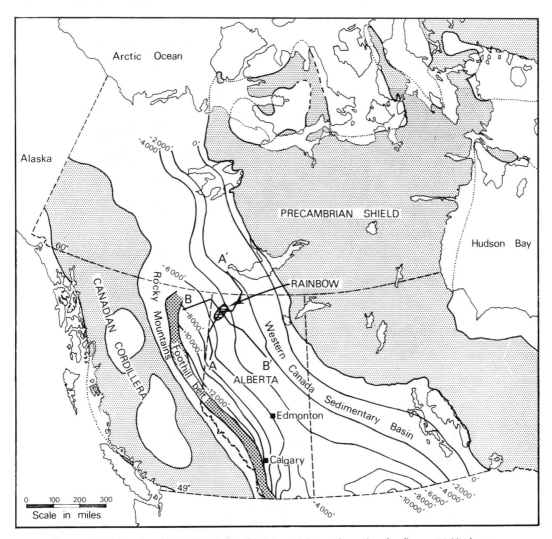

Figure 11/2 Regional Pre-Cambrian Structure. Lines of section for figure 11/4 shown.

In the Rainbow Lake area, basement is generally around 5000 ft below sea-level or at 6300 ft drill depth, and dips are around one quarter of a degree (25 ft per mile). The basement is overlain mainly by Devonian evaporites, carbonates and shales. The Middle Devonian is around 2500 ft thick and restricted almost entirely to carbonates and evaporites. The Upper Devonian has a thick, almost 2000 ft, section of marls and marly shales overlain by a limestone unit, the Wabamun, 1200 ft thick. A major unconformity exists between the latter and the approximately 700 ft thick Lower Mississippian marls and shales and a further unconformity exists below the overlying 1200 ft of Lower Cretaceous shales which are the bedrock of the area. Glacial deposits of extremely variable thickness cover the area and from an operational point of view surface conditions can be extremely difficult due to surface water and muskeg (mossy growth found in high latitudes, usually as uneven topography encouraging standing water and deep and treacherous swamp, bog or marsh). Understandably, both seismic and drilling operations are most efficient during the annual freeze-up (November–April) and indeed in most areas are impossible outside this period.

The main exploration target for the area is the Rainbow Member of the Keg River

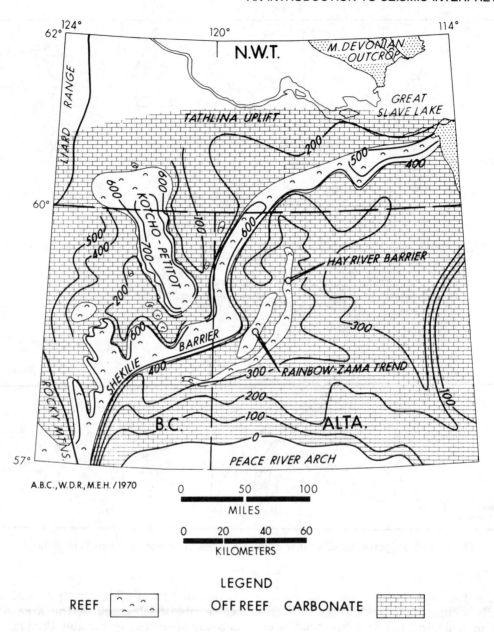

Figure 11/3 Rainbow area, isopach of Keg River Formation showing barrier complexes and reef trend surrounded by off-reef carbonate facies (after Hriskevich, 1970).

Formation where hydrocarbons are trapped in reef structure reservoirs. During the Middle Devonian, the area was a restricted basin in which palaeotemperatures, winds and currents were favourable for the development of reefs which grew up to 800 ft above the main platform of the Lower Member of the Keg River Formation. In an eastward landward direction, the basin was restricted by the Hay River Barrier Reef and a carbonate shelf, and to the south, by the emergent Peace River Arch; to the west and north it was separated from an open marine embayment by the Shekelie Barrier which fringed the Tathlina Arch. The barrier reefs and restrictive structural elements are shown in the map of figure 11/3 and the cross-sections of figure 11/4. The increasing dominance of these barriers resulted eventually in the cessation of reef growth as well as widespread deposition of salt between the reefs, followed by deposition of a continuous anhydrite sequence which serves as the trapping mechanism for the Rainbow Member Reservoirs.

The regional setting is well illustrated by the total intensity magnetic map (figure 11/5). The eastern boundary of the basin is defined by the Hay River Fault which is a major fault system trending NE–SW for hundreds of miles, down-throwing to the northwest, and which is marked by a linear magnetic high of 700 gamma amplitude. In the western part of the map a magnetic low, trending approximately N–S, is probably related to major petrological variations in the basement. As such this zone may have had topographic relief and acted as a hingeline for barrier reef growth.

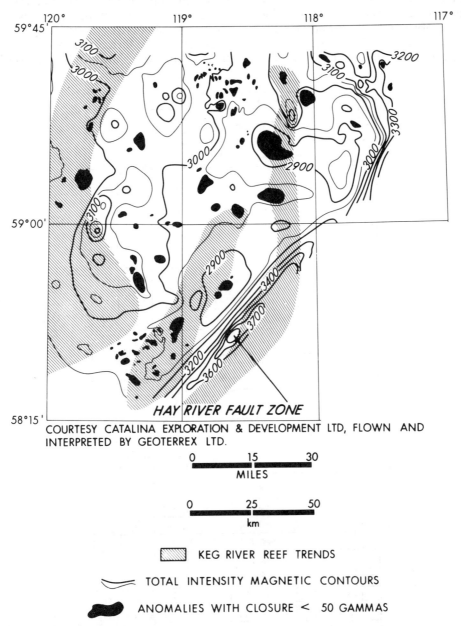

COURTESY CATALINA EXPLORATION & DEVELOPMENT LTD, FLOWN AND INTERPRETED BY GEOTERREX LTD.

Figure 11/5 Central Rainbow area total force intensity aeromagnetic map (after Hriskevich, 1970).

11.3 Seismic interpretation

In the area there are six main reflectors (figure 11/6), and as a means of defining prospective structures it is necessary to interpret and map five of these. These reflectors have been named and can be described geologically as follows:

(1) Wabamun This reflector marks the top of the Wabamun unit where there is a major unconformity between Upper Devonian and Lower Mississippian rocks. A large velocity change exists between the shale and marl above the Wabamun unit (8–10 000ft/s) and the underlying limestone (15–16 000ft/s) providing reflection coefficients in the range 0.27 to 0.39. The Wabamun is a relatively homogeneous unit over 1000 ft thick for which a two-way interval time would be 0.125 s and thus the reflection is unaffected by succeeding reflections. It is accordingly an excellent reflector in all good data sections.

(2) Slave Point This reflector marks the top of the Middle Devonian. Again we have a limestone unit overlain by shale. The shale unit varies in lime content and has velocities in the range 11–15 000ft/s (the velocity contrasts for these and other formations are well illustrated by the sonic logs shown in figure 11/8(b)). With

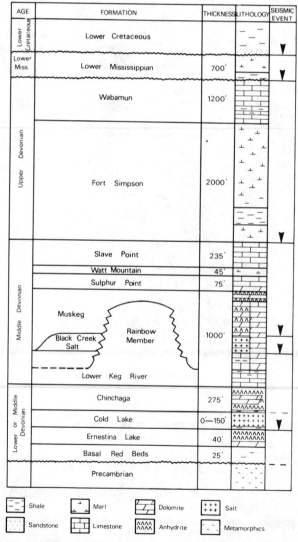

Figure 11/6 Stratigraphic section and main seismic reflectors.

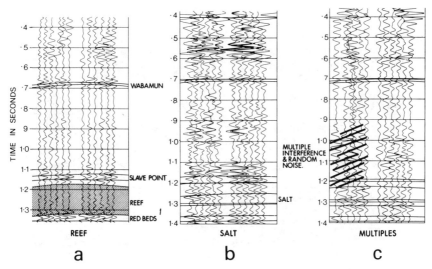

Figure 11/7 Rainbow area seismic field records, 1953 (after Hriskevich, 1970).

the Slave Point Limestone velocity of approximately 18 000ft/s, reflection coefficients vary from 0.14 to 0.31. On multi-fold seismic sections (figure 11/8(a)) the reflection is generally good but it is occasionally degraded by either thickness and velocity variations in the overlying shale units or by thinning of the Slave Point Limestone (maximum thickness 270 ft: equivalent to 0.030 s two-way time), in which case there is interference with reflections from underlying formations of varying thickness and lithology. Reflection quality is poor where multiple suppression is inadequate; first-order Wabamun multiples, and at times the higher-order Cretaceous–Mississippian unconformity multiples, often coincide with the Slave Point reflection. This situation bedevilled earlier single-fold record interpretation (see figure 11/7(c)) and provided one good reason for the application of multi-fold techniques.

(3) & (4) Top and Base Black Creek Salt The salt unit occurs over most of the basin between the reefs, post-dating them, and both top and bottom of this unit provide good reflectors. It lies above the Keg River Formation argillaceous/carbonate units and is overlain by anhydrite and dolomites. Utilising respective velocities of 18 000ft/s, 14 000ft/s and 17 500ft/s and densities of 2.80, 2.25 and 2.60g/cc, we can derive reflection coefficients of −0.23 and +0.18 for the top and bottom reflectors. The Black Creek Salt has a

maximum observed thickness of 271 ft and thins out completely mainly due to solution around the reefs. However, because of the high reflection coefficients, and with suitable broad-band recording and good quality data, it has been possible to pick peak to trough minima as low as 0.007 s (49 ft) for the top and bottom intervals.

(5) Base Cold Lake Salt This salt unit is remarkably consistent throughout the area and is a good reflector. It occurs about 500 ft below the base of the Rainbow Member and is generally about 70 ft above the Pre-Cambrian basement. Thickness varies from 0 to 180 ft. The base provides velocity contrasts from around 14 000 to 18 000ft/s and one suspects that where maximum amplitudes occur, tuning of the top and bottom is taking place. The event may be up to three cycles in length. In some illustrations it is also described as 'Red Beds'.

In 1964, with the poor quality seismic data then available the selection by Banff Oil* of a prospective location was both speculative and

*In 1965 the original participants in the exploration play were Banff Oil Ltd (operator), Aquitaine Company of Canada Ltd (a subsidiary of Société Nationale des Pétroles d'Aquitaine, France) and Socony Mobil Oil of Canada Ltd. The working interests were 5, 45 and 50% respectively. Banff Oil was subsequently merged with Aquitaine in 1970, which itself merged in 1981 with CDC Oil and Gas Ltd, to form Canterra Energy Ltd.

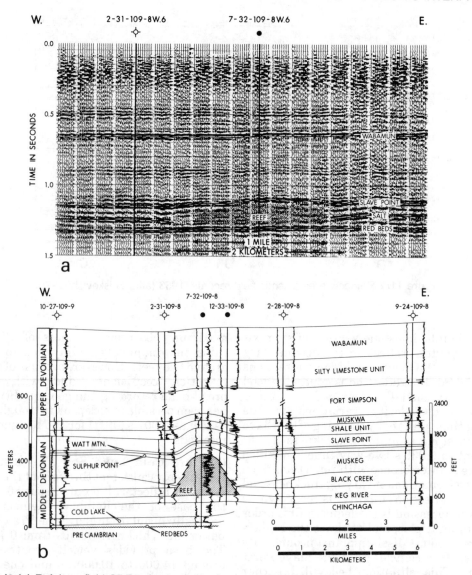

Figure 11/8 (a) Rainbow field CDP seismic section across 'A' discovery pool (after Hriskevich, 1970). (b) Rainbow field geological cross-section across discovery pool showing correlation between wells using gamma ray and sonic logs (after Hriskevich, 1970).

courageous. Only one well had been drilled in the area before; in 1954 the 10-27-109-9 W6M well encountered 245 ft of Black Creek Salt and its complete lack of hydrocarbon indications provided no incentive for a follow-up. According to Hriskevich (1970): 'The seismic data on which the discovery well at Rainbow was based were obtained during 1953–1955 before the use of common depth point techniques . . . seismic interpretation was made extremely difficult by the presence of a severe multiple problem. The

selection of the location involved very close co-ordination between geologists and geophysicists'. Single-fold records of the type referred to above are illustrated in figure 11/7. Figure 11/7(a) shows Wabamun, Slave Point and Cold Lake Salt events quite clearly, plus a supposed reef event. Figure 11/7(b) shows more or less the same section but with a salt event replacing the reef event. Figure 11/7(c) referred to earlier shows poorer data with severe noise including multiples in the Slave Point window. However, despite these difficulties in seismic interpretation, the well (Banff Aquitaine 7-32-109-8 W6M) was successful in discovering both oil and gas in commercial quantities. Results showed a total net pay of 480 ft oil and 291 ft gas and this acted as a great stimulus to obtain the best possible data in new exploration of the area. Common depth point stacking was by then the order of the day, but other improvements rapidly followed:

1. *Survey grid spacing*: Closer density grids were surveyed than previously with E–W lines at quarter-mile spacing and with occasional N–S tie lines over identified anomalies.*
2. *Shot patterns*: In general, single charges of up to 5 lb were used in holes up to 60 ft deep but in glacial drift and other difficult near-surface conditions improvements were achieved by shooting 3-hole patterns.
3. *Spread geometry*: Various spread geometries, split-spread or end-on, were used recording 24 channels with 110–115 ft geophone station spacing. Where necessary, offsets or gaps were utilised to remove ground roll. Shot-hole spacing was 110–150 ft for 12-fold coverage, 220–300 ft for 6-fold.
4. *Geophone arrays*: Generally nine to eighteen phones were used in a 160–170 ft string. Geophones were 14 Hz or less natural frequency.
5. *Recording systems*: The seismic 'boom' at Rainbow coincided with the introduction of

*In the provinces of Canada, much of the Government surveyed areas are divided into Township areas, 6 × 6 miles square, each Township being sub-divided into 36 sections, numbered from 1 in the SE corner alternatively left and right in rows to 36 in the NE corner. Each square mile section is further sub-divided into 16 lsds (legal sub-divisions) numbered in similar fashion from 1 in the SE to 16 in the NE. At the time of the Rainbow discovery, the Alberta Oil and Gas Conservation Board, the Provincial Regulatory Authority, limited oil production wells to within lsds 2, 4, 10 and 12, providing 160 acre per well drainage units, except in special cases. Accordingly it was good practice to shoot seismic lines through the centre of those lsds on a one-half mile spacing. With this type of programme layout, new wells could be staked at a specific shot-point location, as near as possible to the reef crest/maximum pay point. The coverage was subsequently doubled to one-quarter mile spacing to ensure proper delineation of reef anomalies that were sometimes less than three-quarters of a mile across. Thus it can be seen that planning of seismic survey grids in this area was to a large extent influenced by regulatory as well as geological and operational considerations.

the first digital field system, and many crews were equipped with this. Record filter passbands were generally in the range 14 to 65 or 72 Hz. Use of digital systems permitted the application of more advanced processing techniques leading to a significant enhancement of data quality.

6. *Processing*: Analogue processing included the application of weathering corrections, elevation statics, and, optionally, manual statics to artifically 'flatten' the Wabamun or, preferably, the Base Cold Lake Salt reflections. (The Wabamun refracted at about a half-mile offset and consequently only half the traces could be used.) The flattening enhanced the quality of the stacking and aided the interpreter in measuring isochrons on the sections. In addition to the same weathering and elevation statics corrections the use of digital techniques allowed computer centres to provide automatic static-picking routines which could either 'structure-smooth' or 'flatten' a key reflection.

The result of applying these techniques was that data quality was greatly improved and the first-order multiple problem, referred to earlier, was easily resolved.

As exploration drilling accelerated in pace with the expansion of seismic programmes, it became apparent that the reef reflection, as shown on the single-fold record of figure 11/7(a) was not consistent enough to be relied upon as a diagnostic of reef occurrence and could occasionally be confused with the salt reflections. Careful analysis of seismic data did however allow development of an indirect method of reef identification which was based on the following criteria:

1. Thickening of the Slave Point to Base Cold Lake Salt isochron of the order of 0.015+ seconds,
   and/or
2. thinning of the Wabamun to Slave Point isochron of the order of 0.015+ seconds, and
3. absence of the easily identified Black Creek Salt event between the Slave Point and Base Cold Lake Salt events.

Figure 11/8(a) illustrates the principles well. It is an E–W section through the discovery well,

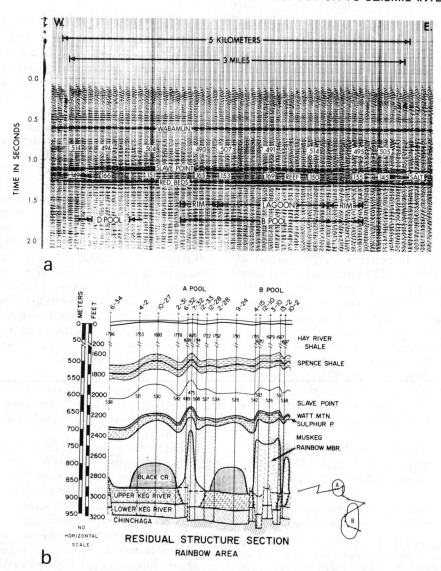

Figure 11/9 (a) Rainbow field seismic section over 'B' pool reef showing outer rim, inner lagoon and 'D' pool pinnacle reef. Note anomalously thin Slave Point to Base Cold Lake Salt (Red Beds) isochron on eastern rim related to faulting of type shown below (after Hriskevich, 1970).
(b) Condensed geological section over 'A' and 'B' pools. Note pronounced faulting in the lowermost section and compare with figure 11/9(a) whose E–W orientation is shown bottom right (after Barss *et al.*, 1970).

Banff Aquitaine Rainbow West 7-32-109-8 W6M. In this section the Red Beds/Base Cold Lake Salt reflection has been artificially 'flattened' (datumised) for maximising stacking efficiency. This and other tying sections on this anomaly comply with the criteria listed above; in particular it should be noted that the prominent Black Creek Salt Reflection disappears from the west one trace-panel east of the 2-31 well and can be seen again five trace-panels from the eastern end of the section (criterion 3). Confirmation of the salt presence to the west is provided by the 2-31-109-8 well which penetrated 80 ft (0.011s) of salt. The geological cross-section in figure 11/8(b) summarises the sub-surface geology. The structural anomaly at the Slave Point level overlies what is known as the 'A' pool; it covers an area of only

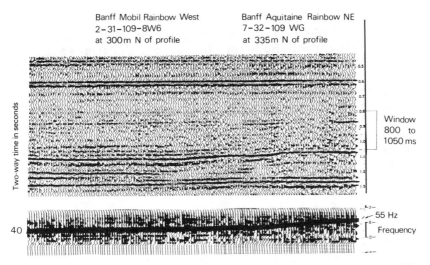

Figure 11/10 Tehze. Frequency analysis of seismic data over 800–1050 ms window. Note shift to high frequency content related to drape, pinching out and compaction resulting in thinning of beds over a reef buildup (Courtesy: Aquitaine and R. O. Lindseth).

985 acres or 1.5 square miles yet the reef is 756 ft high and has 686 gross feet of oil and gas pay and a reservoir of 164 MM bbls.

Reefs at Rainbow come in all shapes and sizes. Figure 11/9(a) is an E–W line over two reefs. The 'D' pool on the left of the section is only 400 acres or 5/8 square miles. The 'B' pool anomaly is large enough, 3640 acres or approximately five square miles, to be termed an atoll reef with outer rim and inner lagoon. All the interpretation criteria are met for both reef structures. However, there are some anomalous isochron values for the 'B' pool related to the rim and lagoonal developments. Basement faults with throws of up to 80 ft are noted in the area, and this would account for the thinning of around 0.009–0.010 s on the upthrown side of the fault seen in figure 11/9(b).

An interesting additional criterion which may be used to aid reef identification is by frequency analysis as described by Lindseth.* In figure 11/10 a frequency analysis of a fixed time-window is provided at the bottom of the seismic section (basically the same line as figure 11/8(a)). A marked frequency shift from 40 to 55 Hz can be noted from left to right over the Keg River Reef Structure. The reason for this frequency shift would appear to be that compaction of the sediments draped over the

*'Recent advances in digital processing', 1968, unpublished report.

reef structure produces thinning of beds, and consequently high frequency 'tuning' effects.

11.4 Interpretation and mapping of Rainbow Reefs

11.4.1 Tehze Pool

Reconnaissance seismic line T108-51 shown in figure 11/11 located a reef anomaly and additional lines T108-44 (figure 11/12) R9-32 (figure 11/13) and R9-33 (figure 11/14) confirmed its areal extent (see colour section and figure 11/15 for locations).

The Wabamun, Slave Point (2nd leg), top and bottom of the seismic events for Black Creek Salt, and the Base Cold Lake Salt are identified. T108-51 (figure 11/11) is, by processing, 'flattened' on the Base Cold Lake Salt event. It shows typical 'dead' reef character, 'roll-over/drape' on the Slave Point of around 0.030 s, and Black Creek Salt character disappearing at shot-point V16567 and reappearing at V16593. The east side of the reef appears to have steeper relief. T108-44 (figure 11/12) is flattened on the Wabamun. Roll-over on the Slave Point is 0.023 to 0.035 s west to east and the smooth Base Cold Lake Salt reflector shows a slight high below the reef. It is apparent that the reef has a steep western margin and that the Black Creek Salt character disappears at shot-point V17362, reappearing at V17395.5. Section R9-32 (figure 11/13) is flattened on the Base Cold Lake Salt. Roll-over is over 0.030 s and the steep relief to the north is accentuated by an apparent fault in the Slave Point at shot-point Z1168 which may be induced by both Salt removal and the flattening of the same fault at Base Cold Lake Salt level. The Black Creek Salt character disappears at that point and reappears at shot-point Z1298. Section R9-33 (figure 11/14) is also flattened on the Base Cold Lake Salt. It crosses the west

flank of the reef. Consequently, the roll-over is in the region of 0.02 s: however, the steep relief to the north is obvious and a fault on the Slave Point is apparent at shot-point Z1118. From a thickness of 0.021 s or 144 ft, at shot-point Z1128, the Black Creek Salt disappears at Z1117 and reappears at Z1091.

Values for the three main reflectors were picked from the four illustrated lines and posted on the two isochron maps (figures 11/15 and 11/16); the lines tie perfectly. Additional values for the surrounding lines were obtained from

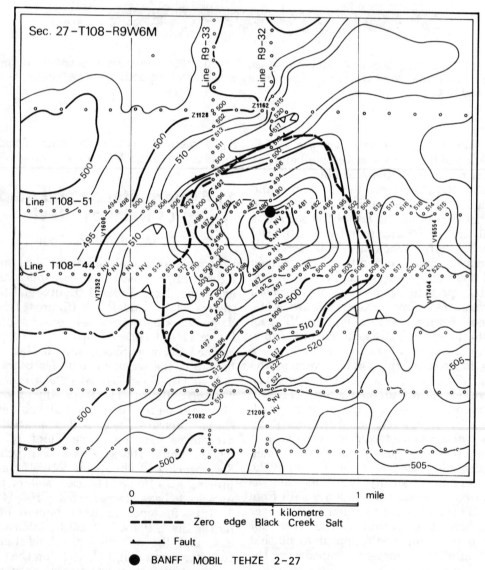

Figure 11/15 Tehze: map of two-way reflection time interval between Wabamun and Slave Point horizons (for clarity, isochron values are posted only for the seismic lines illustrated).

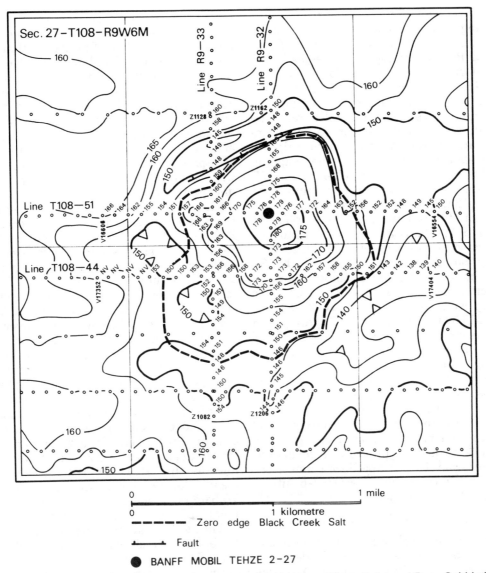

Figure 11/16 Tehze: map of two-way reflection time interval between Slave Point and Base Cold Lake Salt horizons (for clarity, isochron values are posted only for the seismic lines illustrated).

Aquitaine (but not posted for greater map clarity) and the maps contoured. Minimum closures of 0.028 s occur on the Wabamun to Slave Point isochron, and 0.025 s on the Slave Point to Base Cold Lake Salt isochron. As it is known that there is negligible structural relief on either the Wabamun or Base Cold Lake Salt, it can be assumed that the respective but complementary thinning and thickening reflects structure on the Slave Point, with perhaps some compaction-induced interval velocity increase effect in the upper isochron:

evidence from wells shows that the anomaly is almost entirely a drape effect. The third reef criterion, absence of Black Creek Salt, is indicated by a zero edge as identified by the disappearance of the salt seismic character. Outside this sub-circular edge, salt is identified everywhere and there is up to 0.025 s closure over salt anomalies on both isochrons. It can thus be seen that the loss of up to at least 270 ft (0.039 s) of salt by solutioning, and/or partial deposition, contributed the largest part of the drape effect seen on the maps and sections.

From off-reef well isopach comparisons, salt-solutioning has been identified during three periods, Late Middle Devonian, Upper Devonian and Post-Mississippian. It seems likely from uniformity of Wabamun–Base Cold Lake Salt values that salt removal must have occurred prior to the end of the Devonian. The greater minimum and maximum closures of the Wabamun-Slave Point isochron suggest compaction and/or compaction-induced (higher) velocity effects over the reef. The mapping highlights apparent steeper relief on the NE convex side of this pinnacle reef implying a prevailing NE wind. On the western, leeward side, a pair of tails, particularly the larger one trending SW, show an incipient tendency toward crescent atoll development. This reef is an outstanding anomaly: using the 0.150 s Slave Point to Base Cold Lake Salt contour, the maximum diameter is under a mile. The discovery well Banff Mobil Tehze 2-27-108-9 W6M had 958 ft of reef; 595 ft of hydrocarbon pay and reserves of 27 MM bbls oil and 13 BCF natural gas were proven in the reef and overlying carbonate units.

Figure 11/17 (see colour section) is a 1976 reprocessing of Line T108-51. Significant improvement of reflection quality is achieved through more refined normal moveout and static corrections plus deconvolution; the anomaly is further heightened by a condensed horizontal scale (12 traces per inch versus 8 per inch and a 1 trace gap every two shot-points on the original presentation).

11.4.2 'AA' Pool

Ever since the original discoveries at Rainbow, opinion has been divided as to how widespread the deposition of the Black Creek Salt was. One school of thought has it that the salt was deposited everywhere within the Middle Devonian Elk Point Basin that was not emergent, the other school maintains that the salt was not deposited much beyond the lower lying areas, as evidenced by the present day seismic mapping of the formation as shown in figure 11/21, and the isolated occurrences in the Zama area (McCamis and Griffith, 1967). Certainly salt solutioning has been demonstrated by isopach comparisons to take place throughout time from the Middle Devonian to the present (Barss *et al.*, 1970). The disappearance of the salt reflector, combined with the extreme thinning of the seismic isochrons immediately surrounding

reefs, such as at Tehze, confirm this. Either way, the non-deposition or widespread solutioning of the Black Creek Salt means that outside the limits of the Rainbow-Black Creek Salt Basin, around its rim and northwards in the Zama and Shekelie area, the criteria are reduced to two: i.e. mapping of the two isochrons, one above the reef and the other including it. Provided data quality is reasonable, reefs in these areas are still easy to interpret; anomalies can no longer be confused with salt thickenings, but may be with unexpected basement highs.

One such example of a reef occurring outside the limits of present day salt occurrences is the Rainbow 'AA' Pool (see figure 11/21 for location). Seismic sections north–south and east–west through the pool, and mapping of the two main isochrons are shown in figure 11/18(a-d) (colour section). Apart from the absence of a salt reflection in the off-reef section, this reef is similar to the Tehze Reef. However, direct comparison of the seismic sections and the mapping should not necessarily be made because of different sets of instruments and processing parameters, which could give rise to phase, and therefore time shifts. The Base Cold Lake Salt reflection has been picked one cycle, about 0.030 s, lower here, and although when compared with Tehze, the Wabamun to Slave Point isopach and isochron have thinned towards this area, the reduced reef section results in, at the crestal position, thicker Wabamun to Slave Point and thinner Slave Point to Base Cold Lake Salt isochrons.

The two lines are flattened at the base of Cold Lake Salt which both masks a slight low identified by the few wells which penetrated the complete sedimentary section, and also introduces artificial highs and lows at the relatively flat regional Wabamun. Although the lines flank the main crest between the 8 and 9-14 wells, the steepest drop-offs at the Slave Point level are in the north and west, confirming an apparent NW–SE trend which mapping of reef and near-reef structure and isopachs indicate. A bifurcation is provided by a lobe extending north toward the 16-14 well.

The thinning on the Wabamun to Slave Point isochron reflects the underlying structure. Not all of the wells penetrated either the base of the Cold Lake Salt or even the (Lower) Keg River platform, so neither can most of the wells be tied to the Slave Point to the Base of Cold Lake Salt isochron, nor can a complete comparative reef isopach map be derived. Those known reef well isopachs, and estimates for others, have been shown on the second map. On neither the Keg River subsea structure nor the reef isopach is a north lobe established. More likely there is a small embayment immediately NW of the 9-14 well. If this is so, the cardiod type shape of this small reef would suggest main wind and current directions from the SE with the NW embayment being to the leeward. This shape would be similar to that of the Tehze reef and also the larger late Middle Devonian, Judy Creek reef, as described by Klovan and May (1974).

The Rainbow 'AA' Pool although covering a greater area than the Tehze Pool has lesser vertical relief, totalling about 800 ft (700 ft above platform). It is filled almost to the spill point, that is, down to the level of the top of the inter-reef Upper Keg River, 240 ft above the reef platform. This equates with the 0.170 s contour, figure 11/18(d). The gross pay thickness is just over 500 ft, the net (porosity) pay ranging up to 365 ft at the crestal position. The productive pay area covers 1 square mile, providing reserves of 78 MM bbls and 22 BCF of gas, with a primary and enhanced combined recovery factor of 78%. One of the wells initially tested at over 3300 bpd of oil, and to date, over 38 MM bbls and 34 BCF of gas have been produced.

11.5 Regional summary

The spate of discoveries and further government lease sales in the Rainbow Lake region at one time reached a level such that development plans required geophysicists to double as explorationists and productionists. In developing well prognoses for reservoir delineation and production, it soon became evident, on a localised basis, that there were simple empirical relationships between the Slave Point drape in time and depth, and the amount of vertical relief on the Rainbow Member. Careful analysis

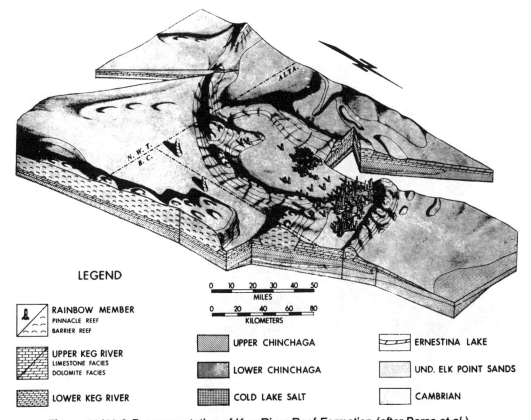

LEGEND

- RAINBOW MEMBER
  PINNACLE REEF
  BARRIER REEF
- UPPER KEG RIVER
  LIMESTONE FACIES
  DOLOMITE FACIES
- LOWER KEG RIVER

0 10 20 30 40 50
MILES
0   20   40   60   80
KILOMETERS

- UPPER CHINCHAGA
- LOWER CHINCHAGA
- COLD LAKE SALT

- ERNESTINA LAKE
- UND. ELK POINT SANDS
- CAMBRIAN

Figure 11/19 3-D representation of Keg River Reef Formation (after Barss *et al.*).

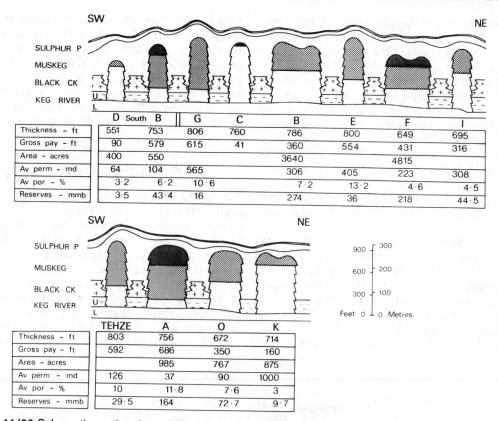

	D South	B	G	C	B	E	F	I
Thickness – ft	551	753	806	760	786	800	649	695
Gross pay – ft	90	579	615	41	360	554	431	316
Area – acres	400	550			3640		4815	
Av perm – md	64	104	565		306	405	223	308
Av por – %	3·2	6·2	10·6		7·2	13·2	4·6	4·5
Reserves – mmb	3·5	43·4	16		274	36	218	44·5

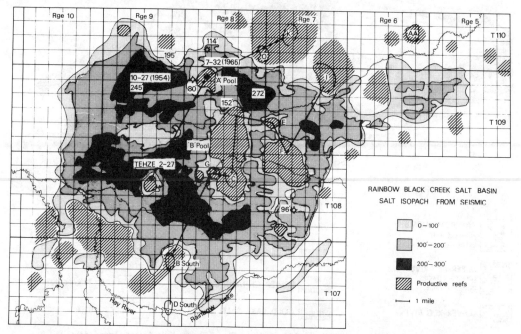

	TEHZE	A	O	K
Thickness – ft	803	756	672	714
Gross pay – ft	592	686	350	160
Area – acres		985	767	875
Av perm – md	126	37	90	1000
Av por – %	10	11·8	7·6	3
Reserves – mmb	29·5	164	72·7	9·7

Figure 11/20 Schematic section through Rainbow area reefs showing comparative data on reservoir characteristics. Lines of section and locations shown in figure 11/21. Gas caps in heavy shading, oil pay in intermediate shading.

Figure 11/21 Salt isopach map of the Rainbow Black Creek Salt Basin (after Barclay, in Mobil internal report, 1967). Lines of section shown for figure 11/20.

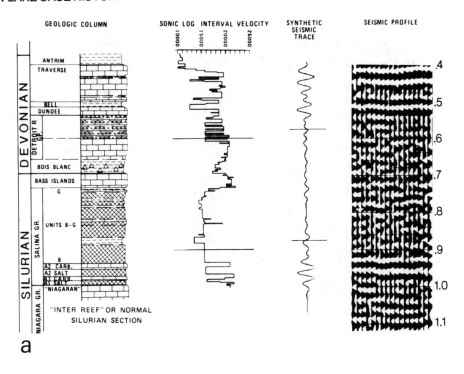

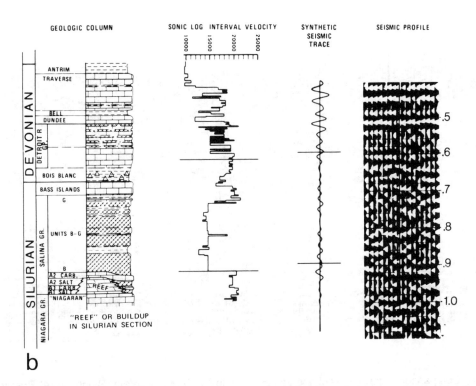

Figure 11/22 Stratigraphic sections in North Michigan compared with sonic logs, synthetic seismic traces and seismic profiles. (a) Inter-reef situation; note thin bed tuning due to alternations of salt and carbonates in lower Salina zone. (b) Reef situation: note lack of reflectivity in lower Salina zone due to loss of salt beds (after Caughlin*et al.*, 1976).

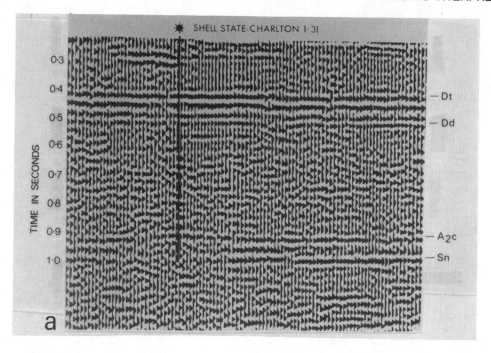

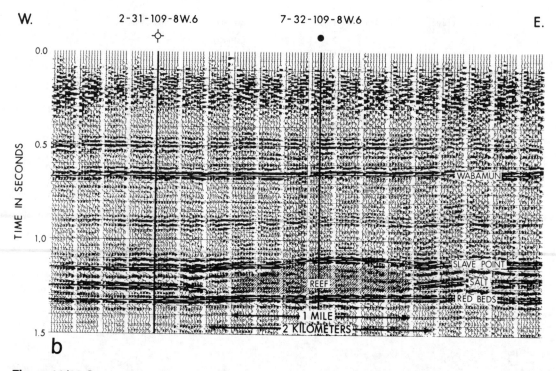

Figure 11/23 Comparison of seismic sections across reefs in North Michigan and Rainbow. (a) Reef in North Michigan: note loss of reflectivity in reef zone beneath well 1–31. (b) Section across Rainbow field discovery pool: note similar loss of reflectivity beneath well 7-32-109-8 W6. (Section (a) after Caughlin *et al.*, 1976; section (b) after Hriskevich, 1970).

of these relationships eventually resulted in prognosis accuracies in the order of 50 ft or less, or within 1% of target drill-depth.

With methodical seismic mapping and tying in of well data, it was eventually possible to prepare maps on the Keg River Reef/Rainbow Member itself. A computer derived isometric projection of a reef map was shown previously in Chapter 6, figure 6/9. A simplified three-dimensional regional view of these reefs is shown in figure 11/19 and in figure 11/20 a schematic section is shown including data on the reservoir characteristics. As mentioned earlier, it was seismically possible to map the top and bottom of the Black Creek Salt (green and blue reflectors, respectively, in figures 11/11 through 11/14), and, given the homogeneous half-velocity of 7000 ft/s, derive the isopach. Using 100 ft salt isopach contours, a seismic-derived map of the complete Rainbow Black Creek Salt Basin is shown in figure 11/21. These last two figures are part of the end product of analysis of a vast amount of data, interpreted by many individuals and a number of companies. The history of exploration and development of the Rainbow area has depended on two main factors. First, the initial successful interpretation of old data which led to the discovery of oil and gas by Banff Aquitaine and second, the fact that this coincided with a great technical advance in seismic exploration methods, common depth point stacking and digital recording. This coincidence led to an immediate and major expansion in Alberta exploration effort, both seismic and drilling. The end result, coming in only five years, was that all the reef reservoirs shown in figure 11/19 were discovered with a 50% drilling success ratio, providing the establishment of reserves of 2.2 billion barrels of oil and 1.5 trillion cubic feet of natural gas in the Rainbow and associated Zama North area.

11.6 Reef interpretation comparison, Rainbow and Michigan

In closing it is interesting to note the similarity of the interpretation criteria used in a more recent successful reef play in Northern Michigan in the USA (Caughlin *et al.*, 1976). Targets are in the Silurian Niagaran reefs. As shown on the stratigraphic columns of figure 11/22, the off-reef section is characterised by the presence of salt. Also, as at Rainbow, a prominent shale/limestone reflection ('Dundee') is used as a datum marker to demonstrate isochron thinning/drape over the reef. The similarities of the interpretation methods are modified due to the thinner salt and carbonate beds in the Niagaran play and are summarised by the comparison table. Seismic sections from Rainbow and Northern Michigan are shown in figure 11/23. Despite the difference in data quality, the similarity is strong: note

Reef presence criteria – comparison between Rainbow and Michigan

	Rainbow area	*Michigan area*
Isochronal Thinning:	Wabamun to Slave Point	Dundee to A2 carbonate
Isochronal Thickening:	Slave Point to Base Cold Lake Salt	A2 carbonate to Niagaran (generally the Niagaran itself is not seen because of the lack of reflectivity: this itself is an alternative criterion).
Reef Zone Character Change:	Disappearance of Black Creek Salt (Main Salt Basin)	Weakening or disappearance of A2 carbonate (the thin salt and carbonate beds encourage a strong 'tuned' event when present: the absence of the salt induces 'detuning').

especially the abrupt change of character going from off-reef to on-reef in both cases.

Acknowledgements

Appreciation is extended to Aquitaine Company of Canada Ltd and Mobil Oil Canada Ltd for provision of previously unpublished seismic sections and maps which were revised for use herein, and to John A. MacDonald, Mobil Oil Canada Ltd, for helpful advice.

References

D. L. Barss, A. B. Copland and W. D. Ritchie, Geology of the Middle Devonian reefs, Rainbow area, Alberta, Canada in: Halbouty, M. T. (ed) *Geology of the Giant Petroleum Fields*. Mem. AAPG, No. 14 (1970), pp. 19–49.

W. G. Caughlin, F. J. Lucia and N. L. McIver, The detection and development of Silurian reefs in Northern Michigan. *Geophysics*, **41** no. 4 (1976), pp. 646–658.

M. E. Hriskevich, Middle Devonian reef production, Rainbow area, Alberta, Canada. *Bull. AAPG*, **54**, no. 12 (1970), pp. 2260–2281.

J. E. Klovan and P. May, The development of Western Canadian Devonian reefs and comparison with Holocene analogues. *Bull. AAPG*, **58** no. 5 (1974), pp. 787–799.

J. G. McCamis and L. S. Griffith, Middle Devonian facies relationships, Zama, Alberta. *Bull. CPG*, **15** no. 4 (1967), pp. 434–467.

Chapter 12

Geophysical Case Study of the Kingfish Oilfield of Southeastern Australia

12.1 Introduction

The Kingfish Oilfield is located in the Gippsland Basin of southeastern Australia. It lies approximately 50 miles offshore in 250 feet of water and is currently Australia's largest producing oilfield with an estimated ultimate recovery of approximately one billion barrels (figure 12/1).

The oil is trapped within a large east–west trending Late Eocene palaeotopographic high. The top of the reservoir is an unconformity surface which is sealed by fine-grained marine clastics of Oligocene age. The reservoir itself consists of Lower Eocene and Upper Palaeocene deltaic and marginal marine sandstones of the Latrobe Group.

Large lateral variations in the average velocity to the top of the reservoir have complicated the seismic mapping of this field. This chapter will discuss the origin of the velocity problems, the magnitude of their effects, and the practical techniques which have been used to obtain velocities for reliable time–depth conversions.

12.2 Regional stratigraphy of the Gippsland Basin

The regional stratigraphy of the Gippsland Basin is summarised in a schematic cross-section which runs from north to south across the Basin (figure 12/2). All the commercial hydrocarbons discovered in the Basin occur in sands within the Upper Cretaceous to Eocene Latrobe Group which is a thick sequence of predominantly non-marine fluvial and deltaic coarse clastics. The overlying Lakes Entrance formation is comprised of calcareous mudstones and shales which were deposited during a widespread marine transgression of the Basin during the Oligocene. Marine sedimentation continued through the Miocene to Recent with the deposition of skeletal limestones, calcarenites and marls which are referred to as the Gippsland Formation.

Sea-level changes and structural movements during Miocene time triggered the development of a complex system of large scale submarine canyons. The sedimentary fill within these channels gives rise to the principal velocity problems experienced in the Basin. At the head of each channel the fill is a mixture of coarse skeletal fragments, but in the middle and distal portions of the channels the infilling sediment is a dense micritic limestone. The interval velocity of this micritic limestone may be as high as 15 000 ft/s, compared to the 8000 to 9000 ft/s interval velocity of the surrounding shales. This large velocity contrast results in significant lateral variations in the average velocity to the underlying seismic markers and these variations severely distort the seismic reflection times to any underlying structure.

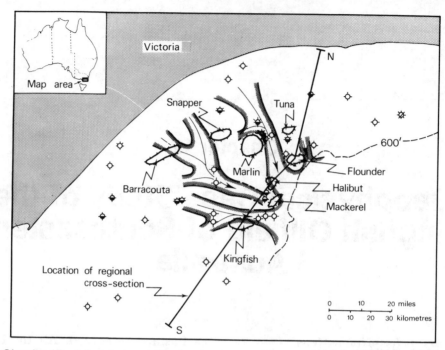

Figure 12/1 Gippsland Basin location map showing the major oil and gas fields and the Miocene Channel systems.

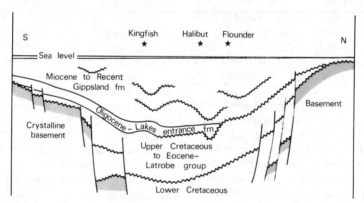

Figure 12/2 Diagrammatic cross-section of the Gippsland Basin along the line S–N of figure 12/1.

Figure 12/1 shows the distribution of the major Miocene channels. They originate in the north-western part of the Basin and trend to the south-east where they coalesce to the east of the Kingfish-Mackerel area. It should be noted that these channels pass across most of the major oil and gas fields which are therefore affected to some degree by the resulting velocity problems.

12.3 Historical résumé

The first seismic survey in the offshore portion of the Gippsland Basin was conducted in 1962 and comprised single-fold, split-spread, analogue reconnaissance data. The Kingfish structure was recognised on two lines from this survey.

Prior to the drilling of Kingfish-1, two subsequent seismic surveys provided eight more lines of seismic control over the structure. These lines consisted of six-fold, split-spread, analogue data and revealed a large channel complex in the Miocene section above the Kingfish structure. A similar Miocene channel system was known to cause lateral velocity variations in the vicinity of the Marlin field, 25

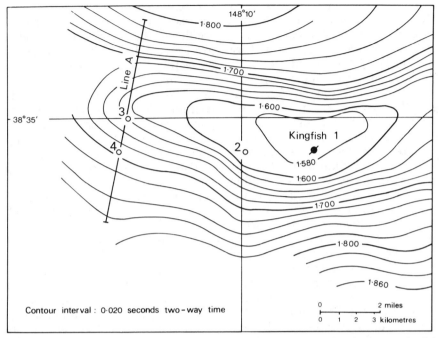

Figure 12/3 Two-way reflection time map of the top of the Latrobe Group at Kingfish.

miles to the north. It was considered, however, that any velocity variation which might be present in the Kingfish area would not be sufficient to cause drastic modification of a structure which was mapped as having 100 ms of closure over 50 square miles. In addition, the accuracy of velocities derived from $T-\Delta T$ analysis of split-spread data from a relatively short cable was suspect and the use of such velocities in depth conversion was considered just as likely to produce depth errors as to correct them.

The first well, Kingfish-1, was therefore drilled on the time crest (figure 12/3). The well intersected 114 ft of oil trapped immediately beneath the top of the Latrobe Group and the velocity information from the well indicated an oil-water contact at a reflection time of 1.590 s. Relating this oil–water contact to the top of the Latrobe time map suggested a low percentage trap fill and an estimated 200 million barrels of oil-in-place. This assessment was disappointing considering the size of the closure evident from the seismic reflection data.

However, the interval velocities derived from the sonic log showed a more severe velocity problem than had been anticipated (figure 12/4). The channel sediments had an interval

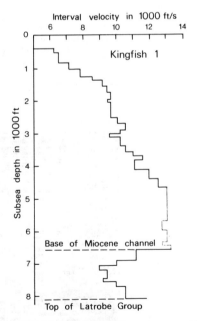

Figure 12/4 Interval velocities measured in the post-Latrobe section of the Kingfish-1 well.

velocity of 13 000 ft/s which was significantly faster than anticipated in the pre-drill analysis. The interval velocity of the underlying and surrounding Lakes Entrance shales was 9000 ft/s or 4000 ft/s slower than the channel fill. It was realised that the reduction in travel time in those areas where there was a thick high velocity channel fill would cause a considerable distortion of the underlying time structure at the top of the Latrobe Group. It was also apparent that a more comprehensive seismic programme with the emphasis on obtaining velocity data was warranted in order to more accurately delineate the Kingfish structure prior to any additional drilling. Consequently, ten digitally recorded six-fold CDP seismic lines were shot across the structure. As suspected, $T-\Delta T$ velocity analysis of this data confirmed the existence of a severe velocity gradient in the area and this gradient conformed closely to the form and trend of the base of the Miocene channel system.

From the velocity data, it became apparent that the Kingfish-1 well had been drilled on the flank of the field only one half mile from the eastern limit of the oil–water contact. In fact, due to the severity of the velocity gradient across the area, the revised crest of the structure was interpreted, at this time, to be two miles west of the time crest.

To verify the revised structural interpretation of the reservoir, Kingfish-2 was drilled in late 1967 at a location 11 000 ft to the west of the Kingfish-1 well. This stepout well, which intersected the top of the Latrobe Group 118 ft shallower than Kingfish-1, penetrated 232 ft of gross oil sand above the same oil–water contact as seen in the discovery well.

A further stepout well was considered necessary to delineate both the structural configuration and stratigraphy of the western portion of the reservoir and in early 1968 Kingfish-3 was drilled at a location some six miles to the west of Kingfish-1.

Following the drilling of Kingfish-3 as a successful confirmation well, it was apparent that the three wells tied to the available seismic grid were adequate to provide reliable definition of an oil reservoir having reserves of approximately one billion barrels. In May 1968, the Kingfish oilfield was declared commercial.

The development of the Kingfish field was based on two 21 conductor platforms. The 'A' platform was located approximately 7000 ft west of Kingfish-2 and the 'B' platform approximately 5500 ft east of Kingfish-2 (see figure 12/7). Development drilling began in March 1970 and continued until October 1971. Both platforms were brought on stream during 1972.

Subsequent to the development, three additional stepout wells have been drilled. Kingfish-4 was drilled near Kingfish-3 in 1973

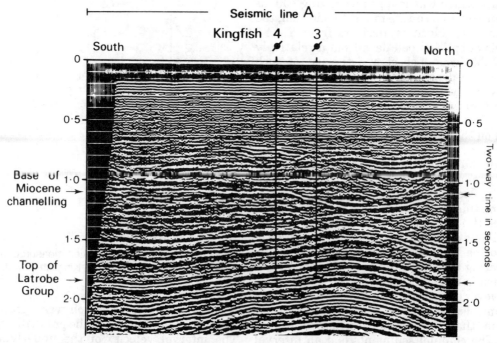

Figure 12/5 Seismic line A (see figure 12/3) across the Kingfish Field.

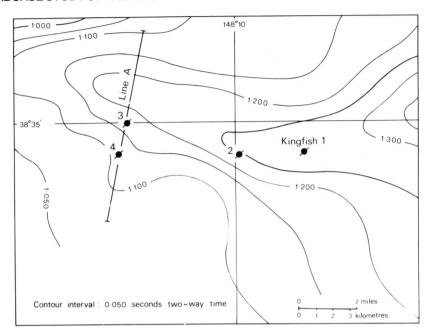

Figure 12/6 Seismic reflection two-way time to the base of the Miocene Channel system at Kingfish.

to clarify the stratigraphy of the western end of the field. This well also confirmed the structural picture of this area.

In late 1974 and early 1975, Kingfish-5 and Kingfish-6 were drilled to the east of Kingfish-1 in an attempt to define the eastern extremity of the oil. Both of these wells failed to encounter hydrocarbons and discredited the detailed structural interpretation of this area of the field. The implication of these East Kingfish results will be discussed later.

12.4 General geometry of the Kingfish Field

As discussed above, the relationship between the time structure at the top of the Latrobe Group and the apparent structure seen on the seismic sections is complicated by variations in the average velocity caused by the high interval velocity of the Miocene channel fill. This effect is illustrated on the seismic line A (figure 12/5) which crosses the western part of the Kingfish field. On this section, the top of the Latrobe time crest of the field can be seen immediately below the axis of the Miocene channel and about one mile north of the Kingfish-3 location. The well control on this line places the true crest two miles south of the time crest, near the Kingfish-4 well. The time crest on this seismic section is in fact about 180 ft downdip from the

true structural crest. Figure 12/6 is a time structure map on the base of the Miocene channel system over Kingfish and shows an east–west axis passing north of the Kingfish-2 and Kingfish-3 wells. This map shows that the relief of the channel in the Kingfish area is about 0.200 s two-way time, or approximately 1200 ft.

The relationship between the axis of the Latrobe time structure and the axis of the Miocene channelling is obvious. The time structure of the top of the Latrobe Group has been severely distorted by the velocity effects of the channel. Correction for the velocity gradient induced by the channel results in the structural interpretation shown in figure 12/7. This map shows that Kingfish-1, in drilling the time crest, was actually 160 ft downdip and five miles east of the true crest of the structure. In fact, the Kingfish-1 well came rather close to completely missing a billion barrel oilfield!

In figure 12/8, the oil–water contact from the Latrobe structure map has been superimposed on the time map and this shows the relationship between the time picture and true structure more clearly. Of particular interest is the extent of the oil–water contact in the southwestern part of the field. If seismic reflection times had been converted to depth using only the Kingfish-1 time-depth information, then the extreme south-western sector of the field would

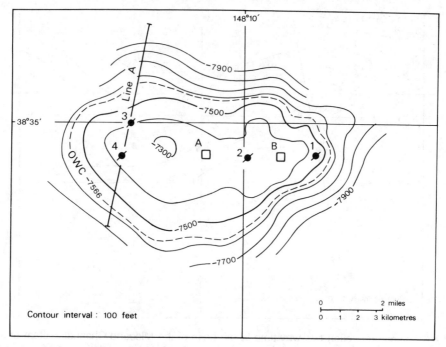

Figure 12/7 Structure map on top of the Latrobe Group at Kingfish. Depth contours in feet.

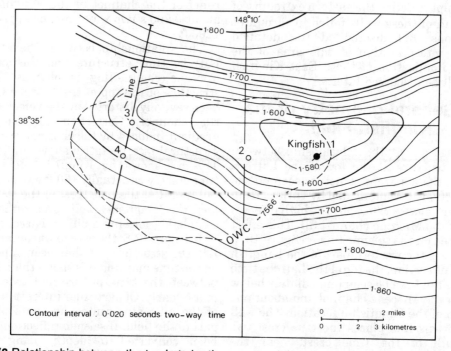

Figure 12/8 Relationship between the top Latrobe time map and the oil–water contact at the Kingfish Field.

have been interpreted as almost 1000 ft below the contact.

The Miocene channel system, with its variable high velocity fill, extends over a large part of the Gippsland Basin, as shown in figure 12/1, and many other areas have the same general velocity problem as Kingfish. Exploration experience in the Basin has shown that some apparent structures seen on seismic reflection data are entirely velocity induced, whereas others, like Kingfish, are highly distorted by anomalous velocity gradients. Some real structures are not apparent at all on the reflection data. Consequently, every prospect evaluation involves the accurate mapping of average velocities to the top of the Latrobe Group in order to obtain a usable structural picture. As there is not enough well control to do this mapping directly, it is necessary to rely upon indirect velocity determinations from common depth point seismic data to provide the bulk of the information.

12.5 Velocity analysis

12.5.1 Normal moveout velocities and scattergrams

In most offshore areas it is now a relatively simple matter to obtain good quality normal moveout velocities from the routine processing of common depth point seismic data. The general technique used by the processing contractors is to apply an assumed normal moveout to a common depth point gather and then to measure the coherency (a measure of correlation between traces). The normal moveout velocities derived in this manner would only correspond to true average velocities in the unlikely circumstance that the earth was homogeneous and had a constant velocity down to the particular reflection of interest, in which case the seismic rays would have travelled along straight ray paths. Although the moveout velocities are precisely what is required to correct for the effects of moveout during processing, these functions are not of particular use, in the raw state, for converting seismic reflection times to geological depths. In order to obtain usable velocities for this purpose, it is necessary to adjust the moveout velocities for the effects of refraction of the seismic rays as they pass through the layered earth. The remainder of this chapter will be directed towards this problem.

Figure 12/9 shows the type of seismic velocity data that is currently being used in the Gippsland Basin. This 'scattergram' is an output from the '700 package' velocity program provided by Geophysical Service International (GSI). In the Gippsland Basin such scattergrams are routinely generated every 600 m on every seismic line giving a high density of velocity control. The velocity program scans for velocities at alternate depth points (every 50 m) along each seismic line and the scattergram output is the averaged velocities from groups of 12 such depth points. A search for continuity in time, amplitude, dip and ΔT of events helps to ensure that the velocities that are averaged come from individual continuous reflections. This statistical approach reduced the effect of the random scatter which is prevalent in single depth point velocity determinations.

The scattergram itself is basically a statistical plot of normal moveout velocity against seismic reflection time. The different symbols in the plot represent a ranking of the data points. It should be noted that the prime ranking hierarchy of these points is not based on correlation amplitude but rather on the number of depth points over which the particular seismic event or 'segment' can be followed by the computer. The circles indicate the highest amplitude correlations within each 100 ms gate.

The seismic display on the left-hand side of the scattergram is a gather from a central depth point. This gather has been corrected for normal moveout by the application of a previously chosen reference velocity function which is shown plotted on the scattergram. The reference display is a useful aid in the interpretation of the scattergram.

The interpretation of the scattergram involves the drawing of a time–velocity curve which is the best fit to those data points which are considered to be reliable. Obviously, data from multiples and reverberations, etc. may appear on the scattergram, and these need to be recognised and discounted in the interpretation. The normal moveout velocity to any particular reflection can be read from the interpreted scattergram and these raw velocities can then be used to determine the average velocities needed to convert seismic reflection times to depths. Specifically, in the Gippsland Basin, it is necessary to determine the average velocity of the top of the Latrobe Group which is the principal objective horizon.

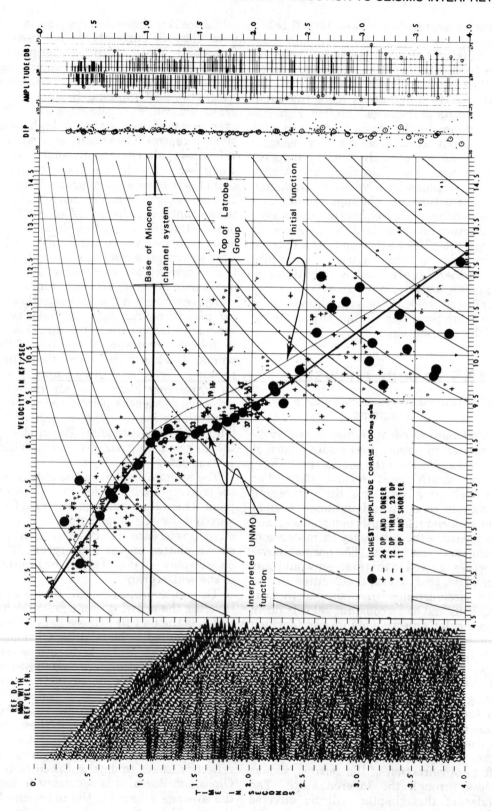

Figure 12/9 Typical scattergram in the Kingfish area.

12.5.2 Conversion factor approach to depth conversion

Prior to 1975, the paucity of reliable velocity picks above the base of the Miocene channelling precluded the use of the Dix Approximation to derive average velocities to the top of the Latrobe Group. Instead, a method for depth conversion was adopted which merely involved the application of a percentage correction to the smoothed moveout velocities to produce 'true' average velocities. The smoothing or editing of the moveout data and the selection of appropriate 'conversion factors' are discussed below.

In situations where the sedimentary sections consist of horizontal layers, it is found that reflections from common depth point gathers are approximately hyperbolae on time versus offset plots. In these cases, moveout velocities have a relatively simple relationship to the true average velocities. However, the channelled areas of the Gippsland Basin represent a situation which departs significantly from the horizontally layered case and the channels can produce large variations in normal moveout velocity which are not related to average velocity changes. The origin of these 'geometric effects' is illustrated in figure 12/10(a). This illustration models a flat Latrobe surface at 5000 ft with an overlying Miocene channel filled with sediments having an interval velocity of 12 000 ft/s. Seismic ray paths have been traced through this model to generate six common depth point 'gathers' and two of these are shown. It can be seen that each of the ray paths contributing to a common depth point set will traverse a different thickness of high velocity material. The moveout curve is significantly non-hyperbolic with the outer rays arriving either too soon or too late depending on the position of the gather relative to the channel geometry. When a distorted moveout curve is forced by a velocity analysis programme to fit a hyperbola, the implied moveout velocity for the central shotpoint of the gather will be either faster or slower than the corresponding moveout velocity for a 'layer-cake' model. It should be noted that such distorted moveout velocities are quite suitable for moveout correction prior to the stacking of common depth point data, but if the distorted velocities were to be used as the basis for depth

conversion, an erroneous picture would result. Figure 12/10(b) illustrates the normal velocity profile for the top of the Latrobe Group that results from this channel distortion. The dashed rms (zero-offset) velocity profile has also been plotted to demonstrate how the average velocity to the Latrobe surface should vary. The diagram shows that the moveout velocity is considerably faster than the rms velocity in the vicinity of the channel edges and is slower beneath the channel axis.

Figure 12/10(c) shows the reflection time profile of the Latrobe surface which is flat in depth but distorted to give an apparent time structure due to the overlying channel. Depth conversion using raw moveout velocities results in the depth profile of the Latrobe surface shown in figure 12/10(d). It should be noted that there is now an apparent structure in depth. It is imperative to avoid drilling such an anomaly in an expensive offshore exploration venture and so it is essential to correct for the distorting effect of the channel geometry before attempting to use moveout derived velocities for depth conversion.

As a practical example consider the seismic section A across the Kingfish Field referred to earlier in the text (figure 12/5). Figure 12/11 shows a normal moveout velocity profile for the top of the Latrobe Group as picked from the scattergrams along this line. The distorting effects of the channel geometry which we saw in the modelled situation are also expected here and they have been allowed for in the smooth line interpretation which shows the normal moveout velocities being too fast at the channel edge and too slow at the channel axis.

The smoothing of the raw moveout profiles to remove the estimated geometric effects of the channels is a key step in the velocity interpretation. The step is highly interpretive and requires considerable experience. Full use must be made at this stage of all local knowledge on velocity gradients, velocity contrasts, data quality and channel base and fill geometry.

Once the profiles have been smoothed the intersections are checked for mis-ties and adjustments are made as necessary. When making these adjustments it should be kept in mind that seismic lines which are more nearly parallel to channel axes will suffer less from the distorting effects of channel irregularities. These lines are used as the key to eliminating or

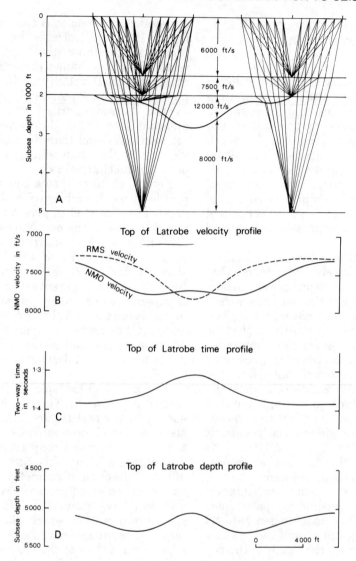

Figure 12/10 A — Ray tracing model used to investigate the distortion of normal moveout hyperbolae beneath channels. B — Top of Latrobe velocity profile beneath the high velocity channel fill. C — Top of Latrobe time profile. D — Top of Latrobe 'depth' profile derived from the distorted moveout velocities.

minimising the effects of ray path distortion on profiles which cut the channels at higher angles.

A map of smoothed moveout velocities is constructed after the smoothed profiles have been tied. Ideally, at any point on this map, the contours represent the moveout velocity which would be obtained for the top of the Latrobe Group at that location if there were 'layer-cake' geology with no channel problems. If done

successfully the smoothing process will have removed the distortion due to the outer rays of the common depth point gathers passing through a significantly different section to the more vertical rays.

The smoothed moveout velocity map must now be converted to an average velocity map by applying a percentage conversion factor. The conversion factors are determined at the well control by comparing the smoothed moveout

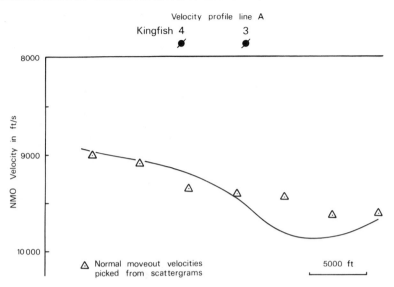

Figure 12/11 Profile showing normal moveout data points and a smooth interpretation in which the distortions due to the geometry of the base of the channel have been removed.

velocity directly with the velocity obtained from the well survey. There is a danger, however, that the smoothing of the velocity profiles is in error and this would obviously lead to an erroneous conversion factor at that well. To guard against this problem, a ray path modelling program which uses the sonic log data is used to predict the normal moveout velocity and the conversion factor at each well. This computed value of the conversion factor is used to verify the smoothing of the profiles.

Away from the well control the prediction of conversion factors is more difficult and relies on an appreciation of the factors which cause the difference between the normal moveout velocities and the true velocities. Raypath model studies at the wells indicate that only three parameters significantly influence conversion factors. These are:

1. The effective offset (maximum source–detector separation after muting).;
2. the depth to the top of the Latrobe Group; and
3. the interval velocity contrasts within the post-Latrobe section.

In the Gippsland Basin it is found that both the effective offset and the depth to the top of the Latrobe Group do not change enough over the area of a prospect to have a significant effect and this leaves the third parameter as the controlling influence in local conversion factor variations.

Experience shows that there is a bigger difference between the normal moveout velocities and the average velocities when the seismic rays travel through a thicker section of high velocity channel fill sediments. The reason for this is related to the greater refraction of the ray paths in the channelled areas, giving a bigger difference between the straight-line and the minimum travel-time paths. Consequently, channel base maps and channel fill isopachs are used to guide the trends of the conversion factor map between the well control. The conversion factor will decrease towards the channel axes where the average velocity is a smaller percentage of the moveout velocity. This effect is illustrated in figure 12/12 where conversion factors have been determined at eight wells in the Kingfish area and verified by raypath modelling. Comparison of this conversion factor map with the channel base map in figure 12/6 shows the predicted decrease in conversion factors towards the channel axis and also the anticipated similarity in form between the two maps.

The conversion factor map shows a total variation of about 1.5% across the Kingfish Field. Although this represents only about a 1.3% variation in the average velocity it corresponds to a difference of about 100 ft in depth. Such a change across the Kingfish Field

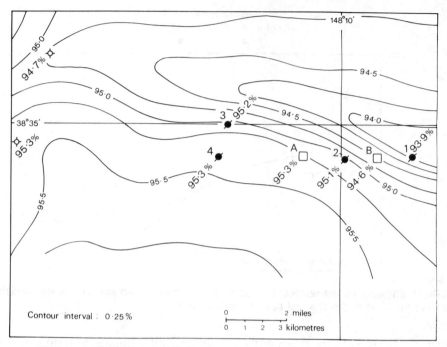

Figure 12/12 Conversion factor map for the top of the Latrobe Group.

could alter the estimated reserves by more than 100 million barrels and so it can be seen that the conversion factor variations can be highly significant.

At this stage in the interpretation procedure the conversion factor map and the smoothed moveout velocity map are combined to give an average velocity map to the top of the Latrobe Group (figure 12/13). This map is then used to produce a depth map to the top of the Latrobe Group from the seismic reflection times.

The velocity analysis technique which has been discussed in detail above was sufficient to define the size and shape of the main Kingfish Field for development purposes.

12.5.3 Recognition of additional velocity problems at East Kingfish

Subsequently, on the basis of this velocity interpretation technique and additional seismic data, the Kingfish-5 well was drilled to test errors in predicting seismic velocities beneath the deepest and most complex Miocene channelling found over the Kingfish structure (figure 12/16). The top of the reservoir was encountered 221 ft low to prediction and 35 ft below the field oil–water contact. Although this well did fail to encounter hydrocarbons, the revised structure map, based on this new well control and an updated velocity picture, showed an extension of the field to the north-east, although not as significant as predicted in the pre-drill analysis.

Later, the Kingfish-6 well was drilled one mile SSW of the Kingfish-5 well to further test the eastern extension of the field. This well also failed to encounter hydrocarbons and intersected the top of the Latrobe reservoir 63 ft deeper than predicted. The depth error in both of these wells was due to an error in predicting average velocity beneath the complex channelling of the Miocene section.

In the light of these disappointing well results it was obvious that a more accurate method of converting seismic reflection times to depths was needed in order to better define both the reserves of the field and the geometry of the flanks. The opportunity for this came in 1974–75 when a seismic survey was shot which gave considerably higher resolution seismic data and, in particular, better quality scattergrams than had previously been available. This improvement was achieved, in part, by placing strict control on several factors affecting normal moveout determination. The gun to cable offset was carefully monitored and

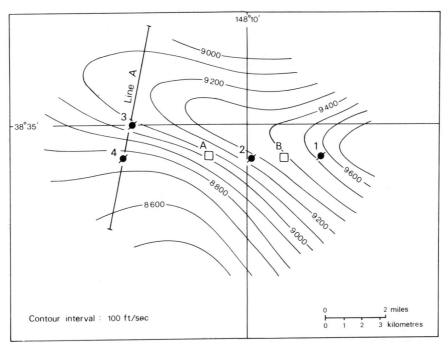

Figure 12/13 Average velocity map to the top of the Latrobe Group at Kingfish.

stringent signal/noise ratio specifications were prescribed. These factors plus the use of a 2 ms sample rate and a relatively short group interval on the cable ensured good definition of the moveout curves.

Additional factors contributing to the improved data quality were the strict control of processing parameters and the monitoring of the data at several stages through the processing stream. Random errors in normal moveout velocity analysis, due to such effects as noise and multiples intersecting the primary events, were again minimised by use of the GSI '700 package' velocity analysis program. The new data gave a marked improvement in normal moveout velocity determinations, particularly from the shallow section. The increase in resolution also allowed a better definition of the complex channel geometries which are the major cause of moveout velocities being unreliable indicators of true velocity. With these improvements it became feasible to make use of a method of depth conversion based on interval velocities.

12.5.4 The interval velocity approach to depth conversion at East Kingfish

At the eastern end of the field, Kingfish-5

encountered thicker channel fill sediments with interval velocities as high as 14 150 ft/s compared with the 9400 ft/s of the adjacent shales and marls. This velocity contrast of almost 5000 ft/s is more severe than seen in the central and western parts of the field and the time–depth conversion problems are correspondingly more complex. A post Kingfish-6 top of the Latrobe depth map produced from a one kilometre grid of 1974 high resolution data using the conventional conversion factor method showed two ridges running east–west along the crest of the structure (figure 12/14). The position of these ridges coincides with shoulders seen on the top of Latrobe reflection on the east Kingfish seismic sections (figure 12/16). These shoulders lie beneath the high velocity channel edges. It was difficult to explain these shoulders as real geological features and suspicion was thrown on the smoothing of the moveout velocity profiles. Consequently, in view of the improved data quality obtained in the 1974 shooting and the errors in depth prediction at Kingfish-5 and Kingfish-6, it was decided to apply a detailed interval velocity analysis over the eastern end of the field.

This interval velocity approach was a

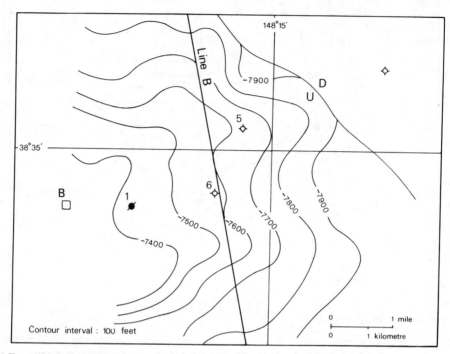

Figure 12/14 East Kingfish: depth to the top of the Latrobe reservoir based on the conversion factor technique.

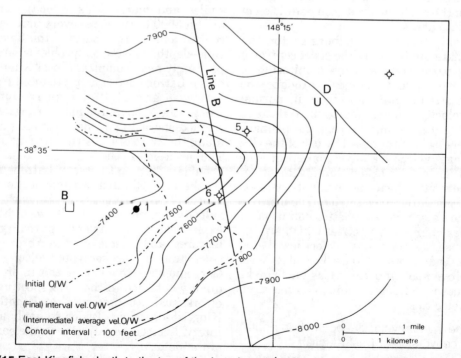

Figure 12/15 East Kingfish: depth to the top of the Latrobe reservoir based on the interval velocity technique.

'layer-cake' method with the post-Latrobe section being divided into some 14 distinct sedimentary layers and wedges using high resolution seismic data such as shown in figure 12/16. The digitised time horizons were then merged with a velocity file made up of interpreted scattergram data and the output was a set of Dix interval velocity profiles for each seismic line. These profiles were smoothed and tied with the cross lines. Once again, lines parallel to the channels were believed in preference to the cross lines where the geometric effects are the greatest. When all the horizons had been tied a series of interval velocity maps were made and a set of isopachs generated. Summing the isopachs produced a depth map to the top of the Latrobe reservoir. This map was slightly too deep because the interval velocities were derived from normal moveout velocities rather than RMS velocities and a small correction factor was needed to tie the well control. The depth structure map on the top of the reservoir produced by this interval velocity approach (figure 12/15) shows a smooth

structure without the anomalous double ridges observed previously.

The depth picture obtained from the interval velocity technique was tested using a ray tracing program. Depth cross-sections together with the smoothed interval velocities were input and the program traced the zero offset ray-paths. The first output was a set of simulated time sections which could be compared with the original data. The effects of migration due to the structure dip of the reflections and also due to refraction across dipping interfaces were apparent at this stage and the depth models were slightly modified to correct for the discrepancies.

The program was then used to simulate scattergrams along the seismic lines. These synthetic scattergrams and the derived normal moveout profiles to the top of the Latrobe Group compared favourably with the actual data and gave confidence to the depth interpretation.

In practice it often happens that the simulated time sections and scattergrams do not match the basic data sufficiently well to

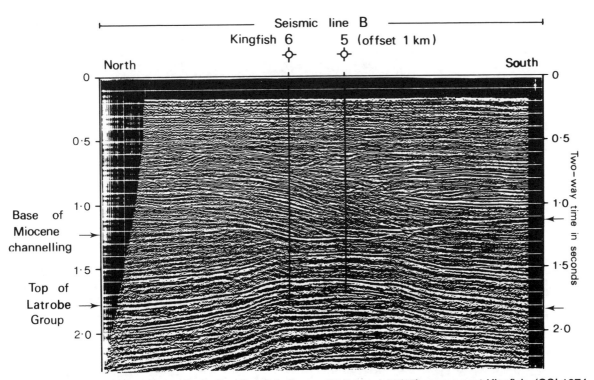

Figure 12/16 High resolution seismic line (line B in figures 12/14 and 12/15) across east Kingfish. (GSI 1974 vintage survey.)

confirm the depth interpretation. In such cases, the parameters used to generate the depth models would be adjusted and the ray tracing analysis repeated until good agreement was obtained.

12.6 Summary

The rational exploration and development of the Gippsland Basin demands that detailed velocity analyses be made over every prospect. The general method that has been adopted for deriving average velocities from seismic velocities is simple, with the most critical steps in the procedure being the smoothing of the moveout velocity profiles and the prediction of conversion factor variations. Both of these steps are highly interpretive and can only be carried out successfully with a sound knowledge of the local geology and an appreciation of the influence that the geology can have on seismic velocities. In particular, this method has been applied in the Kingfish area where it has proved successful in defining the shape of the field sufficiently for proper development and reservoir management.

In areas with complex velocity problems but good seismic velocity data, such as the eastern part of the Kingfish Field, the interval velocity method can be used as an alternative method of depth conversion. This method has the added advantage that ray tracing programs can be used to check the results.

In figure 12/15, the -7566 O/W contact is shown shifted in accordance with the refinements in interpretation. The wells nos. 5 and 6 though dry, contributed data which was significant in the final velocity interpretation. The cost of drilling these wells was probably thus well justified.

Chapter 13

The Hewett Gas Field Case History

In 1959, at Groningen onshore in the northeast Netherlands, NAM, the Shell-Esso consortium, drilled the discovery well on what eventually proved to be a giant gas field with reserves of $1.642 \times 10^{12} \, m^3$ (58 TCF) (see figure 13/1). The productive reservoir was in Rotliegendes (Permian) dune sandstones. Subsequent to the discovery, aeromagnetic surveys were conducted over the North Sea. When integrated with the regional geology, they confirmed predictions, as summarised by Wills (1951), of extension of Permian and younger basins westwards under the North Sea to the English mainland (see figure 13/2).

In 1962, seismic reflection surveys commenced in the southern North Sea; the following year, three separate consortia, headed by the respective operators, Phillips Petroleum, Atlantic Refining Co and Gulf each commissioned three separate surveys on a $\frac{1}{2}° \times 1°$ latitude and longitude grid. They were planned to interlock with each other and after trading each other's data, provided a grid of 20 km square, i.e. approximately subdivided by $10' \times 20'$ areas. A fourth consortium, led by Signal Oil*, joined this arrangement, shooting infill lines in the southern part of the survey area; the final programme grid spacing was on a $5' \times 10'$ grid, or approximately 10 km square

*Subsequently merged into Burmah Oil, the North Sea division of which was later incorporated into BNOC Development Ltd.

(figure 13/3). The 13600 line km were obtained over an area of 114000 km². The data were acquired by three different contractors, Geophysical Service International, Seismograph Service Limited and Robert H. Ray.† Single-fold split-spread profiles were obtained utilising a two-boat operation and dynamite shooting. Navigation and position fixing was by Decca Sea Search involving the use of a master and two slave stations.

From the outset, interpretation revealed the existence of large structures, trending mainly NW–SE, approximately parallel with an ancient high to the southwest, the London-Brabant massif. Correlation of the seismic with onshore well information in the UK, Netherlands and Germany, combined with evidence of extra strong reflectivities and strong absorption led to identification of Permian salt structures. Although these were not as intensely deformed as were the well-documented salt domes and diapirs of Germany, they still showed pronounced relief; in response to rejuvenation of normal faults in the underlying Permian and Carboniferous rocks, the highly plastic halite had flowed, creating folds and salt collapse structures in the overlying Triassic, Jurassic and Cretaceous strata.

From the 1964 Zechstein time structure map shown in figure 13/4 and the accompanying

†Now known as Geosource Petty-Ray

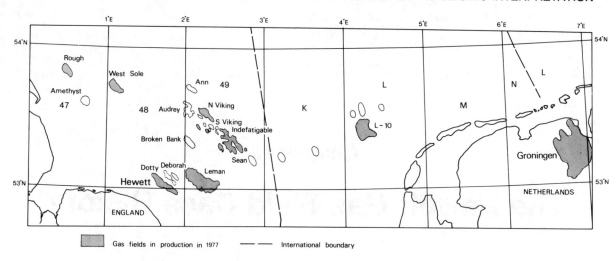

Figure 13/1 Gas fields of the southern North Sea and the Netherlands.

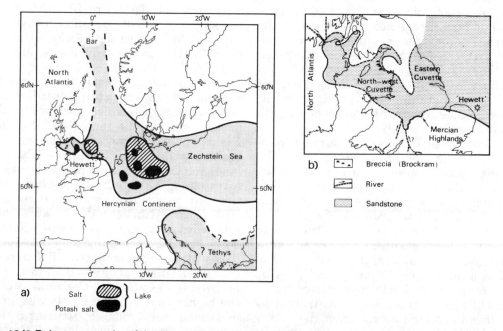

Figure 13/2 Palaeogeography of the Upper Palaeozoic/Lower Mesozoic: (a) Late Permian: the Zechstein Sea extended from E. Europe to Ireland. Fluctuations of evaporite cycles led to precipitation of large deposits of halite and/or potash salts. The Hewett sands were deposited immediately on top of the Zechstein presumably after some rejuvenation of the nearby London-Brabant massif. (b) Lower Trias (Middle Bunter): Lower Bunter sands ('Hewett' sands) have limited distribution, near highland areas; Middle Bunter sands (the 'Bunter' of Hewett) were deposited over a wide area in large delta fans (after Wills, 1951).

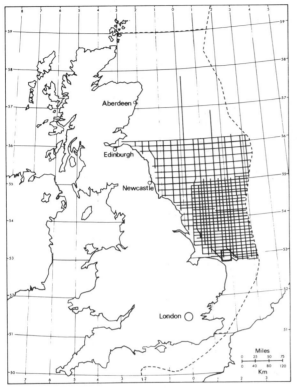

Figure 13/3 Initial seismic reconnaissance of the UK North Sea: the 1963 Phillips/Arpet/Gulf/Signal Group Surveys. The dense grid spacing is 10 km square: the rest is 20 km square. The map area of figure 13/4 is highlighted (after Phillips Petroleum Expl. UK Ltd).

regional geological section, (see figure 13/5) Phillips Petroleum identified one large structure, 15 miles off the coast of East Anglia. In the UK 1964 First Round of Offshore Production Licence awards, the Phillips Consortium were successful in their application for blocks 48/30 and 52/5, over the southeast part of the structure; the Arpet Group became the successful licensees in blocks 48/28 and 29 over the northwest part of the structure. Dip and strike orientated surveys were commissioned in 1964 and 1965. By this time multi-fold or common depth point shooting had 'arrived' in the North Sea and most lines were recorded and processed 3- or 6-fold with a welcome increase in quality. Based on the interpretation of these, Arpet spudded a well on the structure and in late 1966 penetrated significant gas zones in two separate Triassic sandstone reservoirs, with a lesser show in the Permian Hauptdolomit formation. A stepout well, 48/29–2, 2.4 km to the SSE,

found gas in these same three zones, and immediately following this, the Phillips Group made a discovery 13 km to the southeast of the original 48/29–1 well confirming the large areal extent of the reservoirs. The Hewett Field had been discovered.

13.1 Seismic interpretation

Several hundred miles of line were shot over the structure between 1963 and 1968. Initially, data were obtained using a two-boat operation, with dynamite charges up to 110 kg, split-spread cables up to 1200 m either side of the shotpoint, 20–25 hydrophones per station and analogue recording. Gradually, single-fold was superseded by 3-fold and 6-fold stacking, and analogue instruments gave way to digital; new sources evolved improving on production efficiency, signal enhancement and environmental acceptability. The oldest data suffered from very poor normal moveout control over the first 0.5 s and as a result the stacking is not the best. Pre-deconvolution data suffer very badly from water bottom reverberations. This shallow water area (40–20 m over the crest of the structure) is subject to strong tidal variations due to its location near the coastline and the confluence of the southern North Sea with the English Channel, and the Humber, Wash and Thames estuaries. This factor, combined with the limited resolution of the North Sea first phase position-fixing instrumentation, led to navigational and surveying errors which resulted in poor stacking and line misties.

Four lines, two dip-orientated NE–SW, and two strike orientated lines, NW–SE, have been chosen to highlight the seismic interpretation of the Hewett structure; together, they form a closed loop. They are of variable quality, not only because they were shot and processed at different periods during a rapidly changing seismic technology, but also because of their particular orientation relative to the complex-faulted Hewett structure. These seismic sections are reproduced in figures 13/6, 7, 8 and 9.

Of several original reflection events, we will confine ourselves to the three most important (see figure 13/15 for sonic log correlations):

1. *Zechstein*: This is the strongest event on the section. A strong velocity contrast is obtained between the Hewett sandstone with a velocity of around 12 000 ft/s and evaporites, salts and some shales ranging

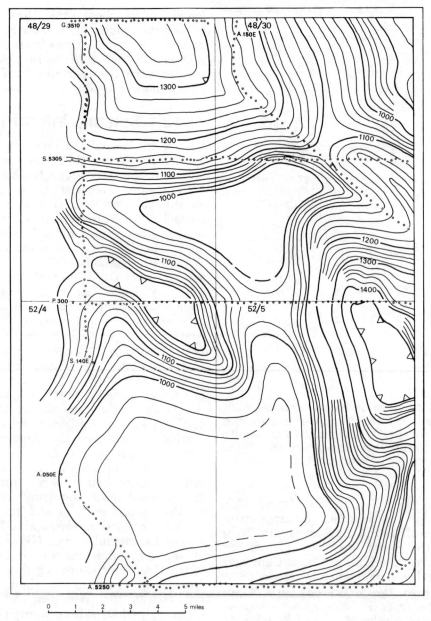

Figure 13/1 Zechstein time structure map from reconnaissance data prepared for the 1964 licence application. Despite the loose (10 km) control, the existence of broad structural highs was outlined by the interpretation. Of these blocks, two each were awarded to the Phillips and Arpet Groups (after Phillips Petroleum Expl. UK Ltd).

down from 16 500 ft/s through 14 000 ft/s to 12–14 000 ft/s, respectively. This event is one cycle above the Z3 or Hauptanhydrit event which is the deepest continuous reflector available for mapping the underlying Rotliegendes reservoirs of the southern North Sea (the Zechstein halite acts as an absorber of seismic energy and underlying

events can only be mapped by undershooting or refraction).

2. *Bunter Sandstone*: This varies in quality but is generally continuous. A velocity reversal obtains between the overlying compact shales and the partly consolidated highly porous sandstone, when gas-filled. When tight or water-filled, the velocity contrast

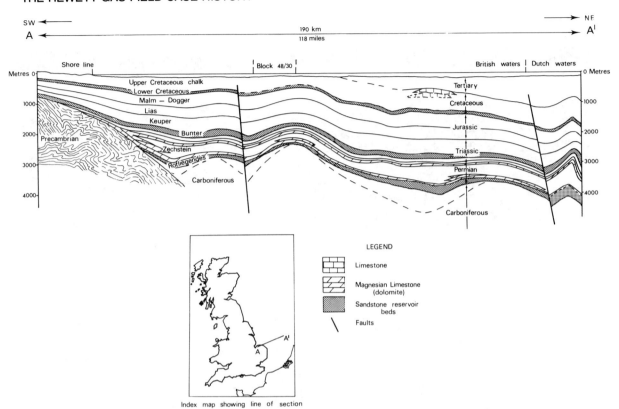

Figure 13/5 NE–SW geological section, offshore East Anglia as prepared in 1964 (after Phillips Petroleum Expl. UK Ltd).

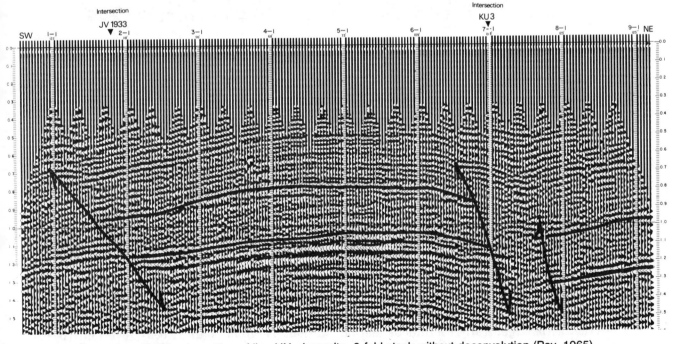

Figure 13/6 Seismic section of line HN, dynamite, 6-fold stack without deconvolution (Ray, 1965).

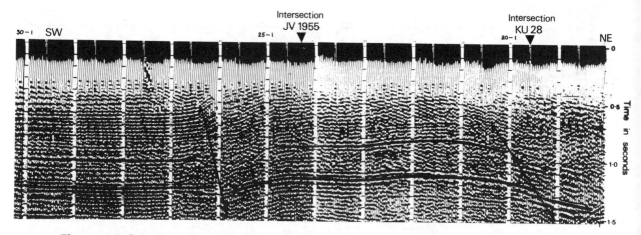

Figure 13/7 Seismic section of line DF, dynamite, 6-fold stack without deconvolution (Ray, 1965).

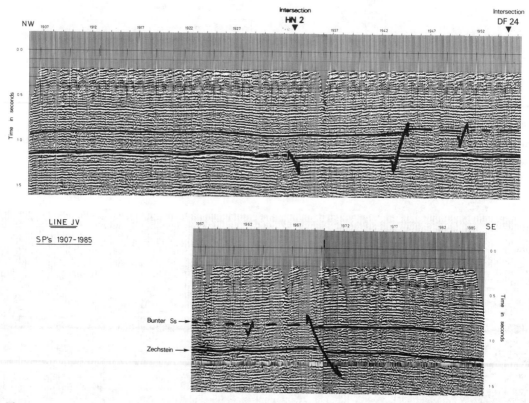

Figure 13/8 Seismic section of line JV, dynamite, 3-fold stack without deconvolution (SSL, 1967).

diminishes and a normal or weak event is obtained.

3. *Keuper*: This an event near the top of the Triassic. It is generally continuous and of good quality throughout the area, except on the older data: it comes in as early as 0.4 s and is often buried in noise on poorly NMO-corrected traces. Reflection generation is from halite beds overlain by mudstones.

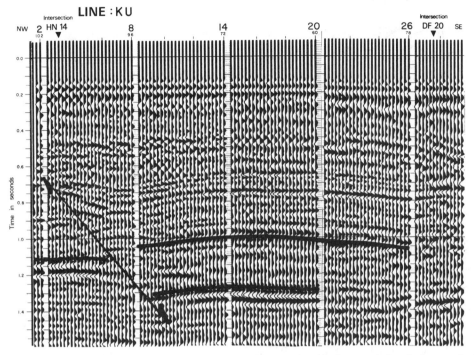

Figure 13/9 Seismic section of line KU, Aquapulse, 12-fold stack with deconvolution before and after stack (Western, 1968).

13.2 Seismic sections

Four interpreted seismic sections are shown in figures 13/6, 7, 8 and 9; locations of seismic lines are shown in figures 13/10 and 13/11.

Although the Keuper reflection is an important event, being the probably Early Cimmerian unconformity and consequently being important in terms of identifying the structural growth history, on most of these sections the reflection is NR and therefore not identified: for identification see the recent sections in figures 13/17(a) and 13/17(b).

HN (Ray, 1965, 6-fold, dynamite, 1680 m split-spread cable, 70 m offset, no deconvolution), figure 13/6.
This NE–SW dip line is located 1 km northwest of the Phillips discovery well 52/5–1, on the apex of both the Hewett and Bunter sandstone reservoirs. The major bounding faults are seen at the Bunter level at SPs 1½ and 7. Correlation of both the Bunter and Zechstein reflections across the southwestern fault cannot be made. Correlation over the northwestern fault is reliable; because of the uncertainty of fault alignment, the two faults shown here are shown

as one broad fault zone on the respective maps. The ties with lines KU and JV are good at the Zechstein level. At the Bunter level, the tie with JV is good but the tie with KU is impossible, due to its occurrence at the coincidence with the major northwestern bounding fault. Reflection quality is good and the picks are reliable over the main structure. Note the occurrence at the 0.8 to 0.9 s levels, between SPs 3½ and 6, of a flat, slightly dipping reflector. As suggested by Cumming and Wyndham (1975), its contrast with the anticlinal curvature of the structure and its flatness suggest it is a gas/water contact.

DF (Ray, 1965, 6-fold, dynamite, 840 m split-spread cable, 35 m offset, no deconvolution), figure 13/7.
This dip line is located to the southeast of the structure, as it plunges in that direction. The major northeastern bounding fault is again apparent at SP 20, at Bunter level. The major bounding fault in the southwest, which was not seen on line HN, occurs at SP 26. The steep dips into the SW bounding fault suggest, as nearby lines show, that a fault occurs around SP 25 throwing down to the southwest. Ties with line JV are good; the ties with line KU again are not

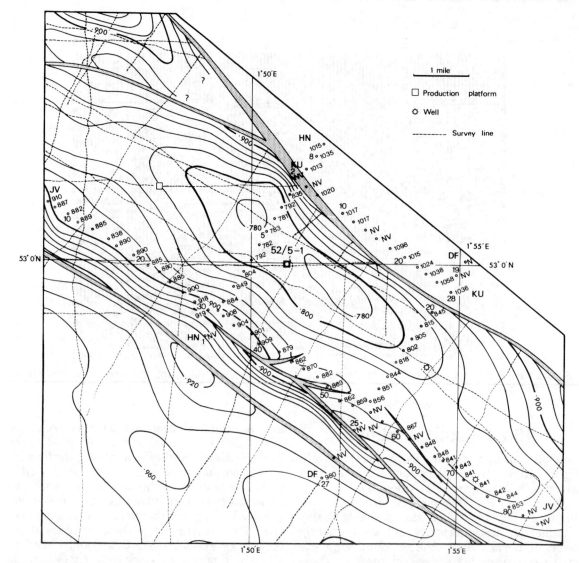

Figure 13/10 Bunter Sandstone time structure map over the SE area of the Hewett Field: SP locations and two-way reflection time values for seismic lines HN, DF, KU and JV are shown; faint dashed lines indicate the location of other lines used for the complete interpretation of this horizon (note the irregularity of many of these lines due to tidal influences and the limited resolution of initial position-fixing systems) (after Phillips Petroleum Expl. UK Ltd).

possible due to the proximity of the major northeastern bounding fault. Reflection quality is fair, and the picks are reliable at both the Bunter and Zechstein levels, especially over the crest of the main structure.

JV (SSL, 1967, 3-fold, dynamite, 1200 m split-spread cable), figure 13/8.

Being a strike line, the structural alignment is relatively uninteresting, running as it does along the southwest flank of the structure, close to the southwestern bounding faults. Ties with lines HN and DF are good at both levels. Several faults are apparent at both the Bunter and Zechstein levels: they are offshoots from the main southwestern bounding fault. Picking is not so reliable as on the dip lines, but the common water-bottom reverberations do not help. The reflection strength of the Bunter is weak or irregular over most of the section; however, in the southwest, it is particularly strong between SPs 1970 and 1980. This is the

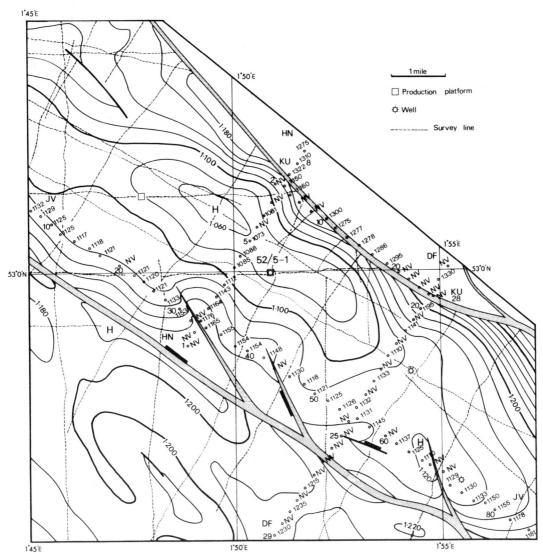

Figure 13/11 Zechstein time structure map, with location of seismic lines HN, DF, JV and KU. The Zechstein immediately underlies the Hewett Sandstone (after Phillips Petroleum Expl. Uk Ltd.)

part of the line closest to the Bunter gas/water interface (see figure 13/12), and the strength of the reflection may be due to 'gas effect', as described later in the discussion of hydrocarbon indicators.

KU (Western, 1968, 12-fold, Aquapulse (4 guns; 4 pops per shotpoint, summed), 1200 m cable, deconvolution before and after stack), figure 13/9.
This NW–SE strike line ties lines DF and HN at their northeastern extremities. Data quality is quite good; however, as this parallels the

NW–SE northeasterly bounding fault diffractions and energy loss make reflection continuity across the faults doubtful, and the line ties difficult. Only one tie is made, with line HN at the Zechstein level, on the upthrown side of the major fault. This, in fact, is the only portion of the main Hewett structure shown on this section. Identification of the Bunter on the upthrown side of the fault is not possible, due to loss of fold and fault effects. Identification of the Bunter and Zechstein on the downthrown side of the fault is by character correlation.

The picking and mapping of lines HN, DF, JV and KU is shown on the two seismic time structure maps, 'Bunter Sandstone' (figure 13/10) and 'Zechstein' (figure 13/11). The remainder of the contouring on these two maps is abstracted from Phillips in-house maps (based on interpretation of the seismic sections shown in faint dashed lines in figures 13/10 and 13/11). Although the mapping of these reflections suggests a simple fold structure, the complexity of the bounding fault relationships is well illustrated by close analysis of these seismic sections. This is particularly so in the southwest, where matching of the fault strikes at the Bunter and Zechstein levels is difficult. At the Bunter level, the main offset from the southwestern bounding fault throws down to the southwest on line DF, creating a small graben, whereas on line HN, it throws down to the northeast, creating a southwesterly dipping block between itself and the main southwestern bounding fault.

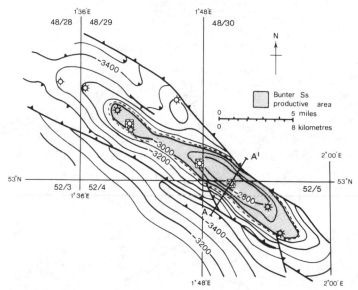

Figure 13/12 Bunter Sandstone depth map (after Cumming and Wyndham, 1975).

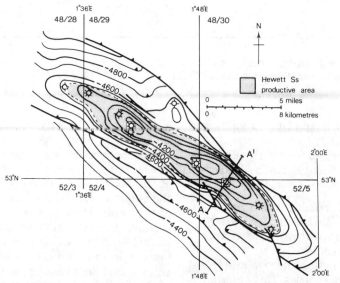

Figure 13/13 Hewett Sandstone depth map (after Cumming and Wyndham, 1975).

When integrated with the valuable well velocity and gas/water interface data, the seismic maps were converted to depth and isopach maps which provided the base for reservoir development studies. The final structure maps for both reservoirs are shown in

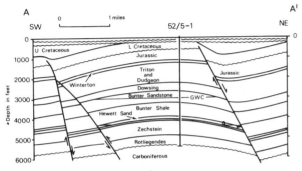

Figure 13/14 Geological dip section A–A', through Phillips discovery well 52/5–1. Line of section is shown in figures 13/12 and 13/13 (after Cumming and Wyndham, 1975).

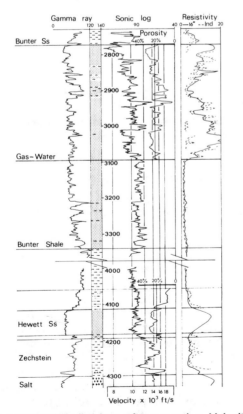

Figure 13/15 Wireline logs of type section. Velocity scales have been superimposed on the sonic log (after Cumming and Wyndham, 1975).

figures 13/12 and 13/13. A geological dip-oriented section of the reservoir, through the Phillips discovery well 52/5–1 and parallel with seismic line HN, is illustrated in figure 13/14. Typical sonic and other wire-line logs are compared in figure 13/15.

The Hewett Gas Field is unique in the southern North Sea in being Triassic; the underlying Rotliegendes which is the reservoir for all other southern North Sea fields, is thought to lack a suitable seal, as the faulting atypically extends up through the Zechstein salt. These same faults probably provided the vertical migration path for the gas which is thought to have the same source as the Rotliegendes fields, i.e. from Carboniferous coal beds; however, it should be noted that the gases in the two reservoirs are different in their chemical composition, the Bunter alone having hydrogen sulphide, and much higher carbon dioxide and nitrogen (Cumming and Wyndham, 1975). Shales and mudstones with anhydrite provide the seals for both the Bunter Sandstone and Hewett Sandstone reservoirs.

Seismic surveys and a total of seven exploration wells established the field outlines as shown in figure 13/16. The lower reservoir extends over 27 km NW/SE, is 4.34 km in maximum width and has an area of approximately 90 km^2 (35 square miles). Total recoverable reserves are in the neighbourhood of 85 $\times$ 10^{9}m^3 (3 TCF).

A unitisation agreement was signed between the Arpet and Phillips Groups in April 1969 and production from the first of three platforms commenced in July of that year. The combined group were successful in their application for Block 52/4 in the 1970 third round of bidding: after the statutory 50 per cent relinquishments, made on the sixth anniversary of the initial awards in the first and third rounds, the field is now completely contained by the part blocks 48/28a, 29a and 30a, 52/4a and 5a, all of which are licensed to the combined Phillips and/or Arpet Group. The complete list of companies is as follows:

Phillips Petroleum Exploration UK Ltd (operator for Phillips Group and Hewett Unit).
Agip (UK) Ltd
Fina Exploration Ltd
Century Power and Light Ltd
Halkyn District United Mines Ltd
Oil Exploration Ltd
Plascom (1909) Ltd

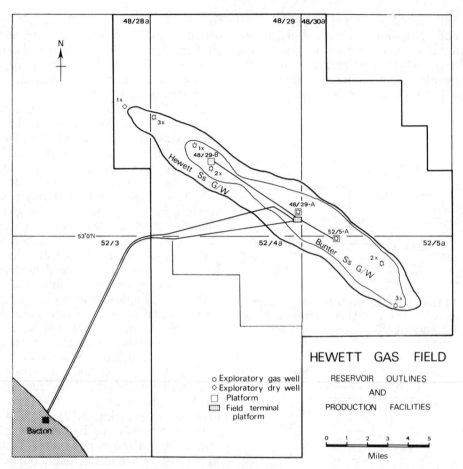

Figure 13/16 The Hewett gas field. The limits of the two Triassic Sandstone reservoirs are shown: the field covers an area of 90 sq km, and is produced from three production platforms (after Phillips Petroleum Expl. UK Ltd.).

Arpet Petroleum Ltd (Operator for Arpet Group).

    British Sun Oil Company Ltd
    North Sea Exploitation and Research Company Ltd
    Superior Oil (UK) Ltd
    Canadian Superior Oil (UK) Ltd
    Sinclair (UK) Oil Company Ltd

13.3 Hydrocarbon indicators

Subsequent to reservoir studies and the completion of development drilling, it was decided to utilise newly developing techniques of direct hydrocarbon detection. In 1974 a short survey was run over the field using the best acquisition and processing parameters available. Two of the lines are shown in figures 13/17(a) and 13/17(b):

Figure 13/17(b) The NE/SW dip line parallels the line HN, discussed earlier. It also passes very close to the well 52/5–1 which was drilled near the crest of both reservoirs. Comparison of the two lines serves to illustrate the dramatic advance in acquisition and processing. Concordant with the increase in quality, there is an increase in frequency content and it is now possible to identify the Hewett Sandstone reflections. In addition, the Zechstein events have acquired high resolution character; what was thought on the early data to be a 0.05 seconds/20 Hz wavelet is now a 0.05 seconds/40 Hz doublet with the null amplitude zone almost certainly defining the absorptive Zechstein halite layer. The faults are now much easier to define. Two hydrocarbon indicator criteria are quite noticeable: a strong black peak, indicating a negative acoustic impedance

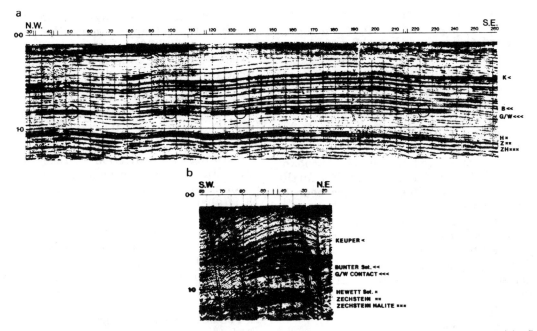

Figure 13/17 Seismic sections derived from 1974 vintage surveys. (a) Strike section along axial crest of the field. Most notable character effects are gas-induced negative reflectivity, amplitude diminishing off-reservoir, gas-water contact reflection and quarter-wavelength 'tuning' which is circled in the figure.
(b) Dip section near 52/5–1. This is a remarkable improvement on the data from line HN which is nearby. The severe faulting at the margins obscures some hydrocarbon criteria; however the gas–water contact in the Bunter Sandstone is apparent, and there is even a suggestion of one in the Hewett Sandstone (both sections after Phillips Petroleum Expl. UK Ltd.)

appears at the top of the gas-laden Bunter Sandstone; as was seen on line HN, but here with much better resolution there is also what appears to be a dipping gas water contact underneath the Bunter Sandstone top. The dip in the contact is probably due to a velocity gradient which follows the dip of the structure.

Figure 13/17(a) This strike line follows the crest of the structure from the northwest of 48/29–2 to southeast of 52/5–3. It is most notable structurally that a saddle separates the major part of the Hewett structure, in the southwest, from the two 'satellite' extensions in the northwest. The superb quality of this section almost makes the necessity for true amplitude sections redundant. Several striking hydrocarbon indicator criteria are apparent:

1. a flat-lying event below the Bunter Sandstone reflector, between shotpoints 145–220,
2. a negative acoustic impedance peak for the Bunter Sandstone reflector, between shotpoints 1–240,

3. a second leg 'bright spot'/thin bed tuning as the negative acoustic impedance peak and the flat-lying event converge, viz, at shotpoints 35–63 and 90–110, 120–140 and 215–235.

13.4 Reflection coefficients and amplitude response analysis

Reflection coefficient analysis of the hydrocarbon indicators mentioned above is hampered by the difficulty in obtaining the sub-surface gas velocities. For the Bunter reservoir, the following averaged (and rounded) parameters have been determined.

V_{bm},	velocity of Bunter Sandstone matrix	= 18 000ft/s
ρ_{bm},	density of Bunter Sandstone matrix	= 2.65g/cc
V_w,	velocity of reservoir water	= 5000ft/s
V_{sh},	velocity of overlying shales	= 10 500ft/s
ρ_{sh},	density of overlying shales	= 2.30g/cc
ϕ,	porosity of the reservoir	= 0.257
S_w,	water saturation of reservoir	= 0.137
V_g,	velocity of Bunter gas	= 1400ft/s
ρ_g,	density of Bunter gas	= 0.07g/cc

Using the Wyllie Time Average Equation (Wyllie *et al.*, 1956 and Chapter 2 p. 00) the velocity of the gas saturated Bunter Sandstone

$$V_{bg} = \text{reciprocal of } \phi\left(\frac{S_w}{V_w} + \frac{\phi(1-S_w)}{V_g} + \frac{(1-\phi)}{V_{bm}}\right)$$

$$= 4840\text{ft/s}$$

This is considerably below the lowest sonic-derived values, which averaged 10 070ft/s from four wells, and which were probably highly affected by mud invasion. For further use, a more compatible value has been extracted from a recent paper on fluid saturated reservoir velocities. It more closely matches the lowest values seen on the sonic logs: it is $V_{bg} = 8500\text{ft/s}$ (Gregory, 1976).

For V_{bw}, velocity of Bunter Sandstone, water-saturated, the Wyllie equation can be used:

$$\frac{1}{V_{bw}} = \frac{\phi}{V_w} + \frac{(1-\phi)}{V_{bm}}$$

$$= (0.257/5000) + (0.743/18000),$$

i.e. $V_{bw} = 10\ 800\text{ft/s}$ (Average sonic velocities of four wells was 11 330ft/s, but again they are probably affected by mud-invasion.)

For ρ_{bw}, reservoir density, water-saturated:

$$\rho_{bw} = \phi\rho_w + (1-\phi)\rho_{bm}$$
$$= 0.257 + (0.743 \times 2.65)$$
$$= 2.23\text{g/cc}$$

Similarly ρ_{bg}, reservoir density, gas-saturated is given by:

$$\rho_{bg} = \phi S_w\rho_w + \phi(1-S_w)\rho_g + (1-\phi)\rho_{bm}$$
$$= (0.257 \times 0.137) + (0.257 \times 0.863 \times 0.07)$$
$$+ (0.743 \times 2.65) = 2.02\text{g/cc}$$

(Average log densities of four wells was 2.08, again probably too high because of mud effect.)

The reflection coefficient for the top of the Bunter gas reservoir,

$$RF_{bg} = (V_{bg}\rho_{bg} - V_{sh}\rho_{sh})/(V_{bg}\rho_{bg} + V_{sh}\rho_{sh})$$
$$= \frac{((85 \times 2.02)-(105 \times 2.30))}{((85 \times 2.02) + (105 \times 2.30))}$$
$$= -0.1689.$$

The reflection coefficient for the gas/water interface,

$$RF_{bw} = (V_{bw}\rho_{bw} - V_{bg}\rho_{bg})/(V_{bw}\rho_{bw} + V_{bg}\rho_{bg})$$
$$= \frac{((108 \times 2.23) - (85 \times 2.02))}{((108 \times 2.23) + (85 \times 2.02))}$$
$$= 0.1676.$$

The values are rather higher than predicted by Gregory, but approximate the results of Toksoz *et al.* (1976).

The reflection coefficient for the off-structure top of Bunter, wet,

$$RF_{sh/bw} = (V_{bw}\rho_{bw} - V_{sh}\rho_{sh})/(V_{bw}\rho_{bw} + V_{sh}\rho_{sh})$$
$$= \frac{((108 \times 2.23)-(105 \times 2.30))}{((108 \times 2.23)+(105 \times 2.30))}$$
$$= -0.0014.$$

Knowing these reflection coefficients we can now draw several conclusions about the meaning of the various events in figures 13/17(a) and (b).

1. Flat-lying events immediately below top Bunter are gas/water interfaces, reflection strength is moderate, but the events are distinctive because of their intrinsic horizontal aspect where other dips are well developed.
2. The negative acoustic impedance for the top of the pay zone is confirmed. Reflection strength is just moderate.
3. The near-zero acoustic impedance calculation for the top of a wet Bunter reservoir suggests that 'dim spots' will be produced from the edges of the gas reservoir, and down-dip. This is exactly what we see from SP 240 and southeastwards in figure 13/17(a).
4. With equal and opposite coefficients the top pay zone and gas/water reflections are ripe for exhibiting tuning or bright spots as the pay thickness tends towards the $\lambda/4$ (one-way time; $\lambda/2$, two-way time) relationship. Analysing the sections, a 0.040 s minimum phase Ricker wavelet seems the most likely choice to synthesise reflection response; with the above velocity $V_{bg} = 8500$ ft/s, the $\lambda/4$ (one-way time) tuning occurs at 0.010 s or 85 ft. By coincidence, the pay thickness at 48/29–2 is 75 ft and if one looks at close-by SP43, in figure 13/17(a) we do get tuning and a bright spot, extending in fact to SP62; this and three other similar effects are circled at SPs 98, 134 and 224. Apart from a good Bunter gas/water contact reflection, figure 13/17(b) fails to show other hydrocarbon criteria due to masking by faulting and diffractions.

A plot of Bunter sandstone amplitude versus time response is shown in figure 13/18. It was derived as follows:

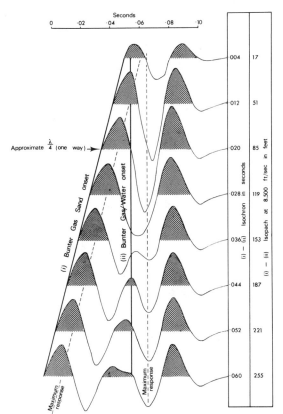

Figure 13/18 Bunter Sandstone wavelet response. Varying the gas pay thickens results in 'tuning' at 85 ft (0.020 s separation) and separation of the shale/gas and gas/water wavelets at 145 ft.

A minimum phase Ricker wavelet of length 0.065 s and effective wavelength of 0.040 s (effective frequency 25 Hz) was used for the top of the pay sand response. By convention it is shown as a peak (negative reflection coefficient). For the gas/water response, the same wavelet was inverted and amplitude modulated to compensate for the partial attenuation of the wavelet at the preceding (gas sand) interface. Starting at 0.004 s wavelet separation (17 ft) with near maximum interference and near-zero amplitude response, the wavelets were separated in increments of 0.008 s (two-way time 34 ft), and their resultant responses plotted. The $\lambda/4$ wavelength (one-way time) reinforcement or tuning is readily apparent at 0.020 s separation, which is equivalent to a pay thickness of 85 ft. Separation of the wavelets occurs around 0.034 s or 145 ft and is complete for a gas sand thickness of 170–264 ft (0.040 to 0.062 s, two-way time delay). The maximum

Bunter gross pay thickness is 323 ft, recorded at the Phillips discovery well 52/5–1. Given enough seismic data of the quality exhibited in figure 13/17(a), and using the criteria developed here, the construction of accurate gas pay maps for the Bunter reservoir should be straightforward.

In this chapter we have used the Hewett gas field case history to briefly trace, over about 15 years, the development of seismic acquisition and interpretation techniques. This history commenced with the early phase of exploration for hydrocarbons in the southern North Sea in 1962, through the discovery phase to that of detailed reservoir studies and the use of modern direct hydrocarbon indicator techniques. As has been discussed in earlier chapters, modern exploration can benefit from what is now considered to be advanced technology even during the earliest exploration phase, but any case history can only relate those events which have occurred over a limited span of years, though in the oil industry a decade is quite a long period. Seismic techniques which were being heralded as revolutionary in the early 1960s are now obsolescent and surely the techniques of the late 1980s will have advanced far beyond those now in current use. Such advances cannot be accurately predicted, but hopefully they will allow the geophysicist to discover and contribute to the exploitation of hydrocarbon reserves which can now be regarded only as unquantified and ill-defined potential resources.

References

A. D. Cumming and C. L. Wyndham, The geology and development of the Hewett gas field: in A. W. Woodland (ed) *Petroleum and the Continental Shelf of North West Europe* Vol. 1 *Geology*, Applied Science Publishers, Barking, England (1975).

A. R. Gregory, Fluid saturation effects on dynamic elastic properties of sedimentary rocks. *Geophysics*, **41** (1976) No. 5, pp. 895–921.

L. J. Wills, *A Palaeogeographucal atlas of the British Isles and adjacent parts of Europe*. Blackie and Son Limited, London (1951).

M. N. Toksoz, C. H. Cheng and A. Timur, Fluid saturation effects on dynamic elastic properties of sedimentary rocks. *Geophysics*, **41** (1976) No. 4, pp. 621–645.

M. R. J. Wyllie, A. R. Gregory and L. W. Gardner. Elastic wave velocities in heterogeneous porous media. *Geophysics*, **21** (1956), pp. 41–70.

Acknowledgements

The use of seismic sections, maps and other illustrations was kindly permitted by the Hewett Operator, Phillips Petroleum and its partners (as listed previously). Various Phillips and Arpet staff are to be thanked for the use of their work, Oil Exploration (Holdings) Ltd for its encouragement, and Dr T. C. Richards for personal communication on some of his work on reflection amplitudes. Unless otherwise stated, all seismic interpretive conclusions are the authors'.

Section 3
Appendices

Appendix 1

Table of Selected Chemical and Physical Properties of Common Rocks, Minerals and Materials*

Note: Throughout the literature, there are many sources of experimental data on the physical and chemical properties of rocks, minerals, etc. This table quotes only one such source and, though incomplete, it is included to give a guide to the range of properties exhibited by such materials.

Material and % porosity	Chemical formula or make-up	Travel time compressional µS/ft	Density, Grain (g/cc)	Density, Observed water-saturated (g/cc)	Bulk modulus dyne/cm²$\times10^{11}$	Rigidity modulus dyne/cm²$\times10^{11}$	Young's modulus dyne/cm²$\times10^{11}$	Poisson's Ratio	Lamé Constant dyne/cm²$\times10^{11}$	Travel Time; Shear µS/ft
Rocks										
Chalk	$CaCO_3$	~75	2.73	2.30–2.50	5.0	2.3	6.0	0.30	3.5	~135
Dolomite, 2%	$CaMg.2CO_3$	43–45	2.87	2.85						~83
Dolomite, 15%	$CaMg.2CO_3$	64–66	2.87	2.62						~120
Limestone, 2%	$CaCO_3$	49–52	2.72	2.69	4.0	2.5	6.2	0.25	2.4	~90
Limestone, 20%	$CaCO_3$	75–78	2.72	2.40						~135
Sandstone, 2%	$SiO_2 + ?$	~59	2.65–2.67	2.57	2.4	1.5–2.0	4.6	0.23	2.0	102
Sandstone, 30%	$SiO_2 + ?$	94–110	2.65–2.67	2.18	1.0–2.0	0.5–1.0				~160
SS, unconsol'd., 40%	$SiO_2 + ?$	>110	2.65–2.67	2.03	~0					~180
Shale	$SiO_2 + ?$	60–170	2.65–2.67	1.90–2.70	0.57	1.5				
Basalt	$Al + Fe + ? + SiO_3$	50–	2.40–2.80		4.5	2.3	6.0	0.28	3.0	92
Gneiss	$Al + Fe + ? + SiO_3$	45–	2.80–3.00		5.7	≈2.7	≈7.0	0.27–0.24	3.5	
Granite	$Al + Fe + ? + SiO_3$	51–	2.50–2.80		2.5	1.5–2.5	4.0–5.5	≈0.20	1.3	~92
Slate	$Al + Fe + ? + SiO_3$	48–	2.60–2.80		6.3	4.7	11.3	0.20	3.1	
Minerals										
Anhydrite	$CaSO_4$	50	2.95–2.99	2.97–3.01	5.6	3.1	7.9	0.26	3.5	90
Calcite, 0%	$CaCO_3$	45–50	2.72	2.72	6.0–7.0	5.0–6.0				~85
Chert, 0%	SiO_2	51	2.65	2.64						~92
Coal (anthracite)	C	90–120	1.40–1.80	1.35–1.80						
Coal (lignite)	$C + O + H$ etc.	140–180	1.20–1.60	1.17–1.50						
Dolomite, 0%	$CaMg.2CaO_3$	41–48	2.97	2.88	8.2	3.0–4.0	7.4	0.30		~78
Gypsum	$CaSO_4.2H_2O$	~53	2.32	2.35	4.7	2.7	6.8	0.26	2.9	
Halite	$NaCl$	66	2.16	2.03	2.3	1.5	3.6	0.24	1.3	113
Illite (dry)	$KAl_2(AlSi_3)O_{10}(OH)_2$		2.60–2.90							
Kaolinite (dry)	$Al_4 Si_4 O_{10}(OH)_8$		2.61–2.68							
Montmorillonite	$(CaMgAl)_2 Si_4 O_{10}(OH)_2 + H_2O$		2.20–3.00							
Quartz, 0%	SiO_2	50–100	2.65	2.65	4.5	3.5				
Fluids										
Air	$O_2 + N_2 + CO_2$ etc.	920	0.00123		0.000014					
Crude Oil	$C + H$	232	0.8–0.9	0.8–0.9	0.08–0.17					
Natural Gas	$C + H$	706	0.00074		0.000013					
Water	H_2O	197	1.00	1.00	0.2					
Miscellaneous										
Cement		≈100	≈2.5							
Iron (casing)	Fe	57	7.76		13.3	8.0	20.0	≈0.30	9.2	
Oil Shale			2.0–2.6							

*After A. T. Hingle, Mobil Log Analysis Chartbook.

Appendix 2

Geophysical survey world-wide average unit costs, listed by survey objective, type and method*

Survey objective, type and method	Total crew-months[a] or (man-months)	Line miles or (logged footage) or [no. of stations]	Acquisition cost (US$)	Average coverage / month or (man-months)	Average cost /month or (man-months)	Average cost /mile or (cost /foot) or [no. of stations]
Land						
Seismic (P-wave)						
Dynamite	5,158.3	254,745	1,333,010,875	49	258,420	5,232
Compressed air	111.4	20,354	34,982,988	182	314,030	1,718
Gas exploder	75.6	3,404	12,796,355	45	169,263	3,759
Weight drop	160.1	15,477	38,030,556	96	237,542	2,457
Solid chemical	87.0	9,400	38,420,000	108	441,609	4,087
Vibratory	3,674.3	224,565	844,096,234	61	229,729	3,758
Other	65.2	2,072	11,237,526	31	172,354	5,423
Seismic (S-wave)						
Dynamite	183.2	9,388	41,500,000	51	226,528	4,420
Compressed air	12.0	382	1,300,000	31	108,333	3,403
Vibratory	50.0	3,111	12,168,371	62	243,367	3,911
Seismic (refraction)						
Dynamite	3.5	186	1,395,000	53	398,571	7,500
Vibratory	0.3	3	71,400	10	238,000	23,800
Other	14.0	24	2,007,046	1	143,360	83,626
Gravity	228.0	56,357	7,462,000	247	32,728	132
Gravity	63.3	[23,719]	6,608,578	(375)	104,400	[279]
Resistivity	1.0	15	10,000	15	10,000	666
Self-potential	(2.0)	5	45,500	(3)	(22,750)	9,100
Induced polarization	60.4	2,160	3,895,728	35	64,498	1,803
Remote sensing other	0.5	50	264,000	100	528,000	5,280
Magnetotelluric	12.0	66	568,000	5	47,333	8,606
Magnetotelluric	8.9	[354]	955,800	(40)	107,393	[2,700]
Marine						
Seismic (P-wave)						
Compressed air	701.1	664,232	418,523,218	947	596,952	630
Electrical	7.2	7,800	2,260,000	1,083	313,888	289
Gas exploder	282.7	199,693	114,389,675	706	404,632	572
Implosive	00.6	64,020	42,159,050	051	423,291	649
Solid chemical	32.0	25,600	14,080,000	800	440,000	550
Other	12.0	25,159	5,705,000	2,096	475,416	226
Seismic (refraction)						
Compressed air	0.5	15	144,000	30	288,000	9,600
Gravity	34.8	43,733	1,032,050	1,256	29,656	23
Magnetic	3.1	3,800	75,002	1,225	24,194	19
Airborne						
Magnetic	108.5	725,391	15,214,330	6,685	140,224	20
R-S side-scan radar	2.3	210,950[c]	1,871,400	91,717	813,652	8
Drill hole						
Sonic/velocity logging	(52.9)	[b]	383,525	[b]	(7,250)	[b]

[a] Total includes government and university activity. [b] Not applicable. [c] Square miles.

*Source: Society of Exploration Geophysicists, Leading Edge, September 1982.

Appendix 3

Some characteristics of a range of marine seismic energy sources*

Source	Description	P_a† (bar-m) (0-125 Hz)	P_a/P_b† (0-125 Hz)	Signature Type	Comments
2000-cu. in. air gun array	31 guns (20–240 cu. in.) 2 ea 150-ft line arrays	69	12	A	
1450-cu. in. air gun array	25 guns (20–100 cu. in.) in 5 subarrays spaced 30 m apart (also used with 15-m and 60-m subarray spacing)	52.6	6.3	A	calculated results, designated as 'long' array
4170-cu. in air gun array	20 guns (60–360 cu. in). 1 gun (1645 cu. in.), 4 ea 15-ft line arrays	49.5	16.7	A	
1450-cu. in. air gun array	25 guns (5–200 cu. in.) 2 ea 80-ft line arrays	48.3	8.4	A	designated as 'standard' array
24-liter (1465 cu. in.) air gun array	24 guns [0.16–2.5 liter (10–150 cu. in.)] in 'U-array'	43	12 (0–248 Hz)	A	
0.59-kg (1.3-lb) maxipulse	0.59-kg nitrocarbonitrate charge fired at 40-ft depth	42	1.4	B	debubble process required, Western Geophysical trademark
3150-cu. in. air gun array	5 subarrays of 5 guns each (630 cu. in) spaced 20–56 m apart	41.5	5.1	A	Shell Canada licensee
1200-cu. in. air gun array	26 guns (10–300 cu. in.), 2 ea 65-ft line arrays	40.5	9.2	A	
Flexichoc	200 cm D × 30 cm H evacuated cavity that implodes when 'fired'	40	2.6	C	IFP trademark
1440-cu. in. water gun array	18 guns (80 cu. in. ea), 3 ea 35-ft line arrays	36	3	C	Calculated, P_a = 2.0 bar-m for one gun
0.27-kg (0.6-lb) maxipulse	0.27-kg nitrocarbonitrate charge fired at 40-ft depth	35	1	B	debubble process required
1222-cu. in. air gun array	7 guns (67–410 cu. in.) in triangularly-shaped array (52-ft base × 45-ft height)	20.1	10.3	A	calculated results
700-cu. in. air gun array	20 guns (10–80 cu. in), 2 ea 65-ft line arrays	18	8	A	P_a/P_b is estimated
1341-cu. in. air gun array	7 guns	12	12	A	gun sizes unknown

*Courtesy: GSI

Source	Description	P_a† (bar-m) (0-125 Hz)	P_a/P_b† (0-125 Hz)	Signature Type	Comments
Vaporchoc II	2-kg (4.4-lb) steam at 60 bar and 400°C exhausted through eight jets	8	5	C	debubble process required, calculated, CGG trademark
Vaporchoc I	same steam conditions as Vaporchoc II exhausted through one jet	8	5	C	debubble process required
one sleeve exploder	propane/oxygen actuated, 1.25 s fill, 11 in. D × 7-ft L	0.74	3.2	A	near field signature from 1969 OTC paper no. 1120

Note: All air guns and water guns charged to 2000 PSI

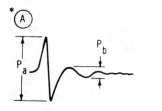
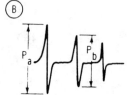
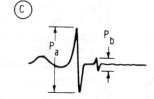

Appendix 4

Interconversion of Units

Length

1 inch = 25.4 millimetres (mm)

1 foot = 12 inches = 0.3048 metre (m)

1 yard = 36 inches = 0.9144 metre (m)

1 mile = 1760 yards = 1.60934 kilometres (km)

1 international nautical mile = 6076.12 feet = 1.852 kilometres (km)

Area

1 acre = 4840 square yards = 4046.86 square metres (m²)

1 hectare = 0.01 square kilometre (km²)

Volume

1 pint = 0.568261 cubic decimetre (dm³)

1 litre = 1 cubic decimetre (dm³)

1 American gallon = 3.78541 cubic decimetres (dm³)
= 0.832674 Imperial gallon
= 0.133681 cubic foot

1 Imperial gallon = 4.54609 cubic decimetres (dm³)
= 0.160544 cubic foot

1 cubic foot = 28.3168 cubic decimetres (dm³)
= 0.0283168 cubic metre (m³)

1 American barrel = 9702 cubic inches
= 158.987 cubic decimetres (dm³)
= 42 American gallons = 34.9726 Imperial gallons
= 5.6146 cubic feet = 0.158987 cubic metre (m³)

1 cubic metre = 1000 litres = 35.3147 cubic feet
= 6.28981 American barrels

Mass

1 ounce = 28.3495 grams (g)

1 pound = 0.453592 kilogram (kg)

1 American (short) ton = 2000 pounds
= 0.907185 tonne

1 tonne (metric ton) = 1000 kilograms (kg)

1 Imperial (long) ton = 2240 pounds
= 1.01605 tonnes

Petroleum specific gravities

Degrees API	25	30	35	40
Specific gravity	0.904	0.876	0.850	0.825

Density

1 gram/cubic centimetre = 1000 kilograms/cubic metre

Magnetic field strength

1 gamma (γ) = 10^{-5} oersted = 1 nanoTesla (nT)

Gravity field strength

1 mGal = 10^{-3} cm/s²

1 gravity unit = 0.1 mGal = 10^{-6} m/s²

Appendix 5

Questions

Included here are a set of questions specifically related to topics covered in this book along with a selection of some questions derived from British University and North American examination papers. The University examination questions in some cases require study of topics not fully covered in this book. It is hoped, however, that students will find it useful to tackle these with the aid of other texts quoted in the reading list.

Review Questions

The reflection of seismic waves is used to delineate sub-surface structure; the reflection of radar waves can be used to detect ships or aircraft. What do you think are the main similarities and differences between these two techniques, and how do the differences influence the methods of data aquisition, processing and display?

Seismic reflection is being increasingly used to study the deep structure of the continental crust, at depths of about 30 km. What modifications to the standard techniques of data acquisition and processing would be desirable in such an application?

Seismic reflection methods have largely been used in prospecting for oil in sedimentary basins. Do you think that there is any hope of using them to help in prospecting for minerals in metamorphic terrain? If so, what modifications of the standard techniques might be needed?

In modern coalmining, it is very desirable to map faults running through a coal-field, even if their throw is as small as 2 m. What problems would be encountered in using reflection seismic methods to detect such small faults in a coal seam at a depth of say, 1 km?

Using the empirical relationships discussed in Section 2.2, estimate the reflection coefficient that might be expected at an unconformity at a depth of 6000 ft, where Tertiary clastics (age 60 m.y.) directly overlie Devonian strata (age 400 m.y.).

In a brine-filled sandstone of porosity 15%, occasional tight streaks of porosity 3% occur. Estimate the reflection coefficient at the top of such a streak, using data and formulae from Section 2.2, and explaining the assumptions you make. Do you think such tight streaks could be detected seismically?

We want to determine the true overburden velocity from analysis of a CDP gather from a dipping reflector. The dip is thought to be about 5°. If the error in deriving an apparent velocity from the CDP gather is negligible, how accurately do we need to know the actual dip if we want 1% accuracy in the true overburden velocity?

A sandstone of velocity 1300 ft/s is encased in a shale such that the reflection coefficients at top and bottom of the sandstone unit are 0.1 and -0.1, respectively. If the seismic source produces a signal which can be approximated as a single cycle of a sine wave of frequency 35Hz, draw the composite reflection waveform that will be observed when the thickness of the sandstone is (a) 400 ft (b) 200 ft (c) 90 ft (d) 40 ft (e) 20 ft. What are the implications for seismic resolution?

In a marine seismic survey, the water depth is 120 m and the rock immediately below the sea bed has a seismic velocity of 2.9 km/s. Assuming a velocity of 1.5 km/s in seawater, at what travel time would we receive refracted sea bed arrivals if the source–receiver offset was (a) 600 m (b) 3000 m?

What is the implication for muting? Should we be concerned about refractions from a deeper horizon, at a depth of 1500 m below surface, where the velocity increases to 3.7 km/s?

For the geometry of figure 4/4, suppose the true static corrections are as follows:

Shot/receiver point 1 2 3 4 5 6 7 8 9 10 11 12 13 14

Static delay ms   0 2 4 6 8 10 30 10 8 6 5 4 2 0 0

Fill in the grid of delay values for each shot–receiver pair. Calculate the values that would be obtained for residual statics at each point, by the method suggested in the text. How do they compare with the true values?

Expand $f(t)=e^{t/T}-(e-1)t/T$ as a Fourier series, valid between $t = 0$ and $t = T$. Draw a graph showing how the function compares with the sum of the first three terms of the series.

A certain seismic source has the signature $(1,1,1,0,0,1,1,0,1)$. If the earth reflection coefficient series were to be $(0,0,1,0,-2,0,0,1)$, what would the reflected signal be? Calculate the cross-correlation of the reflected signal with the source signature, and comment on your results.

Note: here and elsewhere we have taken a reflection coefficient series with terms greater than 1, to simplify the calculation. The results can be simply scaled by division, by say, 10 to give 'realistic' values.

Calculate the autocorrelation coefficients of the source signals $(8,12,6,1)$ and $(2,9,12,4)$, and comment on your results.

Calculate the reflection signal that would be observed from an earth reflectivity series $(1,0,0,-2,1)$ if the source signal were $(6,1,-2)$. Calculate an approximate inverse of length 3 for this wavelet, and apply it to the source signal and to the reflection signal. Comment on your results.

Derive 3-element operators to convert the wavelet $(3,-4,-4)$ into a unit spike at $t = 0$, $t = 1$, and $t = 2$. Which gives the best result?

In performing a velocity analysis on a seismic section, we run out of reliable picks at a TWT of 3.1 s, where the rms velocity is 3.8 km/s. We believe on geological grounds that below this depth there is a very thick sequence of interval velocity about 4.3 km/s. What would be reasonable rms velocities to assume for TWTs of 4 s and 5 s?

Suppose we observe two closely-spaced reflections on a CDP gather. At an offset of 1200 m, they have TWTs of 510 and 530 ms. If the rms velocity is 2.8 km/s at this level, what will the TWTs be after NMO correction? What does this imply for the frequency content of the NMO-corrected data?

Selected examination questions

University of London, Imperial College
Geophysics MSc (1983)

Explain why a 'minimum phase' assumption is required for Wiener predictive deconvolution of seismic traces. What other assumptions does the method require, and how can excessive noise amplification be controlled? What is the minimum phase equivalent of the sampled wavelet 3,7,6,0?

Explain how linear inverse methods can be applied in:

(a) Wavelet shaping filter design.
(b) Elimination of residual statics.

Show how Fourier transform techniques can be used for attenuating multiple reflections during seismic data processing. What precautions are necessary?

A seismic survey is to be performed over deep events dipping at 30° in a medium in which the velocity is 2000 m/s. If the plane wave arrivals are in the frequency band 10 to 50 Hz, what should be the maximum receiver group interval, and why?

University of Leicester
Geology with Geophysics BSc (1982)

Derive the transform relationship between an autocorrelation function and its associated energy spectrum for a finite length, continuous signal. Of what use is the autocorrelation function in geophysics and in particular reflection seismology? Illustrate your answer with a simple numerical example.

Derive the wave equation for a seismic disturbance propagating through a homogeneous, uniform, isotropic medium. Indicate how this equation may be used to predict measurable seismic parameters.

University of Leicester
Geology with Geophysics BSc (1980)

Describe in detail the reasons for the various shot–receiver patterns used in reflection seismology.

In reflection profiling, what fold of CDP cover is possible for a 48 channel receiver if the channel spacing is 10 m and the shot spacing is 60 m?

Describe and explain the reasons for carrying out a 'well-shoot' and a 'sonic log' at a borehole during the course of a seismic reflection survey. What are the problems involved in both types of operation? How is the sonic log calibrated?

University of Aberdeen
Petroleum Exploration Studies MSc (1982)

What is meant by 'normal moveout' on a seismic reflection record? How does it enter into the processing of seismic data and the extraction of information about the rocks?

Various types of 'noise' appear on a raw seismic reflection record. What are they due to and what can be done to eliminate them?

University of Lancaster
Environmental Sciences BSc (1982)

A borehole is logged for seismic velocity and gives the following table of one-way travel times.

Depth (in metres)	One-way time (in seconds)
800	0.4
2600	1.0
3800	1.3
4800	1.5

Plot four small graphs to show the variation of both interval velocity and average velocity with both depth and one-way time. Draw straight line segment graphs as appropriate. Comment on the errors involved in using these representations by comparing the depths computed for a two-way travel time of 2.8 s from both the interval velocity and average velocity graphs.

University of Birmingham
Applied Geophysics MSc (1982)

(a) Define an ideal band-pass filter. To what problems can it be applied? Show, with the aid of diagrams, how a digital time-domain band-pass filter of finite length can be developed.

(b) What criterion does matched filtering seek to achieve? Write down the matched filter corresponding to a general sampled input signal, $s(t)$, of length N, and hence demonstrate the significance of autocorrelation in matched filtering. What considerations determine the design of $s(t)$ in the 'Vibroseis' system?

University of Birmingham
Applied Geophysics MSc (1983)

A limestone horizon at an average depth of 100 m over an area 2 km $\times$ 2 km is to be mapped beneath overlying homogeneous shales. Sandstones up to 20 m in thickness and of velocity intermediate between those of shale and limestone are known to fill broad channels which have been cut into the limestone. Discuss the problems which might arise in the execution and interpretation of both refraction and reflection surveys, indicating how you would choose between these two methods. How would your choice be affected if the area were covered by 10 m of water?

Discuss the different approaches to seismic data acquisition and interpretation that are required in

(a) the broad assessment of the petroleum potential of a relatively unexplored basin,
(b) the detailed exploration of the structure of a proven offshore oilfield, and
(c) the evaluation of the suitability of a proposed route for a pipeline to the shore from the field.

University of Leeds
Geophysics MSc (1981)

Describe the ways in which seismic reflection signals are enhanced with respect to seismic noise in the acquisition of land and marine reflection data.

Describe the various seismic reflection interpretation methods used to investigate structural and lithological hydrocarbon traps.

University of Leeds
Geophysics MSc (1982)

Review the processing of multichannel seismic reflection data both before and after velocity analysis. Clearly indicate, giving reasons, the parts of the processing sequence that are optional.

Describe the design features and survey operation of a multichannel marine seismic reflection system, used to collect multifold common depth point data, having an air gun array as its seismic source. What are seabed reverberations and how are they removed from marine reflection data?

University of Leeds
Geophysics MSc (1983)

Derive expressions for the reflection travel time from an inclined reflector to show that for a given trace on a seismic reflection section:

(a) migration time is greater than unmigrated time, and
(b) migration always increases dip.

Briefly describe one computer method of migration.

Briefly review the various marine seismic energy sources currently in use, indicating, for each, how the effect of the bubble pulse phenomenon is minimised.

Describe the design features and operation of a tuned air gun array and explain the reasons why this acoustic energy source has become so widely used.

University of Wales
Geology BSc (1981)

The proving and exploitation of an offshore oil/gas field includes the use of a wide range of seismic reflection techniques. Describe these techniques.

University College of North Wales
Marine Geotechnics MSc (1980)

What is meant by the bubble pulse in marine seismics, what effect does it have on the seismic record, what are the factors which control it, and how can it be removed (wholly or partially) from the record?

Chelsea College
Applied Geophysics BSc (1982)

Explain the difference between the following in relation to seismic reflection surveying:

(a) average velocity
(b) stacking velocity
(c) rms velocity
(d) interval velocity.

Discuss the application of velocity analysis in the preparation of CDP stacked record sections and the value of the velocity information thus obtained in interpretation.

Give an account of the approach to seismic interpretation in terms of seismic stratigraphy.

Chelsea College
Applied Geophysics BSc (1981)

Give a brief account of each of the following in the context of the processing of sesmic reflection data:

(a) anti-alias filter
(b) deghosting filter
(c) velocity filter.

Discuss the problems involved in the interpretation of unmigrated CDP seismic record sections in terms of geological structures conducive to the accumulation of hydrocarbons.

Imperial College of Science and Technology
Petroleum Geology MSc (1983)

As an oil company geologist, what types of information would you expect to obtain primarily from seismic data interpretation? Discuss how the information you specify may be deduced.

Discuss the standard processing sequence for seismic data, making specific mention of those stages where interpretive decisions need to be made.

Indiana University – Purdue University
Principles of Geophysics Test no. 2 Fall 1980

A set of high resolution seismic instruments is used to record the following two events. Plot the data on graph paper and answer the questions below.

Seismometer Distance (feet) X	Times (seconds)	
	T_1	T_2
0	0.000	0.100
100	.020	.102
200	.040	.108
300	.060	.177
400	.080	.128
500	.100	.141

| Seismometer Distance (feet) | Times (seconds) | |
X	T_1	T_2
600	.120	.156
700	.140	.172
800	.147	.189
900	.154	.206
1000	.161	.224
1100	.167	.242
1200	.174	.260

(a) Is T_1 a refracted or reflected event?
Is T_2 a refracted or reflected event?

(b) Assume that the shot point and seismometers are on a level surface and that the velocity interfaces have no dip.
What are the velocities of the two layers?

(c) From the velocities found above, what rock types would you expect to encounter in a drill hole?

(d) From the reflection data, what is the depth to the top of the reflecting layer?

(e) What is the critical distance? Sketch a diagram showing ray paths out to the critical distance.

(f) The depth, z, to the interface can be shown by refraction theory to be:

$$ z = \frac{T_i}{2} \left(\frac{V_2 - V_1}{V_2 + V_1} \right)^{\frac{1}{2}} $$

From your graph, what are T_1, V_1 and V_2?

Indiana University – Purdue University
Interpretation of Seismic Data Lecture Test

Indicate if the following statements are true or false.

(a) Reflected amplitudes from a rock unit depend only on the acoustic impedance of that particular rock type.

(b) Perfectly rigid rocks overlain by soft rocks will have a positive reflection coefficient.

(c) Reflected energy is usually less than the transmitted energy within any rock layer.

(d) Weathering correction is necessary for interpretation of seismic data.

(e) Seismic velocities generally decrease with increasing depth.

University College at Potsdam, New York
Geophysics Exam. Introduction to Seismic Prospecting

Determine the sampling frequency, Nyquist frequency and anti-alias filter cut-off frequency for a sampling rate of 2 milliseconds.

Using a 24 channel cable with 100 metres between channels towed behind a ship so that the nearest phone is 200 metres behind the ship and the source 100 metres behind the ship. Design a shooting program for 12-fold coverage. Determine how frequently the shots should be fired if the ship is travelling 10,000 metres/hr (5 knots).

What is cross-correlation? Cross-correlate the following source signal with the returned signal.

(a) −1, 2, −1
(b) 2, 0 −1, 2

A fault has a displacement of 30 metres near the surface. Suppose the velocity of shallow reflections is 2 km/s and their frequency is 50 Hz.

(a) What is the wavelength of the seismic waves?

(b) Will this fault be seen? Explain.

For greater depths the velocity increases to 5 km/s and the frequency is 20 Hz.

(c) Will the fault be seen at greater depth? Explain.

(d) What is the resolving power at greater depth?

Queens University, Kingston, Canada
Geology Final Examinations (1970–1982)

Sketch the expected time–distance graph of the first seismic wave arrival for a simple two layer earth (e.g. overburden over bedrock). How would you use this graph to interpret the thickness of the top layer? Indicate how the graph will vary if the interface between the two layers is not horizontal.

If seismic wave velocity is inversely proportional to density[1] by the wave equation, why do seismic velocities generally increase with depth? What are some of the reasons that seismic velocity does not uniformly increase with depth?

Discuss the relationship between seismic wave propagation and the elastic properties of an isotropic material, noting in particular the relationship between seismic velocities and the elastic constants. What is the relationship between the density of the material and the velocity of seismic waves? What causes the attenuation and dispersion of a seismic wave as it travels through an isotropic elastic medium?

List and *briefly* explain all the variable parameters which must be considered when designing a seismic reflection field acquisition program.

Discuss the problem of sample aliasing from both the time domain and frequency domain points of view. What can be done during recording to eliminate this problem?

Describe the seismic processing procedure of deconvolution with reference to both the time domain and the frequency domain.
What is the minimum phase wavelet and why is it an important consideration in deconvolution? Give an example of where deconvolution is particularly useful, showing in a sketch the expected results.

List and briefly describe the various types of seismic energy sources used in land and marine seismic reflection. What are the relative advantages and disadvantages of these sources and which are the most popular (based on present line mileage use).

Appendix 6

Further Reading

H. N. Al-Sadi. *Seismic Exploration.* Birkhauser, Basel (1980).

A. Ameely, H. A. K. Edelman and J. Fertig. *How do Shear Waves Affect Normal P-Waves?* Technical Paper, Preussag and Prakla-Seismos, Hanover.

N. Anstey. *Seismic Interpretation. The Physical Aspects.* International Human Resources Development Corporation, Boston (1977).

N. Anstey. *Seismic Exploration for Sandstone Reservoirs.* International Human Resources Development Corporation, Boston (1980).

Applied Automation Inc. *Opseis 5500 System. Digital Radio Telemetry Seismic Recording System.*

A. J. Berkhout. Wave Field Extrapolation Techniques in Seismic Migration: A Tutorial. *Geophysics,* **46**, no. 12 (1981), pp. 1638–1656.

A. J. Berkhout. *Seismic Migration. Developments in Solid Earth Geophysics no. 14A.* Elsevier, Amsterdam (1982).

R. B. Blackman and J. W. Tukey. *The Measurement of Power Spectra.* Dover, New York (1958).

M. Born and E. Wolf. *Principles of Optics.* Pergamon, New York (1959).

R. E. Chapman. *Petroleum Geology, a Concise Study.* Elsevier, Amsterdam (1973).

M. B. Dobrin. Seismic Exploration for Stratigraphic Traps. *In: Seismic Stratigraphy: applications to hydrocarbon exploration.* American Association of Petroleum Geologists Memoir no. 26 (1977).

W. M. Ewing, W. S. Jordetsky and F. Press. *Elastic Waves in Layered Media.* McGraw-Hill, New York (1957).

A. A. Fitch. *Seismic Reflection Interpretation.* Gebruder Borntraeger, Berlin (1976).

R. Garotta. Land Seismic Shear Waves. *CGG Technical Series,* No. 507.78.05 (1978).

R. Garotta. Comparison between P, SH, SV and Converted Waves. *CGG Technical Series,* No. 521.82.08 (1982).

T. S. Gaskell and K. V. Lacklock. Position Fixing for North Sea Oil. *Journal of Navigation,* **27**, no. 2 (April 1974).

F. S. Grant and G. F. West. *Interpretation Theory in Applied Geophysics.* McGraw-Hill, New York (1965).

R. A. Hardage. *Vertical Seismic Profiles, Part A: Principles.* Geophysical Press, Amsterdam (1983).

E. S. Hills. *Elements of Structural Geology.* Chapman and Hall, London (1972).

G. R. Hoar. Satellite Surveying. *Magnavox Technical Paper Series* MX-TM-3346-81, no. 1005 (1981).

P. Hood. Finite-difference and Wave-number Migration. *Geophysical Prospecting,* **26**, no. 4 (1978), pp. 773–790.

P. Hood. Migration. *In:* A. A. Fitch (ed.) *Developments in Geophysical Exploration 2.* Applied Science Publishers, London (1981).

A. M. Kleyn. *Seismic Reflection Interpretation.* Applied Science Publishers, London (1981).

R. McQuillin and D. A. Ardus. *Exploring the Geology of Shelf Seas.* Graham and Trotman, London (1977).

D. Michou. Vertical Seismic Profile. *CGG Technical Series,* No. 231-7605.07.

A. W. Musgrave (ed.) *Seismic Refraction Prospecting.* Society of Exploration Geophysicists, USA (1967).

L. L. Nettleton. *Gravity and Magnetics in Oil Prospecting.* McGraw-Hill, New York (1976).

D. S. Parasnis. *Principles of Applied Geophysics.* Methuen, London (1962).

C. Powell. An Informal Review of Shore-based Radio Position Fixing Systems. *H.S.S.U.T.* (6 September 1973).

Prakla-Seismos. Migration of Reflection Time Maps. *Prakla-Seismos Information Bulletin,* no. 1. Prakla-Seismos, Hanover (1977).

G. R. Ramsayer. *Seismic Stratigraphy, A Fundamental Exploration Tool.* Offshore Technology Conference Houston 1979, paper no. 3568 (1979).

E. A. Robinson. Physical Applications of Stationary Time Series. Charles Griffin,

London (1980).

E. A. Robinson. *Geophysical Signal Analysis.* Prentice Hall, Englewood Cliffs, N.J. (1980).

E. A. Robinson. Predictive Deconvolution. *In:* A. A. Fitch (ed.) *Developments in Geophysical Exploration 2.* Applied Science Publishers, London (1981).

E. A. Robinson. *Migration of Geophysical Data.* International Human Resources Development Corporation, Boston (1983).

Schlumberger Ltd. *Log Interpretation Principles.* Schlumberger, New York (1972).

W. Schneider. Integral Formulation for Migration in 2- and 3-Dimensions. *Geophysics,* **43,** no. 1 (1978).

Seismograph Service Corporation. *The Robinson-Treitel Reader.* SSC. Tulsa, Oklahoma (1973).

R. E. Sheriff. *Encyclopaedic Dictionary of Exploration Geophysics.* USA (1973).

R. E. Sheriff. *A First Course in Geophysical Exploration and Interpretation.* International Human Resources Development Corporation, Boston (1978).

R. E. Sheriff. *Seismic Stratigraphy.* International Human Resources Development Corporation, Boston (1980).

R. E. Sheriff. *Structural Interpretation of Seismic Data.* AAPG Geologists Education Course Note Series no. 23 (1982).

R. E. Sheriff and L. P. Geldart. *Exploration Seismology (vol. 1 history, theory, data acquisition).* Cambridge University Press (1982).

M. T. Silvia and E. A. Robinson. *Deconvolution of Geophysical Time Series in the Exploration for Oil and Natural Gas. Developments in Petroleum Science,* no. 10. Elsevier, Amsterdam (1979).

D. G. Stone and H. B. Evans. Extrapolating Logs Run in Exploration or Development Wells using Seismic Data. *Society of Professional Well Log Analysts,* 21st Symposium (1980).

W. M. Telford, L. P. Geldart, R. E. Sheriff and D. A. Keys. *Applied Geophysics.* Cambridge University Press (1976).

P. M. Tucker and M. J. Yorston. Pitfalls in Seismic Interpretation. *Society of Exploration Geophysicists,* monograph no. 2 (1973).

K. H. Waters. *Exploration Seismology* (2nd edition). Wiley, New York (1981).

D. G. A. Whitton and J. R. V. Brooks. *A Dictionary of Geology.* Penguin, Harmondsworth (1972).

Appendix 7

Moray Firth Seismic Interpretation Training Package

Data Package for Training in Reflection Seismic Interpretation

The British Geological Survey wishes to announce the availability for purchase of a data package derived from a seismic survey of the Inner Moray Firth. The package is available to any educational or training establishment, including those within commercial organisations, subject to the condition that the data must not be used for any commercial purpose, either for consultation with hydrocarbon exploration companies or for incorporation into any commercial report. All purchasers will be required to sign an agreement for provision of the data in advance of completion of sale.

Contents of the Data Package

The package aims to provide sufficient data for a short course in seismic interpretation in a form such that all relevant maps and sections can be easily copied for students to perform a number of exercises. The data set consists of the following:

1. An explanatory report
   5 copies
2. 1:100 000 scale shot-point map showing entire coverage of the BGS seismic survey of the Inner Moray Firth
   1 copy on transparent film
3. Full scale seismic sections for lines (see map enclosed)*:
   5, 6 and 7 — In total
   10   — SPs 0–270
   11   — SPs 0–560
   12   — SPs 950–1548
   1 copy each on transparent film
4. Well data: Hamilton well no. 12/26-1.
   Composite log
   Sonic log
   Density log
   1 copy each on transparent film

*Figure 10/6 in this book.

5. UV camera monitor record for one location on line 6.
   5 copies
6. Velocity analyses. The type available is the Seiscom Velocity Spectrum analysis. Three analyses are supplied for locations along line 6 and two from line 7, these being either side of well 12/26-1.
   1 copy each on transparent film

Contents of the Explanatory Report

The report briefly describes the data set and by way of introduction gives a short exploration history of the Moray Firth basin. There follows a chapter on data acquisition and processing with sections on specification of the survey, description of the seismic source and cable configuration, recording techniques, position-fixing, quality control and a summary of the processing including a processing flow diagram. A final chapter gives an elementary guide to the interpretation of the data commencing with establishing a tie to well 12/26-1, interpretation of the sections, production of time contour maps, some notes on features revealed by the seismic sections and finally a suggested course structure for a training scheme using the data.

Included with each report, as well as simplified two-way time maps derived from a BGS interpretation of the data, are a set of colour plates showing an interpretation of three reflecting horizons for all the sections provided and coloured geological map of the area.

Purchasing Arrangements

The cost of the complete data set is £90 and the supply of this package is deemed to be zero rated for VAT purposes. Extra copies of the explanatory report may be purchased at £5 per copy. The purchasing procedure will be as follows:

1. An order should be submitted to:
   Marine Geophysics Programme Manager
   British Geological Survey
   Murchison House
   West Mains Road
   Edinburgh EH9 3LA
   Telephone: 031-667 1000
   Telex: 727343

2. On receipt of an order, an 'Agreement' form will be returned for signature to the purchaser along with an invoice. A remittance should be returned to Marine Geophysics Programme Manager, cheque made payable to the Natural Environment Research Council along with a signed copy of the Agreement. On receipt of the remittance data sets will be prepared for despatch to purchasers, but it should be noted that depending on demand this may take a few weeks. All orders will be met in strict rotation according to date of receipt of remittance.

3. Orders received from abroad will be despatched by AIR PARCEL. An additional remittance of £10 should be added whenever delivery outside Great Britain is requested.

Appendix 8

Geological Column

This table shows the subdivision of the geological column into systems, series, and stages. Dates, in millions of years, are taken from the detailed geological time table by Van Eysinga (1975). Names given are those commonly in use for marine rocks in northwest Europe. The fifteen stages of the Quaternary and the eight of the Neogene are omitted for lack of space, as are the modern stage names now accepted for parts of the Ordovician, Silurian, and Carboniferous. In these Palaeozoic systems the more familiar 'series' are listed. The subdivisions of the Precambrian are based entirely on isotopic dates and include both stratigraphic rock units and metamorphic complexes. Neither the names nor the dates on this table will be acceptable to every geologist, as there are areas of disagreement affecting nearly every part of it. The steady accumulation of new data slowly reduces the size of these areas. (Adapted from J. Thackray, *The Age of the Earth*. HMSO, London. Published courtesy British Geological Survey.)

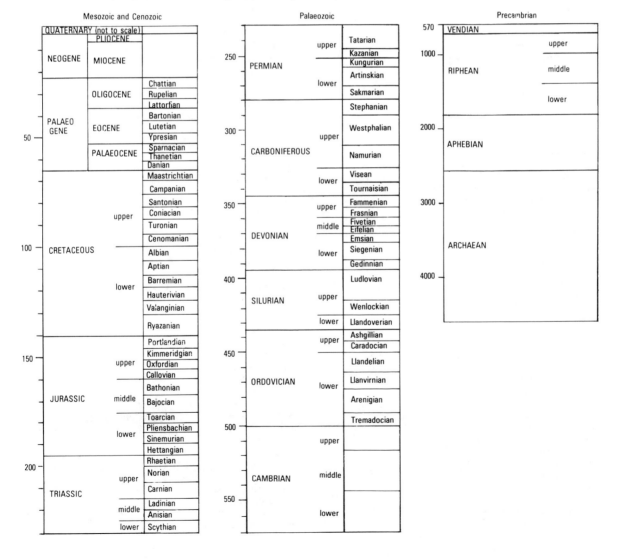

INDEX